LIVER REGENERATION

Edited by **Pedro M. Baptista**

INTECHOPEN.COM

Liver Regeneration
Edited by Pedro M. Baptista

Published by InTech

Janeza Trdine 9, 51000 Rijeka, Croatia

Publishing Process Manager Molly Kaliman

Technical Editor Teodora Smiljanic

Cover Designer InTech Design Team

Additional hard copies can be obtained from orders@intechopen.com

Liver Regeneration, Edited by Pedro M. Baptista
p. cm.
ISBN 978-953-51-0622-7

For EU product safety concerns:
IN TECH d.o.o., Prolaz Marije Krucifikse Kozulić 3, 51000 Rijeka, Croatia,
info@intechopen.com or visit our website at intechopen.com.

Meet the editor

Dr. Pedro M. Baptista, Pharm.D., Ph.D., is currently a researcher at the Wake Forest Institute for Regenerative Medicine in Winston-Salem, NC, USA. This is a world-renown institution in the field of tissue and organ engineering, and the first to accomplish a transplantation of an in vitro-grown organ, the bladder, into a human patient. Dr. Baptista is originally from Portugal, where he obtained his Pharmacy degree at the School of Pharmacy of the University of Lisbon. He has worked in clinical research for the pharmaceutical industry before engaging in his doctoral dissertation project in liver bioengineering and regeneration. His doctoral work resulted in the development of the first human liver ever made in a laboratory, impacting the scientific community at large and widely broadcasted in the media all around the world. He has more than 20 scientific publications including a book, book chapters and several papers and reviews published in prestigious international journals. He is also a regular contributor to the portuguese pharmaceutical magazine "Pharmacy and Distribution" where he publishes a monthly chronicle. Dr. Baptista's current research interests focus on investigating methods to expand stem/progenitor cell numbers required for organ bioengineering and in making the long term transplantation of these lab-grown organs a reality.

Contents

To my family

Preface

This book focuses on the current knowledge regarding the physiologic processes that are triggered after hepatic injury and ultimately lead to liver regeneration. Some of these mechanisms are common to other tissues/organs, but the quickness, precision and effectiveness of liver regeneration in completely restoring its initial physiological function after injury is quite remarkable and unique among all the solid organs. Thus, the knowledge of these specific molecular and cellular mechanisms is crucial for the improvement of the current therapies and ultimately, complete recovery from liver disease.

Hence, the first section of the book comprises multiple chapters that detail the mechanisms of molecular and cellular liver regeneration. Then, the second section describes different animal models used in this field of research, highlighting their significance and contribution to the study of liver regeneration. Finally, the last section presents a chapter on the gold standard for end-stage liver disease, liver transplantation, followed by numerous approaches and strategies for liver regeneration that rely on different cell therapies. The last chapter of this book describes some of the new approaches being developed that rely on tissue and organ bioengineering.

It is then my hope as the book editor that this book will be able to help as many professionals and curious minds as possible, working in or out of the liver field, and that it can shed some light in the intricate mechanisms of organ regeneration.

Pedro M. Baptista, Pharm.D., Ph.D.
Researcher,
Wake Forest Institute for Regenerative Medicine,
USA

Section 1

Cellular and Molecular Mechanisms of Regeneration

1

Hepatocytes and Progenitor – Stem Cells in Regeneration and Therapy

Laura Amicone, Franca Citarella, Marco Tripodi and Carla Cicchini
Dept. Cellular Biotechnology and Hematology,
"Sapienza" University of Rome,
Italy

1. Introduction

The liver is a highly specialized detoxifying organ involved in: i) glucose homeostasis; ii) lipid homeostasis and ketone bodies production; iii) metabolism of amino acids. Most of the liver functions are carried out by the hepatocytes (about 70-75% of hepatic cells) that, together with cholangiocytes (10-5 % of hepatic cells), are of endodermal derivation and constitute the hepatic parenchyma.

The liver has a peculiar and fascinating ability: it is able to regenerate itself after loss of parenchyma for surgical resection or injuries caused by drugs, toxins or acute viral diseases. The ancient myth of Prometheus highlighted this capability: the Titan Prometheus was bound for ever to a rock as punishment by Zeus for his theft of the fire; each day a great eagle ate his liver and each night the liver was regenerated, only to be eaten again the next day.

The liver compensatory regeneration is a rapid and tightly orchestrated phenomenon efficiently ensuring the reacquisition of the original tissue mass and its functionality. Primarily, it involves the re-entry into cell cycle of parenchymal hepatocytes which are able to completely recover the original liver mass (Fausto, 2000). The liver anatomical and functional units reconstitution also requires non-parenchymal cells (endothelial cells, cholangiocytes, Kupffer cells, stellate cells). It is yet not clear if each cell histotype is involved in the proliferative process or if the regeneration requires the activity of a cell with multiple differentiation potential. Recently, the bipotentiality of the hepatocytes, able to divide giving rise to both hepatocytes and cholangiocytes, has been suggested. Furthermore, when injury is severe or the hepatocytes can no longer proliferate a progenitor cell population, normally a quiescent compartment is activated. A population of small portal cells named oval cells was first identified in 1978 by Shinozuka and colleagues (Shinozuka et al., 1978). Now as "oval cells" is indicated a heterogeneous population of bipotent transient amplifying cells, originating from the Canal of Hering (Dabeva & Shafritz, 1993). These cells are normally quiescent but, after injury, rapidly and extensively proliferate and differentiate in hepatocytes and cholangiocytes (Yovchev et al., 2008).

The observation that oval cells are a mixed precursor population suggests their differentiation from liver stem cells (Theise et al., 1999). Since the hepatocytes are able to regenerate themself to compensate liver mass loss, the existence of a liver stem cell, able to drive regeneration in conditions of extreme toxicity affecting the same hepatocytes, has long been debated. Today, there is growing evidence that the liver stem cell exists and its isolation from the organ, its numerical expansion in vitro and its characterization are joint efforts in many laboratories around the world. The interest of the scientific community in the identification, isolation and manipulation of the hepatic stem cell also depends on the fact that the great hopes placed in the use of mature hepatocytes in cell transplantation protocols for the treatment of liver diseases have been disappointed. The basis of these unsatisfactory therapeutic approaches lie in the paradox, not yet resolved, of the inability of hepatocytes, which show *in vivo* a virtually unlimited proliferative potential, to grow *in vitro* to quantitatively and qualitatively amount suitable for cell transplantation in adults.

2. Hepatocyte and regeneration

Regeneration of the original liver mass after damage has been extensively studied in rodents after two-thirds partial hepatectomy (PH) (Bucher, 1963). Regeneration of the liver depends on both hyperplasia and hypertrophy of the hepatocytes, cells that in a normal adult liver exhibit a quiescent phenotype. Hypertrophy begins within hours after PH then hyperplasia follows (Taub, 2004). This occurs first in the periportal region of the liver lobule then spreads toward the pericentral region (Fausto & Campbell 2003).

The restoration of liver volume depends on three steps involving the hepatocytes: i) initiation, ii) proliferation and iii) termination phases.

The initiation step depends on the "priming" of parenchymal cells, mainly via the signaling pathways triggered by the cytokines IL-6 and TNF-α secreted by Kupffer cells, rendering the hepatocytes sensitive to growth factors and competent to replication.

After the G0/G1 transition in the initiation phase, the hepatocytes will enter into the cell cycle (Taub, 2004). Growth factors, primarily HGF, epidermal growth factor (EGF) and TGF-α, are responsible of this second step of regeneration in which the hepatocytes both proliferate and grow in cell size, activating the IL-6/STAT-3 and the PI3K/PDK1/Akt pathways respectively. The first signaling cascade regulates the cyclin D1/p21 and also protects against cell death, for example by up-regulating FLIP, Bcl2 and Bcl-xL. The latter pathway regulates cell size via mammalian target of rapamycin (mTOR) (Fausto, 2000; Serandour et al., 2005; Pahlavan et al., 2006; Fujiyoshi & Ozaki 2011). Numerous growth factors (for example HGF, TGF-α, EGF, glucagon, insulin and cytokines like TNF, IL-1 and -6 and somatostatin (SOM)) are implicated in the regeneration process.

The HGF is a potent growth factor mainly acting on hepatocytes in a paracrine manner binding to its specific trans-membrane receptor tyrosine kinase c-met. HGF is secreted as an inactive precursor and stored in the extracellular matrix (ECM), then activated by the fibrinolytic system (Kim et al., 1997). Plasmin and metalloproteinases (MMPs) degrade the ECM and release pro-HGF that, in turn, is cleaved into an activated form by the urokinase-type plasminogen activator (u-PA)(Kim et al., 1997). The HGF/met signaling is transduced to its downstream mediators, i.e. the Ras-Raf-MEK, ERK1/2 (Borowiak et al., 2004), PI3K/PDK1/Akt (Okano et al., 2003) and mTOR/S6 kinase pathways, resulting in cell cycle progression.

TGF- α is another growth factor relevant in liver regeneration (Tomiya et al., 2000). It belongs to the EGF family, of which all members (EGF, heparin binding EGF-like factor and amphiregulin) transduce trough the common receptor EGF receptor (EGFR) and exert overlapping functions (Fausto 2004). This factor acts in autocrine and paracrine fashions and its production and secretion are induced by HGF.

IL-6 induces mitotic signals in hepatocytes through the activation of STAT-3 (Cressman et al., 1996). The IL-6/STAT-3 signaling involves several proteins: the IL-6 receptor, gp130, receptor-associated Janus kinase (Jak) and STAT-3. The IL-6 receptor is in a complex with gp130, which, after recognition by IL-6, transmits the signal. Jak is responsible of gp130 and STAT-3 activation after IL-6 binding. The STAT-3 form released by gp130 dimerizes and translocates to the nucleus to activate the transcription. STAT3 controls cell cycle progression from G1 to S phase regulating the expression of cyclin D1. In fact, in the liver-specific STAT3-KO model mice, mitotic activity of hepatocytes after PH is reduced significantly (Li et al., 2002).

The PIK/PDK1/Akt signaling pathways are activated by receptor tyrosine kinases or receptors coupled with G proteins by IL-6, TNF-α, HGF, EGF, TGF-α and others (Desmots et al., 2002) (Koniaris et al., 2003). An important downstream molecule of Akt for cell growth is mTOR (Fingar et al., 2002). The activation of this pathway coexists with STAT-3 signaling. In STAT-3-KO mice no significant differences were observed macroscopically in liver regeneration in comparison to control animals, reaching the liver of these mice after PH an equal size. This observation may be explained considering the increase in size of the hepatocytes. Increase in cell size corresponds to marked phosphorylation of Akt and its downstream molecules $p70^{S6K}$, mTOR and GSK3beta (Haga et al., 2005).

The third phase in liver regeneration is the termination step. A stop signal is necessary to avoid an inappropriate liver functional size but the molecular pathways involved in this phenomenon are not yet clear. A key role is exerted by the cytokine TGF-β, secreted by hepatocytes and platelets, that inhibits DNA synthesis (Nishikawa et al., 1998). In fact, within 2-6 hours after PH, the insulin growth factor (IGF) binding protein-1 (IGFBP-1) is produced to counteract its inhibitor effects (Ujike et al., 2000).

3. Liver progenitor cells and regeneration

When liver parenchyma damage is particularly serious and hepatocytes are no longer able to proliferate, liver regeneration can occur through the intervention of bipotent progenitor cells that can proliferate and differentiate into hepatocytes and bile duct cells. It was 1950 when Wilson and Leduc, studying the regeneration of rat liver after severe nutritional damage, observed for the first time these particular cells, located within or immediately adjacent to the Canal of Hering, and their differentiation into two histological types of liver epithelial cells (Wilson & Leduc, 1950). In 1956 Faber called these cells, which are found in the liver of mice treated with carcinogens (Farber 1956), "oval cells" for their morphology.

The first characterization of oval cells has shown the simultaneous expression of bile ducts (CK-7, CK-19 and OV-6) and hepatocytes (alpha-fetoprotein and albumin) markers (Lazaro et al., 1998). Subsequent studies have shown the activation, during oval cell compartment proliferation, of stem cell genes such as c-kit (Fujio et al., 1994), CD34 (Omori et al., 1997) and LIF (Omori et al., 1996) .

Stable lines of oval cells, useful for *in vitro* and *in vivo* studies of differentiation and of liver colonization, were obtained from normal rat liver F-334 (Hixson et al., 1990), or from rats fed with DL-ethionine (Sells et al., 1981) or treated with allyl alcohol (Yin et al., 1999). In addition, these precursors were stabilized starting from liver explants of animal models of Wilson disease (Yasui et al., 1997) of transgenic mice expressing Ras (Braun, et al., 1987) of p53 knockout mice fed with choline-free diet and finally of human liver (Dumble et al., 2002).

The oval cell is currently the best characterized liver progenitor cell although several studies have demonstrated the presence of precursors/stem cells either residing in the liver or coming from blood.

Regardless of the species in which were observed and the name that was given to them, the progenitor cells of the liver have common characteristics:

- they are very few and hardly recognizable in the healthy liver, but clearly evident as a result of chronic liver injury near the terminal trait of biliary duct;
- they express cholangiocyte and hepatocyte markers;
- they are basophilic, with a high ratio of nucleus/cytoplasm and are smaller than mature hepatocytes (10 µM in diameter compared to 50 of hepatocytes);
- they are immature and have a great proliferative capacity.

Further than oval cells, other bipotential precursor cells able to differentiate and colonize diseased liver in animal models have been isolated from rodent and human livers, allowing the study of molecular mechanisms triggering their differentiation. The identification and characterization of an immortalized bipotent precursor cell was firstly described by Spagnoli and coworkers (Spagnoli et al., 1998) in MMH cell lines. MMHs (Met Murine Hepatocyte) are immortalized cell lines derived from explants of embryonic, fetal and new-born livers derived from transgenic mice expressing a constitutively active truncated human Met receptor (cyto-Met) (Amicone et al., 1997). All of the MMH lines are not tumorigenic and show a differentiated phenotype judging from the retention of epithelial cell polarity and the expression of liver enriched transcriptional factors (LETF). In addition, many of them express hepatic functions. MMHs have been found to contain a cell subpopulation constituted by fibroblastoid cells, called "palmate cells" for their morphology, showing characteristics of a bipotent progenitor. The palmate cells are not polarized, do not express liver specific transcription factors or liver products, but retain the ability to divide and differentiate into hepatocytes and bile duct cells. Unequivocal demonstration that palmate cells can give rise to epithelial-hepatocytes is provided by cloning of individually fished cells and characterization of their progeny. Moreover, as true stem cells, palmate cells are diploid whereas their epithelial progeny is hypotetraploid. All of these findings demonstrate that palmate cells are the precursors of hepatocytes in MMH cell lines. These bipotential liver cells are also able to *in vivo* differentiate into hepatocytes and colonize diseased livers in mice (Spagnoli et al., 1998). Using the same methods of isolation and selection, Strick-Marchand and Weiss subsequently isolated, from mouse embryos wild-type, bipotent cells able to regenerate livers of mice uPA/SCID mice (Strick-Marchand & Weiss 2002). Bipotent progenitors were isolated and stabilized also from pig liver (Strick-Marchand et al., 2004), monkey (Talbot et al., 1994) and human fetal liver (Allain et al., 2002).

The identification of precursor cells has increasingly strengthened the idea that in the liver there are also real stem cells with a wide differentiation potential (capable of explaining many processes not yet fully understood such as liver development and regeneration) and which may give rise, by asymmetric division, to the same bipotent precursor cells.

The immunophenotypic characterization of the heterogeneous oval cell population, in which there are cells expressing hematopoietic stem cells (HSC) (eg, c-kit, CD34 and Thy-1) markers, had initially led to believe that oval cells could originate from the recruitment and differentiation of circulating HSC. In fact, many studies have demonstrated the ability of HSCs to differentiate into hepatocytes *in vitro* and their mobilization from the marrow and recruitment in the liver during regeneration. Two independent works (Wang et al., 2003; Vassilopoulos et al., 2003) however, have shown that stem cells derived from murine bone marrow and transplanted in FAH-/- mice, were involved in the regeneration of the damaged liver tissue through a process of cell fusion with endogenous hepatocytes rather than through a trans-differentiation process. The new hepatocytes in fact had both host and donor genetic markers. The events of trans-differentiation of HSC precursors into oval cells or hepatocytes documented to date are in fact extremely rare (Menthena et al., 2004; Grompe, 2003; Fausto, 2004; Thorgeirsson & Grisham, 2006). Mesenchymal-like cell population, depicting high level of proliferation and possessing a broad differentiation potential, has been isolated from adult human liver (Herrera et al., 2006; Najimi et al., 2007).

The efforts of different research groups is still directed towards the identification and isolation of a cell "resident" in the liver with stem cell characteristics, namely the ability to regenerate itself (self-renewal) and, more importantly, to divide asymmetrically, generating a cell identical to itself and a bipotent progenitor.

Reid and colleagues focused on human hepatic stem cells and highlighted as liver is comprised of different maturational lineages of cells both intrahepatically in periportal zone by the portal triads and extrahepatically in the hepato-pancreatic common duct (Turner et al., 2011). More in dectail, the intrahepatic stem cell niches have been located in the canals of Hering (for pediatric and adult livers) and in the ductal plates (for fetal and neonatal livers) (Schmelzer et al., 2007; Turner et al., 2011; Zhou et al., 2007). The extrahepatic niche was recently unveiled by the Reid's research group that demonstrated the presence of multipotent stem/progenitors in human peribiliary glands, deep within the duct walls, of the extrahepatic biliary trees (Cardinale et al., 2011 and 2012). These cells, which self-replicate, are positive for transcriptional factors typical of endoderm and surface markers typical of stem/progenitors and may express genes of liver, bile duct and pancreatic genes.

Conigliaro and colleagues recently reported the identification, the isolation from fetal and neonatal murine livers, the characterization and the reproducible establishment in line of a non-tumorigenic "liver resident stem cell" (RLSC), that proved to be a useful tool to study liver stem cell biology (Conigliaro et al., 2008). The immunophenotype of this cell (CD34- and CD45-) indicates a not hematopoietic origin and the transcriptional profile highlights the expression of a broad spectrum of 'plasticity-related genes' and 'developmental genes', indicating a multi-differentiation potential. Indeed, RLSCs not only differentiate spontaneously into hepatocytes and cholangiocytes (suggesting their partial endodermal determination), but can be induced *in vitro* to differentiate into osteocytes, chondrocytes and cells of neuroectodermal derivation (astrocytes, neurons). The ability of RLSCs to differentiate spontaneously in hepatocytes, the lack of albumin and the wide differentiation potential place these liver stem cells at the pre-hepatoblast/liver precursor hierarchical position. Notably, RLSCs are also a model to *in vitro* study liver zonation. This term indicates the typical distribution into hepatic lobule of several functions. Most of the main metabolisms of the liver, in fact, are not uniformly distributed over the hepatic lobule but follow gradients of enzymatic activities along the centrolobular/portal axis. Coherently, adult hepatocytes undergo into a post-differentiation patterning resulting into a zonal heterogeneity of gene expression and functions defined "metabolic zonation". Specific enzymatic/metabolic activities, i.e. carbohydrate metabolism, ammonia detoxification, bile formation/transport/secretion and drug biotransformation, are confined to the perivenular (PV, i.e. near the centrolobular vein) or periportal (PP, i.e. near the portal vein) zones of the hepatic lobule (Gebhardt, 1992). The elucidation of the mechanisms responsible for induction and maintenance of the hepatocyte heterogeneity remains one of challenge in experimental hepatology. Intriguingly, inversion of the blood flow direction changes the enzymatic gradients and, consequently, the zonation of some, but not all, the liver metabolisms, thus revealing the influence exerted by the oxygen and circulating molecules on this phenomenon (Kinugasa & Thurman, 1986). For the bloodstream independent gradients, cell-cell and cell-extracellular matrix interactions and paracrine signaling have been suggested as instructive stimuli (Gebhardt & Reichen, 1994). Recently, concerning soluble factors, a key role of the Wnt/β-catenin pathway has been unveiled. Within the hepatic lobuli, Wnt signaling has been proposed to originate from endothelial cells of the central vein and follows a stable gradient that decrease toward the PV–PP axis. In the liver, Benhamouche and collaborators observed a mutually exclusive localization of activated β-catenin and its negative regulator APC in the PV and in PP hepatocytes, respectively. Moreover, these authors demonstrated that genetic manipulation of APC expression and adenoviral delivery of the extracellular antagonist of Wnts DKK allowed to switch the phenotype from PP into PV and vice versa (Benhamouche et al., 2006).

A second key element in controlling hepatic zonation was identified in the transcriptional factor HNF4α: Stanulovic and colleagues have recently shown that this orphan nuclear receptor regulates the zonal expression of some genes, including Cyp7, UDP-glucuronyltransferase and apolipoprotein E (Stanulovic et al., 2007). Their analysis of HNF4α knock-out mice revealed in PV hepatocytes a maintenance of PV genes expression and in PP hepatocytes the inhibition of a PP gene (PEPCK) coupled to the activation of PV genes. These observations led to the conclusion that HNF4α exerts a dual role of activator of PP genes and inhibitor of PV genes in PP hepatocytes. In frame with these observations Colletti and colleagues showed as RLSCs spontaneously differentiate into periportal hepatocytes that, following Wnt pathway activation, switch into perivenular hepatocytes. Moreover, they gathered evidences showing a direct convergence of the canonical Wnt signaling pathway and HNF4α in controlling the hepatocyte heterogeneity. HNF4α and Wnt signaling pathway have been proposed as active members of the same machinery that controls the transcription of differentially zonated HNF4–dependent genes (Colletti et al., 2009).

In conclusion we can say that there are no more doubts about the existence of liver stem cells residing in the liver although there is still much to do especially with regard to the identification and characterization of specific microenvironments able to define the corresponding tissue stem- niche.

4. Molecular mechanisms controlling liver stem cell fate

A stem cell "niche" is believed to maintain the liver progenitor cells in a native state and allows their activation when required. It is conceived as a restricted area in an adult organ that regulates, by means of micro-environmental signaling, stem cell maintenance and differentiation. Stem cell behavior, in particular the balance between self-renewal and differentiation, is ultimately controlled by the integration of autocrine and paracrine factors supplied by the surrounding microenvironment. Stem cells respond to these instructive signals from the niche by changing their expression profile in a reversible manner. In particular, instructive signals received from the niche influence the so-called stem cell "metastable" phenotype. The metastability, currently considered a common characteristic of embryonic and adult stem cells and a manifestation of cell plasticity (McConnell &Kaznowski, 1991; Hay, 1995; Thomson et al., 1998; Blau et al., 2001; Burdon et al., 2002; Reddy et al., 2002; Prindull & Zipori, 2004), consists essentially in the cell capability to change the expression profile in a reversible manner and it is characterized by the co-expression of both epithelial and mesenchymal traits. This highly dynamic cell state may be considered as a balance between epithelial-mesenchymal and mesenchymal-epithelial transitions (EMT/MET). Both the EMT and the reverse process MET are typical events of development, tissue repair and tumor progression. The EMT is the process by which polarized cells, closely attached to each other, gradually lose epithelial features and acquire mesenchymal characteristics, including invasiveness and motility (Thiery et al., 2009). MET refers to the reverse phenomenon often occurring in a secondary site, by which the epithelia-derived mesenchymal cells reacquire their epithelial phenotype.

The observation that a number of stem cells are restricted to a specific differentiation fate suggests that elements pivotal for their metastability and for the coordinated execution of opposite processes, such as self-renewal and differentiation, may be tissue specific. A simple and direct molecular mini-circuitry of master elements of mutually exclusive biological processes, able also to reciprocally influence their own expression, may provide the best device to trigger such complex phenomena.

The availability of a stable stem cell line executing specific differentiation programs discloses a unique possibility to investigate mechanisms regulating alternative cellular choices.

Recently, RLSCs and hepatocytes derived from their differentiation (RLSCdH) permitted to identify a simple cross-regulatory circuitry between HNF4α (master regulator of hepatocyte differentiation and MET inducer) and Snail (master regulator of the EMT), whose expression is mutually exclusive due to their direct reciprocal transcriptional repression (Cicchini et al., 2006; Santangelo et al., 2011). In particular, Cicchini and co-workers showed that Snail represses the HNF4α transcription through the direct binding to its promoter (Cicchini et al., 2006) and that Snail over-expression is sufficient i) to induce EMT in hepatocytes with change of morphology, down-regulation of several epithelial adhesion molecules, reduction of proliferation and induction of matrix metalloproteinase 2 expression and, ii) most relevantly, to directly repress the transcription of the HNF4α gene. These findings demonstrated that Snail is at the crossroads of the regulation of EMT in hepatocytes by a dual control of epithelial morphogenesis and differentiation. More recently, Santangelo and colleagues collected evidence that HNF4α has a direct master role in the MET process of the hepatocyte and that its differentiation role is intrinsically linked to an active repression of mesenchymal program expression (Santangelo et al., 2011). Their data highlight as both, key EMT regulators (Snail and Slug) and mesenchymal genes, have to be included among the target genes relevant for HNF4α1 master function in controlling epithelial phenotype. Their main finding was to ascribe to HNF4α1 a general "anti-EMT" role through the orchestrated repression of both master EMT regulators and mesenchymal markers. HNF4α-mediated repression of mesenchymal gene program, moreover, is executed not only in the dynamic EMT/MET processes but also in the stable maintenance of the hepatocyte epithelial phenotype. In fact, they found that: in dedifferentiated hepatomas HNF4α1 ectopic expression was sufficient to down-regulate Snail, Slug, HMGA2, Vimentin and Fibronectin genes. In addition, in differentiated hepatocytes, HNF4α1 was found stably recruited to the promoters of EMT inducers and its knockdown caused the upregulation of these genes.

Consistent with these observations Garibaldi and colleagues (Garibaldi et al., 2011)Garibaldi et al., in press demonstrated that the same molecular players in an epistatic mini-circuitry are pivotal for the RLSC maintenance. In particular they observed that hepatic stem cells constitutively express Snail and that their spontaneous differentiation into hepatocytes is underlined by negative regulation of Snail expression. Snail silencing causes down-regulation of stemness markers and its ectopic expression in hepatocytes is sufficient to restore their expression. In RLSC Snail stably represses HNF4 and miR-200a-b-c and miR-34a, known as stemness inhibiting microRNAs and distinctive of epithelial cells. This latter activity is probably due to a direct mechanism as suggested by the binding of endogenous Snail to miR-200c and 34a promoters in RLSC. In terms of conceptual advances, these data allow to extend the role of Snail from EMT inducer to stemness stabilizer.

In the light of the previously demonstrated reciprocal repression between Snail and HNF4α these observations have been extended: Garibaldi and colleagues described that HNF4α is required for miR-200a-b-c, and miR-34a expression in hepatocytes and that HNF4α silencing in hepatocytes and its targeting in KO mouse models correlates with a strong down-regulation of their expression. This is probably due to a direct mechanism as suggested by the fact that endogenous HNF4α was found recruited on miR-200a-b, miR-200c and miR-34a promoters in both differentiated hepatocytes and mouse liver. Notably, in HNF4 KO mouse models miRs down-regulation correlates to a strong up-regulation of the stemness markers SCA1 and FOXA1. Thus HNF4α, first identified as a positive regulator of hepatocyte differentiation and recently located at the crossroad of other cellular functional categories (i.e. cell cycle, apoptosis, stress response) appears to participate also in the active repression of stemness.

The proposed mechanism implies that the execution of a stemness program requires the active repression of a differentiation program while the maintenance of the hepatocyte one requires the active repression of stemness traits. These observations, focusing on epithelial differentiation, are centered on a HNF4α/Snail/epithelial-miRs circuitry, however may be conceivable that other differentiation pathways could be regulated by similar mechanisms. In this light Snail can probably be considered as a general factor counteracting (and counteracted by) tissue-specific regulators. This is further suggested by studies indicating that Snail family members repress the expression of tissue-specific inducers as the pro-neural genes sim and rho (Xu et al., 2010) and the skeletal muscle master regulator MyoD (Kosman et al., 1991).

5. Hepatocyte transplantation in cell-based therapeutic

Animal models in which transplanted cells show a selective advantage over resident hepatocytes have been used to study transplantation, proliferation and reconstitution potential of the hepatocytes. Liver animal models belong to three groups (Palmes & Spiegel 2004): i) hepatotoxin-induced models; ii) surgical models; iii) animal models of hereditary liver defects.

Normal adult hepatocytes can be serially transplanted and single hepatocyte can be clonally amplified, showing stem-like properties, and serially passaged to repopulate almost 70% of the liver of (Fah)-deficient mice (Overturf et al., 1999). Excellent results have been obtained by using transgenic *Rag2*$^{-/-}$/*Il2rg*$^{-/-}$ mice (deficient for the recombinant activation gene-2 and the common γ-chain of the interleukin receptor) (Traggiai et al., 2004) or the *Alb-uPA(tg(+/-)* mice (expressing the uroplasminogen activator (uPA) under the transcriptional control of the albumin promoter) (Sandgren et al., 1991)) or mice obtained by the crossing of the above reported genotypes (Haridass et al., 2009; Azuma et al., 2007).

Hepatocyte transplantation protocols in humans have been proposed as an alternative to orthotropic liver transplantation in patients and used for some metabolic disorders i.e. familial hypercholesterolemia, glycogen storage disease type 1a, urea cycle defects and congenital deficiency of coagulation factors (Quaglia et al., 2008). Currently, the liver transplantation is the treatment of choice for acute and chronic end-stage liver failure and for diseases refractory to other treatments; but the limited availability of donor organs is the major limiting factor in this therapeutic procedure. Although different techniques of implants using either complete liver, liver reduced or hyper-reduced "split liver" (liver for two) have tried to overcome the shortage of organs, liver transplantation remain an unsufficient approach to satisfy the needs of patients with liver disease.

In recent years, hepatocyte transplantation has emerged as a potential alternative or complementary procedure to liver transplantation, at least in certain circumstances. The application of this therapeutic modality is based on the concept that cell transplantation would replace the function of the affected organ, either temporarily, allowing the recovery of the organ functionality or the availability of a liver for the transplant, or permanently, preventing need for this last procedure.

The development of this therapeutic approach could provide a new opportunity for patients with liver disease, particularly for children suffering from some metabolic diseases, with certain advantages over liver transplantation. In fact it is a less invasive and risky procedure and it has a lower cost. There is also a greater availability of material to be transplanted and that could be used as a source of cells (organs considered "marginal", material resulting from organ reductions, from partial hepatectomy and cadaveric livers unsuitable for transplantation) and the possibility of using a donor to several recipients.

Despite these advantages, a number of critical issues are still unresolved: the rejection of transplanted hepatocytes, their correct localization and functionality and, mostly, cells availability at the right time. The latter remains a problem that would be definitively solved with the cultivation and the preservation of large scale culture of hepatocytes. Nevertheless, these cells in culture, contrary to what happens *in vivo* during liver regeneration, have a very low proliferative potential and quickly lose their differentiated characteristics.

This implies that cell therapy can be carried out only with freshly isolated cells, not expanded *in vitro*. The number of cells that can be achieved with this approach is usually not sufficient to colonize adult livers, while there is more chance of success in pediatric patients with metabolic diseases of genetic origin since they can be treated with a limited number of hepatocytes.

6. Conclusion

Liver stem cells may represent an important tool for the treatment of the liver diseases. They could be an alternative source of functional hepatocytes aimed at cell transplantation, tissue engineering and bio-artificial liver. Manipulation of stem cells will be more efficient since we know the factors controlling their biology. Only by dissecting the molecular events underlying the stemness, the differentiation choice and the maintenance of the differentiated phenotype can we control stem cell behavior for therapeutic purposes. The translation of *in vitro* studies in *in vivo* experimental models and, finally, in humans is one of the major challenges of experimental hepatology. Moreover, better understanding the mechanisms that control the proliferation of stem and progenitor cells will shed new light on the molecular and cellular basis of liver cancer.

7. References

Allain, J. E., I. Dagher, et al. (2002). "Immortalization of a primate bipotent epithelial liver stem cell." Proceedings of the National Academy of Sciences of the United States of America 99(6): 3639-3644.

Amicone, L., F. M. Spagnoli, et al. (1997). "Transgenic expression in the liver of truncated Met blocks apoptosis and permits immortalization of hepatocytes." EMBO J 16(3): 495-503.

Azuma, H., N. Paulk, et al. (2007). "Robust expansion of human hepatocytes in Fah-/-/Rag2-/-/Il2rg-/- mice." Nature biotechnology 25(8): 903-910.

Benhamouche, S., Decaens, T., et al. (2006). "Apc tumor suppressor gene is the "zonation-keeper" of mouse liver". Dev Cell;10:759-70.

Blau, H. M., T. R. Brazelton, et al. (2001). "The evolving concept of a stem cell: entity or function?" Cell 105(7): 829-841.

Borowiak, M., A. N. Garratt, et al. (2004). "Met provides essential signals for liver regeneration." Proc Natl Acad Sci U S A 101(29): 10608-10613.

Braun, L., M. Goyette, et al. (1987). "Growth in culture and tumorigenicity after transfection with the ras oncogene of liver epithelial cells from carcinogen-treated rats." Cancer research 47(15): 4116-4124.

Bucher, N. L. (1963). "Regeneration of Mammalian Liver." International review of cytology 15: 245-300.

Burdon, T., A. Smith, et al. (2002). "Signalling, cell cycle and pluripotency in embryonic stem cells." Trends Cell Biol 12(9): 432-438.

Cardinale, V., Y. Wang, et al. (2011). "Multipotent stem/progenitor cells in human biliary tree give rise to hepatocytes, cholangiocytes, and pancreatic islets" Hepatology 54:2159-2172.

Cardinale, V., Y Wang, et al. (2012). "The biliary tree-a reservoir of multipotent stem cells".Nature reviews Gastroenterology&Hepatology, advance online publication 28 february.

Cicchini, C., D. Filippini, et al. (2006). "Snail controls differentiation of hepatocytes by repressing HNF4alpha expression." J Cell Physiol 209(1): 230-238.

Colletti, M., C. Cicchini, et al. (2009). "Convergence of Wnt signaling on the HNF4alpha-driven transcription in controlling liver zonation." Gastroenterology 137(2): 660-672.

Conigliaro, A., M. Colletti, et al. (2008). "Isolation and characterization of a murine resident liver stem cell." Cell death and differentiation 15(1): 123-133.

Cressman, D. E., L. E. Greenbaum, et al. (1996). "Liver failure and defective hepatocyte regeneration in interleukin-6-deficient mice." Science 274(5291): 1379-1383.

Dabeva, M. D. and D. A. Shafritz (1993). "Activation, proliferation, and differentiation of progenitor cells into hepatocytes in the D-galactosamine model of liver regeneration." Am J Pathol 143(6): 1606-1620.

Desmots, F., M. Rissel, et al. (2002). "Pro-inflammatory cytokines tumor necrosis factor alpha and interleukin-6 and survival factor epidermal growth factor positively regulate the murine GSTA4 enzyme in hepatocytes." J Biol Chem 277(20): 17892-17900.

Dumble, M. L., E. J. Croager, et al. (2002). "Generation and characterization of p53 null transformed hepatic progenitor cells: oval cells give rise to hepatocellular carcinoma." Carcinogenesis 23(3): 435-445.

Farber, E. (1956). "Similarities in the sequence of early histological changes induced in the liver of the rat by ethionine, 2-acetylamino-fluorene, and 3'-methyl-4-dimethylaminoazobenzene." Cancer Res 16(2): 142-148.

Fausto, N. (2000). "Liver regeneration." J Hepatol 32(1 Suppl): 19-31.

Fausto, N. (2004). "Liver regeneration and repair: hepatocytes, progenitor cells, and stem cells." Hepatology 39(6): 1477-1487.

Fausto, N. and J. S. Campbell (2003). "The role of hepatocytes and oval cells in liver regeneration and repopulation." Mech Dev 120(1): 117-130.

Fingar, D. C., S. Salama, et al. (2002). "Mammalian cell size is controlled by mTOR and its downstream targets S6K1 and 4EBP1/eIF4E." Genes Dev 16(12): 1472-1487.

Fujio, K., R. P. Evarts, et al. (1994). "Expression of stem cell factor and its receptor, c-kit, during liver regeneration from putative stem cells in adult rat." Laboratory investigation; a journal of technical methods and pathology 70(4): 511-516.

Fujiyoshi, M. and M. Ozaki (2011). "Molecular mechanisms of liver regeneration and protection for treatment of liver dysfunction and diseases." J Hepatobiliary Pancreat Sci 18(1): 13-22.

Garibaldi, F., Cicchini, C., et al. (in press). "An epistatic mini-circuitry between the transcription factors Snail and HNF4alpha controls liver stem cell and hepatocyte features exhorting opposite regulation on stemness-inhibiting microRNAs". Cell Death and Differentiation 2011 Dec 2. doi: 10.1038/cdd.2011.175. [Epub ahead of print]

Gebhardt, R. (1992). "Metabolic zonation of the liver: regulation and implications for liver function". Pharmacol Ther; 53:275-354.

Gebhardt, R.& Reichen J. (1994). " Changes in distribution and activity of glutamine synthetase in carbon tetrachloride-induced cirrhosis in the rat: potential role in hyperammonemia". Hepatology;20:684-91

Grompe, M. (2003). "The role of bone marrow stem cells in liver regeneration." Seminars in liver disease 23(4): 363-372.

Haga, S., W. Ogawa, et al. (2005). "Compensatory recovery of liver mass by Akt-mediated hepatocellular hypertrophy in liver-specific STAT3-deficient mice." J Hepatol 43(5): 799-807.

Hailfinger, S., M. Jaworski, et al. (2006). "Zonal gene expression in murine liver: lessons from tumors." Hepatology 43(3): 407-414.

Haridass, D., Q. Yuan, et al. (2009). "Repopulation efficiencies of adult hepatocytes, fetal liver progenitor cells, and embryonic stem cell-derived hepatic cells in albumin-promoter-enhancer urokinase-type plasminogen activator mice." The American journal of pathology 175(4): 1483-1492.

Hay, E. D. (1995). "An overview of epithelio-mesenchymal transformation." Acta Anat (Basel) 154(1): 8-20.

Herrera, M. B., S. Bruno, et al. (2006). "Isolation and characterization of a stem cell population from adult human liver." Stem Cells 24(12): 2840-2850.

Hixson, D. C., R. A. Faris, et al. (1990). "An antigenic portrait of the liver during carcinogenesis." Pathobiology : journal of immunopathology, molecular and cellular biology 58(2): 65-77.

Kim, T. H., W. M. Mars, et al. (1997). "Extracellular matrix remodeling at the early stages of liver regeneration in the rat." Hepatology 26(4): 896-904.

Kinugasa, A. & Thurman, R.G. (1986)."Differential effect of glucagon on gluconeogenesis in periportal and pericentral regions of the liver lobule". Biochem J;236:425-30.

Koniaris, L. G., I. H. McKillop, et al. (2003). "Liver regeneration." J Am Coll Surg 197(4): 634-659.

Kosman, D., et al. (1991) "Establishment of the mesoderm-neuroectoderm boundary in the Drosophila embryo". Science 254, 118-122.

Lazaro, C. A., J. A. Rhim, et al. (1998). "Generation of hepatocytes from oval cell precursors in culture." Cancer research 58(23): 5514-5522.

Li, W., X. Liang, et al. (2002). "STAT3 contributes to the mitogenic response of hepatocytes during liver regeneration." J Biol Chem 277(32): 28411-28417.

McConnell, S. K. and C. E. Kaznowski (1991). "Cell cycle dependence of laminar determination in developing neocortex." Science 254(5029): 282-285.

Menthena, A., N. Deb, et al. (2004). "Bone marrow progenitors are not the source of expanding oval cells in injured liver." Stem Cells 22(6): 1049-1061.

Najimi, M., D. N. Khuu, et al. (2007). "Adult-derived human liver mesenchymal-like cells as a potential progenitor reservoir of hepatocytes?" Cell transplantation 16(7): 717-728.

Nishikawa, Y., M. Wang, et al. (1998). "Changes in TGF-beta receptors of rat hepatocytes during primary culture and liver regeneration: increased expression of TGF-beta receptors associated with increased sensitivity to TGF-beta-mediated growth inhibition." J Cell Physiol 176(3): 612-623.

Okano, J., G. Shiota, et al. (2003). "Hepatocyte growth factor exerts a proliferative effect on oval cells through the PI3K/AKT signaling pathway." Biochem Biophys Res Commun 309(2): 298-304.

Omori, M., R. P. Evarts, et al. (1997). "Expression of alpha-fetoprotein and stem cell factor/c-kit system in bile duct ligated young rats." Hepatology 25(5): 1115-1122.

Omori, N., R. P. Evarts, et al. (1996). "Expression of leukemia inhibitory factor and its receptor during liver regeneration in the adult rat." Laboratory investigation; a journal of technical methods and pathology 75(1): 15-24.
Overturf, K., M. Al-Dhalimy, et al. (1999). "The repopulation potential of hepatocyte populations differing in size and prior mitotic expansion." The American journal of pathology 155(6): 2135-2143.
Pahlavan, P. S., R. E. Feldmann, Jr., et al. (2006). "Prometheus' challenge: molecular, cellular and systemic aspects of liver regeneration." J Surg Res 134(2): 238-251.
Palmes, D. and H. U. Spiegel (2004). "Animal models of liver regeneration." Biomaterials 25(9): 1601-1611.
Pelletier, L., S. Rebouissou, et al. (2011). "HNF1alpha inhibition triggers epithelial-mesenchymal transition in human liver cancer cell lines." BMC cancer 11(1): 427.
Prindull, G. and D. Zipori (2004). "Environmental guidance of normal and tumor cell plasticity: epithelial mesenchymal transitions as a paradigm." Blood 103(8): 2892-2899.
Quaglia, A., S. C. Lehec, et al. (2008). "Liver after hepatocyte transplantation for liver-based metabolic disorders in children." Cell transplantation 17(12): 1403-1414.
Reddy, G. P., C. I. McAuliffe, et al. (2002). "Cytokine receptor repertoire and cytokine responsiveness of Ho(dull)/Rh(dull) stem cells with differing potentials for G1/S phase progression." Exp Hematol 30(7): 792-800.
Sandgren, E. P., R. D. Palmiter, et al. (1991). "Complete hepatic regeneration after somatic deletion of an albumin-plasminogen activator transgene." Cell 66(2): 245-256.
Santangelo, L., A. Marchetti, et al. (2011). "The stable repression of mesenchymal program is required for hepatocyte identity: A novel role for hepatocyte nuclear factor 4alpha." Hepatology.
Sells, M. A., S. L. Katyal, et al. (1981). "Isolation of oval cells and transitional cells from the livers of rats fed the carcinogen DL-ethionine." Journal of the National Cancer Institute 66(2): 355-362.
Serandour, A. L., P. Loyer, et al. (2005). "TNFalpha-mediated extracellular matrix remodeling is required for multiple division cycles in rat hepatocytes." Hepatology 41(3): 478-486.
Shinozuka, H., B. Lombardi, et al. (1978). "Early histological and functional alterations of ethionine liver carcinogenesis in rats fed a choline-deficient diet." Cancer research 38(4): 1092-1098.
Schmelzer, E., L. Zhang, et al. (2007). "Human hepatic stem cells from fetal and postnatal donors". J Exp Med 204:1973-1987
Spagnoli, F. M., L. Amicone, et al. (1998). "Identification of a bipotential precursor cell in hepatic cell lines derived from transgenic mice expressing cyto-Met in the liver." J Cell Biol 143(4): 1101-1112.
Stanulovic ,V.S., Kyrmizi, I., et al. (2007) "Hepatic HNF4alpha deficiency induces periportal expression of glutamine synthetase and other pericentral enzymes". Hepatology;45:433-44
Strick-Marchand, H., S. Morosan, et al. (2004). "Bipotential mouse embryonic liver stem cell lines contribute to liver regeneration and differentiate as bile ducts and hepatocytes." Proceedings of the National Academy of Sciences of the United States of America 101(22): 8360-8365.

Strick-Marchand, H. and M. C. Weiss (2002). "Inducible differentiation and morphogenesis of bipotential liver cell lines from wild-type mouse embryos." Hepatology 36(4 Pt 1): 794-804.

Talbot, N. C., V. G. Pursel, et al. (1994). "Colony isolation and secondary culture of fetal porcine hepatocytes on STO feeder cells." In vitro cellular & developmental biology. Animal 30A(12): 851-858.

Taub, R. (2004). "Liver regeneration: from myth to mechanism." Nat Rev Mol Cell Biol 5(10): 836-847.

Theise, N. D., R. Saxena, et al. (1999). "The canals of Hering and hepatic stem cells in humans." Hepatology 30(6): 1425-1433.

Thiery, J. P., H. Acloque, et al. (2009). "Epithelial-mesenchymal transitions in development and disease." Cell 139(5): 871-890.

Thomson, J. A., J. Itskovitz-Eldor, et al. (1998). "Embryonic stem cell lines derived from human blastocysts." Science 282(5391): 1145-1147.

Thorgeirsson, S. S. and J. W. Grisham (2006). "Hematopoietic cells as hepatocyte stem cells: a critical review of the evidence." Hepatology 43(1): 2-8.

Tomiya, T., I. Ogata, et al. (2000). "The mitogenic activity of hepatocyte growth factor on rat hepatocytes is dependent upon endogenous transforming growth factor-alpha." Am J Pathol 157(5): 1693-1701.

Turner, R., O. Lozoya, et al. (2011). "Human hepatic stem cell and maturational liver lineage biology". Hepatology 53(3):1035-45.Ujike, K., T. Shinji, et al. (2000). "Kinetics of expression of connective tissue growth factor gene during liver regeneration after partial hepatectomy and D-galactosamine-induced liver injury in rats." Biochem Biophys Res Commun 277(2): 448-454.

Vassilopoulos, G., P. R. Wang, et al. (2003). "Transplanted bone marrow regenerates liver by cell fusion." Nature 422(6934): 901-904.

Wang, X., M. Foster, et al. (2003). "The origin and liver repopulating capacity of murine oval cells." Proceedings of the National Academy of Sciences of the United States of America 100 Suppl 1: 11881-11888.

Wilson J.W., Leduc E.H., (1950). "Abnormal mitosis in mouse liver". *Am J Anat*. Jan;8 6(1):51-73.

Xu, H. et al. (2010) "Liver-enriched transcription factors regulate microRNA-122 that targets CUTL1 during liver development". *Hepatology* 52, 1431-1442, doi:10.1002/hep.23818.

Yasui, O., N. Miura, et al. (1997). "Isolation of oval cells from Long-Evans Cinnamon rats and their transformation into hepatocytes in vivo in the rat liver." *Hepatology* 25(2): 329-334.

Yin, L., D. Lynch, et al. (1999). "Participation of different cell types in the restitutive response of the rat liver to periportal injury induced by allyl alcohol." *Journal of hepatology* 31(3): 497-507.

Yovchev, M. I., P. N. Grozdanov, et al. (2008). "Identification of adult hepatic progenitor cells capable of repopulating injured rat liver." *Hepatology* 47(2): 636-647.

Zhou, H., L.E. Rogler, et al. (2007). "Identification of hepatocytic and bile ductular cell lineages and candidate stem cells in bipolar ductular reactions in cirrhotic human liver" *Hepatology* 45: 716-724.

2

Liver Progenitor Cells, Cancer Stem Cells and Hepatocellular Carcinoma

Janina E.E. Tirnitz-Parker[1,2], George C.T. Yeoh[2] and John K. Olynyk[1,2]
[1]*Curtin University,*
[2]*Western Australian Institute for Medical Research, University of Western Australia*
Australia

1. Introduction

There is great interest in the biology of liver progenitor cells (LPCs) because of their stem cell-like ability to regenerate the liver when the hepatocyte pool is exhausted. Barely detectable in healthy tissue, they emerge upon chronic insult in periportal regions, proliferate and migrate to injury sites in the parenchyma and eventually differentiate into hepatocytes and cholangiocytes to restore liver mass, morphology and function. The increasing worldwide shortage of livers for orthotopic transplantation means LPCs have assumed more prominence as candidates for cell therapy as an alternative therapeutic approach for the treatment of various liver diseases. However, an LPC response is usually seen in pre-cancerous liver pathologies and their high proliferation potential makes them possible transformation targets; associations that overshadow their restorative capability. This mandates that we continue to investigate the factors that govern their activation, proliferation and especially their differentiation into mature, functional cells to effectively direct transplanted cells towards regeneration and not tumorigenicity.

2. Normal liver tissue turnover

Tissue regeneration and maintenance in healthy intestine and skin is achieved within days and weeks respectively. In contrast healthy liver has a very slow cell turnover rate and the vast majority of hepatocytes is considered to be in the quiescent, non-proliferative G_0 phase of the cell cycle. It has been estimated that at any one time only 1 in 20,000 to 40,000 hepatocytes is undergoing mitotic cell division with an average life span of 200 to 300 days (Bucher & Malt, 1971).

The mechanisms by which hepatic cells are replaced in healthy liver are controversial. An early model, the "streaming liver" hypothesis is based on the metabolic zonation and differential gene expression patterns of periportal compared to pericentral hepatocytes. Periportal cells were proposed to proliferate and migrate ("stream") towards the central area with maturation during the journey and terminal differentiation achieved when the cells reached the central zone (Zajicek *et al.*, 1985; Arber *et al.*, 1988; Sigal *et al.*, 1992). However there is no convincing evidence for a periportal to pericentral differentiation gradient and while hepatocytes in opposing lobular areas are responsible for different metabolic functions, cells in either location are considered to be fully differentiated. By reversing the blood flow in the liver, Thurman and Kauffman demonstrated that this lobular zonation is not dependent on hepatocyte lineage progression but rather due to metabolite-induced gene regulation (Thurman & Kauffman, 1985). Retroviral marking studies provided additional evidence against the "streaming liver" model since transplanted cells, traceable by β-galactosidase expression, remained in the original location for 15 months (Bralet *et al.*, 1994). Furthermore, experiments performed with mosaic livers of chimeric rats (Ng & Iannaccone, 1992) as well as approaches using transgenic hAAT/β-gal mice (Kennedy *et al.*, 1995) demonstrated that hepatocytes proliferate clonally during normal tissue renewal throughout the whole liver lobule. Collectively, these findings led to the conclusion that normal liver cell plates lack the existence of a main proliferative compartment and instead randomly distributed hepatocytes mediate normal liver turnover by slow clonal expansion without involvement of a liver stem cell (Ponder, 1996).

3. Liver regeneration

The liver has an enormous capacity to regenerate by (1) replication of remaining, healthy hepatocytes, (2) activation, expansion and differentiation of a stem cell compartment, or (3) by a combination of these processes. Which pathway is employed depends on the nature of the injury, its severity and duration. This is discussed in greater detail in the sections to follow.

3.1 Hepatocyte-mediated regeneration

The hepatic regenerative capacity is most clearly seen after surgical removal of liver mass. This model, referred to as partial hepatectomy (PHx), was introduced by Higgins and Anderson (Higgins & Anderson, 1931) and it is unquestionably the best studied liver regeneration model due to its simplicity of design and reproducibility. In the rat two-thirds PHx is performed, whereas in the mouse usually only the left lobe is removed due to technical difficulties in the performance of two-thirds PHx surgery in mice, with resultant high mortality (Fausto *et al.*, 2006). The removed lobes do not re-grow. Instead there is compensatory, hyperplastic growth of all residual cellular populations until the size of the organ achieves proportionality to the body size, as determined by metabolic demands of the organism (Kawasaki *et al.*, 1992; Starzl *et al.*, 1993). The different liver cell types do not divide simultaneously but show different kinetics in DNA synthesis. Periportal hepatocytes, with a presumably shorter G_1 phase than pericentral cells (Rabes, 1976), are the first to undergo a wave of mitosis but DNA synthesis progresses to eventually involve the whole lobule with the exception of a few glutamine synthetase-positive, pericentral cells (Gebhardt, 1988). The proliferating hepatocytes are thought to provide mitogenic stimuli for the other hepatic cell populations. Biliary ductular cells, Kupffer and hepatic stellate cells (HSCs) and finally sinusoidal endothelial cells enter DNA synthesis about 24 hours later (Michalopoulos & DeFrances, 1997) with synchronised proliferation of each cell type for at least the first wave of replication. The greatest increase in liver mass can be seen by 72 hours with complete mass restoration after about one week (Grisham, 1962).

Although it was known from early experiments that repeated PHx does not exhaust hepatocyte growth (Simpson & Finck, 1963), the enormous proliferative capacity of adult hepatocytes has previously been underestimated. Rhim *et al.* showed that newborn uPA overexpressing mice with continuous hepatocytic necrosis could be rescued by transplantation of a small number of hepatocytes that required between 10 to 15 rounds of replication to generate sufficient liver mass (Rhim *et al.*, 1994; Rhim *et al.*, 1995). In addition, serial transplantation experiments performed in tyrosinemic mice caused by a deficiency for fumarylacetoacetate hydrolase (FAH) revealed that hepatocytes are capable of undergoing more than 70 cell doublings without loss of functionality (Overturf *et al.*, 1997). Conversely there is also recent evidence that hepatocytes might reach a state of "replicative senescence" under certain chronic conditions such as advanced cirrhosis, perhaps due to telomere shortening (Paradis *et al.*, 2001; Wiemann *et al.*, 2002).

3.2 Liver progenitor cell-mediated regeneration

Repeated replication of healthy hepatocytes is the most efficient way to restore liver mass and function during normal tissue renewal and repair. If this process is inhibited or blocked during chronic chemical or carcinogenic hepatocyte insult, the liver relies on stem cell-like LPCs for its restoration. These cells are also referred to as "oval cells" in rodents (Fausto & Campbell, 2003) and the "Ductular Reaction" in humans due to their rather ductular phenotype in most human chronic liver diseases (Roskams & Desmet, 1998; Theise *et al.*, 1999).

3.2.1 History, origin and features of liver progenitor cells

The appearance of oval-like cells in the livers of rats treated with the azo dye "Butter Yellow" was originally reported in 1937 (Kinosita, 1937). Two decades later, Farber introduced the term "oval cell" for this population after observing small ovoid cells with a scant basophilic cytoplasm and a high nuclear to cytoplasmic ratio following treatment of rats with carcinogenic agents (Farber, 1956a, 1956b). Shortly after, Wilson and Leduc documented the proliferation of ductular cells that gave rise to hepatocytes and possibly new interlobular bile ducts in mice fed a methionine-rich, bentonite-supplemented diet and they were the first to suggest the existence of a bipotential liver progenitor or stem cell (Wilson & Leduc, 1958). Many experimental models involving toxins and carcinogens, alone or in combination with other surgical or dietary regimes, have since been developed and these facilitated the study of these progenitor cells, which are now widely accepted to represent adult LPCs; the progeny of hepatic stem-like cells.

The precise origin of LPCs remains uncertain, even though many researchers have addressed this question. The lack of definite evidence regarding the cellular source of LPCs may reflect differences in the models used to induce them and has also been hampered by a lack of specific LPC markers. Lenzi *et al.* suggested bile ducts as the structure of origin and argued that LPCs express biliary markers such as cytokeratin (CK) 7 and CK19 and lack expression of the mesenchymal cell markers vimentin and desmin. Additionally, the degree of LPC proliferation during early ethionine-induced carcinogenesis was found to be proportional to the increase in biliary tree volume and the authors claimed that LPCs are simply part of spatially expanded cholangioles (Lenzi *et al.*, 1992).

Other investigators have proposed an extrahepatic origin for LPCs. After it became apparent that some LPCs share c-kit, CD34 and Thy-1 expression with haematopoietic stem cells

(Fujio *et al.*, 1994; Omori *et al.*, 1997; Petersen *et al.*, 1998a), Petersen *et al.* were the first to suggest that LPCs could be derived from epithelial precursors in the bone marrow (Petersen *et al.*, 1999). Bone marrow-derived cells that potentially contribute to liver regeneration would enter via the portal vasculature and locate adjacent to the ducts in the periportal region, which is why Sell extended the preceding proposition by suggesting the periductular LPC as the candidate cell for an extrahepatic, bone marrow-derived stem cell in the liver (Sell, 2001). To test the hypothesis that cells from the bone marrow contribute to the formation of LPCs and hepatocytes, several investigators performed cell transplantation studies. They generally followed the fate of male bone marrow cells or purified haematopoietic stem cells transplanted into lethally irradiated female recipients that were in most cases subjected to liver injury. It was demonstrated that very minor fractions of LPCs or hepatocytes were donor-derived in both healthy and diseased livers (Petersen *et al.*, 1999; Theise *et al.*, 2000a, 2000b; Krause *et al.*, 2001; Wang *et al.*, 2002). The responsible population in the bone marrow capable of repopulating the liver was thought to be of c-kit^{high}Thy^{low}Lin^{neg}Sca-1^{pos} phenotype (Lagasse *et al.*, 2000). Soon after, the bone marrow was found to contain another stem cell subpopulation, the multipotent adult progenitor cell (MAPC), which can be induced to express hepatocyte phenotype and functions *in vitro* (Schwartz *et al.*, 2002) and is capable of differentiating into hepatocyte-like cells when transplanted into the liver (Jiang *et al.*, 2002). When donor-derived hepatocytes were examined genotypically, it was noted that they contained both donor and host genetic markers, indicating cell fusion as the likely mechanism by which hepatocytes are generated from bone marrow and not by transdifferentiation of haematopoietic stem cells (Vassilopoulos *et al.*, 2003; Wang *et al.*, 2003b). On the other hand, haematopoietic stem cells co-cultured with injured liver tissue separated by a trans-well membrane were shown to convert to a hepatocyte phenotype without fusion due to humoral factors released from the liver tissue. When engrafted into injured liver the haematopoietic stem cells differentiated into functional hepatocytes and their plasticity was proposed to facilitate the conversion, rather than the rare cell fusion event that was only seen at later stages of the experiment (Jang *et al.*, 2004). Recently it was demonstrated by transplantation of lacZ-transgenic bone marrow into virally or steatotically challenged mice that the contribution of extrahepatic cells to LPC-generated hepatocytes is minimal (Tonkin *et al.*, 2008). Collectively, these experiments show that some bone marrow cells are capable of producing hepatocytes (with or without fusion, depending on the model and cell population used) to restore injured liver. However, it occurs at a low frequency and efficiency unless a strong selective pressure is applied (Thorgeirsson & Grisham, 2006). It is likely that the more significant role of bone marrow cells is to generate non-parenchymal cells during liver regeneration (Forbes *et al.*, 2004). The usual regeneration processes after acute and chronic liver injuries appear to rely predominantly on intrahepatic cells.

The most widely accepted view is that LPCs originate from liver-resident precursor or stem cells, which lie dormant and present in such low numbers as to be undetectable in normal liver. However, they can be activated to proliferate under certain pathological conditions (Fig. 1). Evidence from experiments showing that LPCs always emerge from periportal liver zones and the fact that selective periportal damage inhibits the LPC response (Petersen *et al.*, 1998b) have led to the conclusion that the precursor cell likely resides somewhere in the vicinity of the portal triad. Grisham and Porta found ductular proliferation in carcinogen-treated rats that they attributed to activated stem-like cells from the Canals of Hering, the anatomical boundary between terminal bile ducts and the most distal hepatocytes of the hepatic plate (Grisham and Porta, 1964). Microscopic studies of early histological changes in rats following 2-acetylaminofluorene (2-AAF)/partial hepatectomy (PHx) treatment also show elongated ductular branches that are formed by proliferating LPCs, which originate from a stem cell compartment located in these canalicular-ductular junctions. The newly formed biliary structures represent cellular extensions of the Canals of Hering and remain connected to the terminal biliary ductules by a continuous basement membrane (Paku *et al.*, 2001). Reid and colleagues suggested epithelial cell adhesion molecule (EpCAM) as a suitable marker for isolation and study of these Canals of Hering-derived LPCs (Schmelzer *et al.*, 2007). Lineage tracing of Sry (sex determining region Y)-box 9 (Sox9)-expressing cells supports the hypothesis that LPCs derive from the epithelial lining of bile ducts (Furuyama *et al.*, 2011). Theise *et al.* conducted studies comparing normal with acetaminophen-induced necrotic liver and identified the human equivalent to the rodent Canals of Hering, a niche which is similarly thought to harbour stem-like cells that give rise to LPCs or the Ductular Reaction (Theise *et al.*, 1999).

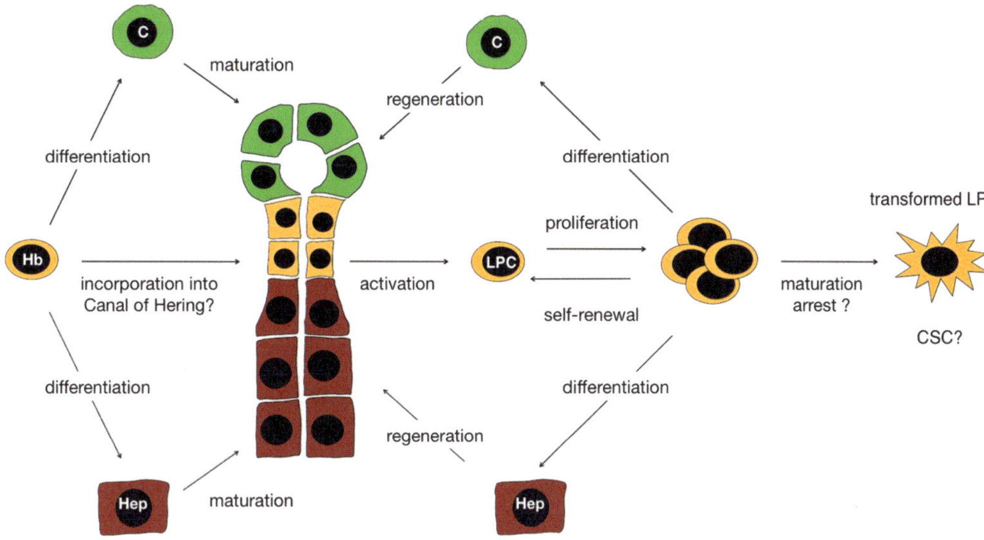

Fig. 1. LPC ontogeny. During liver development hepatoblasts (Hb) differentiate into cholangiocytes (C) and hepatocytes (Hep) and might be incorporated into the Canals of Hering to serve as a stem cell compartment during chronic liver injury. Activated liver progenitor cells (LPC) proliferate after appropriate stimuli, are capable of self-renewal and later commit towards either the cholangiocytic or hepatocytic lineage to regenerate the liver. If kept in a proliferative state, LPCs are likely candidates for transformation and might represent cancer stem cells (CSCs).

LPCs are a heterogeneous cell population and immature as well as intermediate phenotypes are observed before cells that express a differentiated phenotype are identified. Importantly, from activation to differentiation or transformation, they continuously change their morphology, phenotype and accordingly marker expression. LPCs express different combinations of phenotypic markers from both the hepatocytic and biliary lineage (Fig. 2) and also share epitopes with haematopoietic cells and cancer stem cells (CSCs; see table 1).

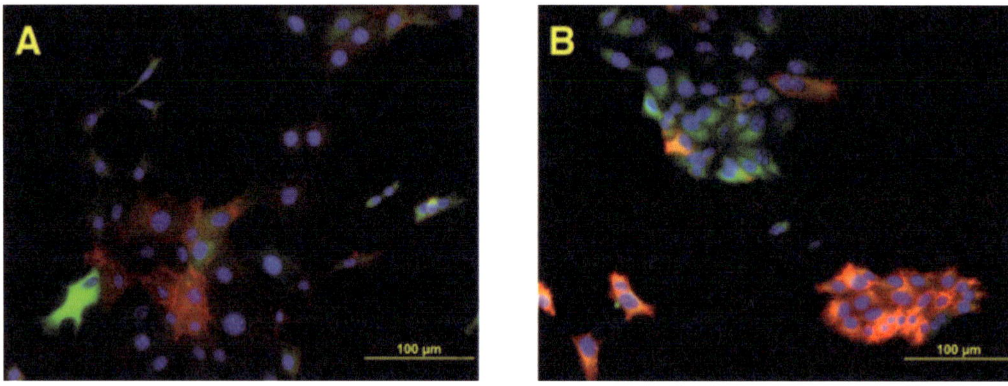

Fig. 2. Bipotentiality of LPCs. Immunofluorescent characterisation of the clonally established LPC line BMOL (Tirnitz-Parker *et al.*, 2007) demonstrates the cells' bipotentiality. Immature BMOL cells co-express the hepatocytic markers muscle 2-pyruvate kinase (A, green) and transferrin (B, green) with the biliary markers A6 (A, red) and CK19 (B, red).

Marker	Hepatocytes	Cholangiocytes	LPCs	CSCs	References
A6	-	+	+	?	Engelhardt *et al.*, 1993, Tirnitz-Parker *et al.*, 2007
AFP	fetal	-	+	(+)	Sell, 1978; Evarts *et al.*, 1987; Smith *et al.*, 1996; Ishii *et al.*, 2010
Alb	+	-	+	(+)	Sell, 1978; Tian *et al.*, 1997; Yamashita *et al.*, 2010
CD24	?	(+)	+	+	Lee *et al.*, 2011; Qiu *et al.*, 2011
CD34	-	+	+	(+)	Omori *et al.*, 1997; Petersen *et al.*, 2003
CD133	?	?	+	+	Suetsugu *et al.*, 2006; Ma *et al.*, 2007; You *et al.*, 2010; Colombo *et al.*, 2011
CK7	-	+	+	?	Golding *et al.*, 1995; Clouston *et al.*, 2005
CK8	+	(+)	+	?	Sarraf *et al.*, 1994; Golding *et al.*, 1995; Sasaki *et al.*, 2008
CK14	-	(+)	(+)	(+)	Bisgaard *et al.*, 1993; Rogler, 1997; Zhang *et al.*, 2010
CK18	+	(+)	+	(+)	Golding *et al.*, 1995; Zhang *et al.*, 2010
CK19	-	+	+	?	Sarraf *et al.*, 1994; Golding *et al.*, 1995; Colombo *et al.*, 2011
c-kit	-	(+)	+	(+)	Fujio *et al.*, 1994; Fujio *et al.*, 1996; Knight *et al.*, 2008
Cx32	+	-	(+)	(+)	Zhang & Thorgeirsson, 1994; Paku *et al.*, 2004; Kawasaki *et al.*, 2011
Cx43	-	+	+	?	Zhang & Thorgeirsson, 1994; Paku *et al.*, 2004
Dlk	-	-	+	(+)	Jensen *et al.*, 2004; Yanai *et al.*, 2010
E-cad	low	high	high	?	Tirnitz-Parker *et al.*, 2007; Ueberham *et al.*, 2007; Van Hul *et al.*, 2009
EpCAM	(+)	fetal	+	+	Schmelzer *et al.*, 2007; Yamashita *et al.*, 2009; Okabe *et al.*, 2009; Yoon *et al.*, 2011
GGT IV	-	+	+	?	Petersen *et al.*, 1998a; Holic *et al.*, 2000
M_2PK	fetal	+	+	?	Tian *et al.*, 1997; Lowes *et al.*, 1999; Tirnitz-Parker *et al.*, 2007
OV-6	-	+	+	?	Dunsford *et al.*, 1989; Zhang *et al.*, 2010; Cao *et al.*, 2011
π-GST	fetal	-	+	?	Tee *et al.*, 1992; Lowes *et al.*, 1999; Oliva *et al.*, 2010
Sca-1	-	-	+	+	Petersen *et al.*, 2003; Qiu *et al.*, 2011
Thy-1	-	-	+	+	Petersen *et al.*, 1998a; Colombo *et al.*, 2011

Table 1. Marker expression by adult liver cells. A6, murine marker, epitope unknown; AFP, α-fetoprotein; Alb, albumin; CD, cluster of differentiation; CK, cytokeratin; c-kit, CD117, stem cell factor receptor; Cx, connexin; Dlk, delta-like protein; E-cad, E-cadherin; EpCAM, epithelial cell adhesion molecule; GGT IV, γ-glutamyl transpeptidase IV; M_2PK, muscle 2-pyruvate kinase ; OV-6, rat and human marker, epitope shared by CK14 and 19; π-GST, pi-glutatione-S-transferase; Sca-1, stem cell antigen 1; Thy-1, thymocyte differentiation antigen 1.

They have been shown to differentiate at least bipotentially into hepatocytes and cholangiocytes (Tirnitz-Parker *et al.*, 2007), and in some models display multipotentiality, also producing intestinal and pancreatic lineages (Tatematsu *et al.*, 1985; Yang *et al.*, 2002; Leite *et al.*, 2007). Hence it is not surprising that there is still not a single LPC-specific marker available and a combination of phenotypic markers is required for their identification or isolation.

LPCs infiltrate the parenchyma in close spatial and temporal association with hepatic stellate cells (HSCs). Following activation, HSCs differentiate from quiescent, vitamin A-rich cells into α-smooth muscle actin-positive myofibroblastic cells, which are capable of matrix degradation to generate space for cell migration as well as fibrogenesis and collagen deposition to provide chronically injured liver with architectural support. The activation, proliferation, migration and differentiation status of LPCs and HSCs, as well as their beneficial as opposed to pathological contributions, are controlled by key cytokines. LPCs and HSCs have been reported to influence each other's behaviour through paracrine signalling. LPCs produce a range of cytokines, including lymphotoxin β (LTβ), which signals via the LTβ receptor on HSCs to activate the NFkB pathway. This results in production of intercellular adhesion molecule 1 (ICAM-1) and regulated upon activation, normal T-cell expressed and secreted (RANTES), which then act as chemotactic agents for LPCs and inflammatory cells involved in the wound healing response to chronic liver injury (Ruddell *et al.*, 2009). Several other factors mediating the LPC response have been identified, including tumour necrosis factor (TNF), TNF-like weak inducer of apoptosis (TWEAK), interferon gamma (IFNγ), and transforming growth factor beta (TGFβ) among others (Knight *et al.*, 2000; Akhurst *et al.*, 2005; Knight *et al.*, 2005; Knight & Yeoh, 2005; Knight *et al.*, 2007; Tirnitz-Parker *et al.*, 2010). Abrogation of these key signalling pathways inhibits the LPC response to injury and prevents or diminishes liver fibrosis in animal models (Davies *et al.*, 2006; Lim *et al.*, 2006; Knight *et al.*, 2008). In the setting of impaired wound healing combined with chronic inflammation, the regenerative fibrotic response turns into pathological fibrogenesis, which can progress to cirrhosis and eventually hepatocellular carcinoma (HCC).

3.2.2 Rodent liver progenitor cell induction models

The majority of commonly used LPC induction models was originally developed to study the process of hepatocarcinogenesis. They generally combine an injuring mitotic stimulus, usually in the form of functional liver mass loss (chemically or otherwise-induced), with a manipulation that chronically damages hepatocytes or blocks their ability to divide and prevents them from contributing to the liver regeneration process. Described below are four examples of the most commonly used regimens.

3.2.2.1 D-galactosamine

This model is mainly used to induce liver injury in the rat. Administration of D-galactosamine inhibits RNA and protein synthesis in centrilobular hepatocytes by trapping and depleting uridine-nucleotides and UDP-glucose (Decker & Keppler, 1972), leading to acute necrosis. Hepatocyte replication is not fully blocked in this model; the response is only delayed. LPCs are resistant to the chemical as they do not metabolise D-galactosamine and are induced to proliferate within 48 hours after injury. They migrate into the parenchyma, where they generate both ductular cells and small hepatocytes (Lemire *et al.*, 1991; Dabeva & Shafritz, 1993).

3.2.2.2 Solt-Farber model and the modified 2-AAF/PHx regime

In this model, which is commonly used in rats and only rarely in mice, injection of the ethylating hepatocarcinogen diethylnitrosamine (DEN) is followed two weeks later by a two-week treatment with 2-AAF and PHx one week into 2-AAF feeding (Solt & Farber, 1976). The most commonly used regimen is a modification to the original Solt-Farber protocol, in which the "initiation" step of DEN injection is omitted and 2-AAF is administered four days before and after PHx, the 2-AAF/PHx regime (Tatematsu *et al.*, 1984). Both models induce proliferation of ductular or periductular LPCs, which accelerates when 2-AAF feeding is terminated, indicating that not only hepatocytes are growth-inhibited by 2-AAF but also LPCs, although to a lesser extent. LPCs differentiate more efficiently into hepatocytes at low doses of 2-AAF, whereas they tend to undergo apoptosis at higher dosages (Alison *et al.*, 1997). As a consequence, the rate at which LPCs differentiate into hepatocytes can easily be controlled through variation of the 2-AAF dose (Paku *et al.*, 2004).

3.2.2.3 Choline-deficient, ethionine supplemented diet (CDE diet)

A dietary deficiency of the lipotrope choline is known to induce hepatic steatosis (Lombardi *et al.*, 1966; Lombardi *et al.*, 1968). This pathology reflects an impaired release of triglycerides in the form of very low-density lipoprotein (VLDL) from hepatocytes, leading to intracytoplasmic deposition of fat vacuoles within a few hours of choline withdrawal. Choline-deficiency has also been reported to induce hepatocarcinogenesis (Ghoshal & Farber, 1984; Yokoyama *et al.*, 1985; Locker *et al.*, 1986). Similar effects were shown for another well-known carcinogen, DL-ethionine. Administered alone, ethionine is an antagonist of methionine and as such an inhibitor of *de novo* choline-biosynthesis thus induces fatty liver (Farber, 1967) and also leads to HCC (Farber, 1956a). When tested in combination with choline-deficiency, ethionine enhances the formation of liver tumours (Shinozuka *et al.*, 1978b), yet surprisingly diminishes the formation of fatty liver during choline-deficiency (Sidransky & Verney, 1969).

An interesting observation during early choline-deficient, ethionine-supplemented (CDE) diet-induced hepatocarcinogenesis studies in rats was the massive proliferation of α-fetoprotein-positive LPCs in the liver (Shinozuka *et al.*, 1978a). Numerous studies using this model to provoke an LPC response in rats were subsequently described. Due to the extensive availability of genetically engineered mouse strains, it became desirable to apply this regimen to mice. The conventional CDE diet used in rats however caused high mortality in mice and was therefore modified to a CD diet with separate administration of 0.165% DL-ethionine in the drinking water. This customised CDE diet (Akhurst *et al.*, 2001) reliably induces the proliferation of LPCs (Fig. 3, Tirnitz-Parker *et al.*, 2007; Tirnitz-Parker *et al.*, 2010) as well as inflammatory cells (Knight *et al.*, 2005) and serves as a murine model of hepatic fibrogenesis (Ruddell *et al.*, 2009; Van Hul *et al.*, 2009) and tumorigenesis following prolonged CDE diet exposure (Knight *et al.*, 2000; Knight *et al.*, 2008).

3.2.2.4 3,5-diethoxycarbonyl-1,4-dihydro-collidine diet (DDC diet)

The hepatotoxin 3,5-diethoxycarbonyl-1,4-dihydro-collidine is also an effective inducer of LPCs as it causes extensive and prolonged liver damage while the diet is administered (Jakubowski *et al.*, 2005). However, in contrast to the CDE diet (see above), a fraction of hepatocytes continue to proliferate for the duration of diet administration (Wang *et al.*, 2003a). Thus the model is unusual in that liver regeneration is accomplished by both hepatocytes and LPCs. It offers an alternate model to investigate mechanisms that regulate LPC proliferation and differentiation. In the context of liver cancer, the DDC model has been used extensively to demonstrate a link between LPCs and HCC. LPCs isolated from p53 null mice subjected to a DDC diet are able to generate both hepatocarcinomas and cholangiocarcinomas following transplantation into immunodeficient mice (Suzuki *et al.*, 2008). By placing a Hepatitis B Virus X transgenic mouse on a DDC diet, Wang and colleagues were able to show that LPCs overexpressing HBx were tumorigenic (Wang *et al.*, 2012). Interestingly, over the same period of seven months, DDC treatment did not induce tumours in wild type mice. In another study, the importance of the Hippo-Salvador pathway, working through inhibition of the yes-associated protein YAP, was shown by subjecting mice with liver-specific ablation of WW45 (drosophila homolog of Salvador and adaptor for the Hippo kinase) to a DDC diet. These mice displayed liver tissue overgrowth, an enhanced LPC response and they developed liver tumours with HCC as well as cholangiocarcinoma characteristics that appeared to be LPC-derived (Lee *et al.*, 2010).

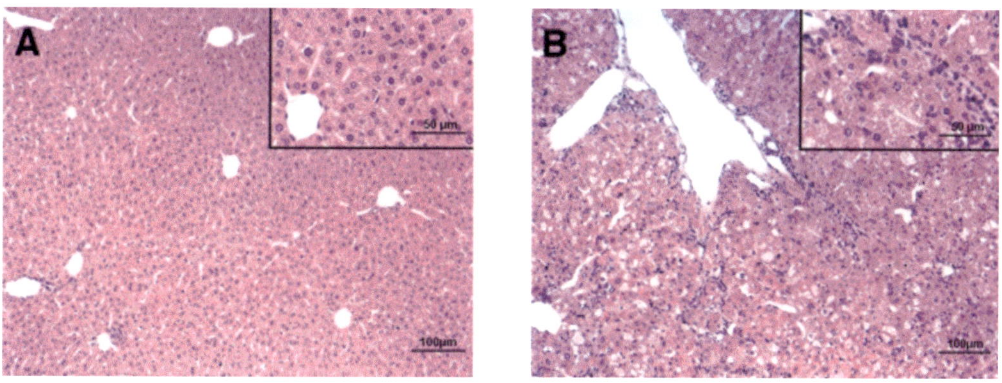

Fig. 3. Histology of normal and chronically injured liver. Adult mice on a control diet display normal liver architecture with orderly cords of hepatocytes and sinusoidal structures in-between the plates (A). On day 21 of the CDE diet, the liver architecture is highly disrupted by steatosis, scattered aggregates of infiltrated inflammatory cells and proliferating LPCs (B).

3.2.3 Liver progenitor cells in human pathologies

LPCs have been identified in a variety of human liver pathologies and are activated like their rodent counterparts to regenerate chronically injured liver (Haque *et al.*, 1996; Theise *et al.*, 1999). Like oval cells in rodents, human LPCs are usually associated with prolonged fibrosis, hepatocellular necrosis, cirrhosis and chronic inflammatory liver diseases. Hence, their proliferation is frequently seen in patients with hereditary haemochromatosis, chronic hepatitis B or C infection, alcoholic liver disease (ALD) and non-alcoholic fatty liver disease (NAFLD) when hepatocytes are inhibited by DNA-damaging oxidative stress (Lowes *et al.*, 1999; Roskams *et al.*, 2003a; Clouston *et al.*, 2005). The degree of stem cell activation and the number of proliferating LPCs in these pathologies was demonstrated to correlate with the progression and severity of the underlying liver disease (Lowes *et al.*, 1999). The activation of human LPCs is characterised by the appearance of reactive ductules, also referred to as Ductular Reaction. Cirrhotic livers have been shown to contain nodules that are usually in close contact with reactive ductules and consist entirely of intermediate hepatocytes, which strongly suggests they originate from LPCs (Roskams *et al.*, 2003a; Roskams *et al.*, 2003b, Falkowski *et al.*, 2003). LPCs always emerge in pathologies with a predisposition to cancer and their proliferation in an environment rich in inflammatory mediators, growth factors or reactive oxygen species renders them likely targets for transformation. Furthermore, inhibition of the LPC response has been demonstrated to reduce the formation of cancerous lesions, strongly supporting a role for LPCs in hepatocarcinogenesis (Davies *et al.*, 2006; Knight *et al.*, 2005; Knight *et al.*, 2008). Very recently LPCs have not only been discussed as cellular precursors for liver cancer but also as potential liver cancer stem cells, which could be responsible for tumour maintenance and recurrence (Marquardt *et al.*, 2011; Rountree *et al.*, 2012).

4. Cancer stem cells

The similarities between adult stem cells and CSC have led to confusion regarding their identity and it has often not been clear in the literature whether CSCs represent transformed progenitor or stem cells or whether both cell types are distinct cell populations that only share the expression of certain cell markers and display a similar biology. Adult tissue stem cells and CSCs are both defined by (i) highly efficient self-renewing ability through asymmetrical cell division and (ii) differentiation capacity along at least two if not more cell lineages. CSCs manifest the additional property of tumour initiation and/or maintenance. Nowadays the consensus is that the term CSC simply describes a cell's potential for self-renewal and ability to give rise to the hierarchic organisation of the heterogeneous lineages of cancer cells that constitute the tumour and does not consider the cell's origin. CSCs may arise from the differentiation arrest and transformation of a normal adult stem cell through oncogenic and/or epigenetic aberrations or the dedifferentiation of a mature cell that subsequently acquires self-renewing capacity. The CSC concept has been debated for many decades and compelling evidence of their existence has only emerged in the past decade.

4.1 Haematopoietic cancer stem cells

The existence of CSCs was first demonstrated in the haematological malignancy acute myeloid leukaemia (AML). Dick and colleagues isolated human AML cells from peripheral blood and transplanted them into non-obese diabetic/severe combined immunodeficient (NOD/SCID) mice. The vast majority of cells was unable to induce leukaemia, even when transplanted in larger numbers, despite the fact that they displayed a leukaemic blast phenotype such as the $CD34^+/CD38^+$ subpopulation. Only 0.01-1% of all AML cells, the $CD34^+CD38^-$ fraction, initiated AML and gave rise to a heterogeneous leukaemia tumour cell mass, classifying them as CSCs. The $CD34^+CD38^-$ cells could be serially transplanted and reliably developed AML with the same morphology and cell surface marker expression as the original tumour (Bonnet & Dick, 1997). Additional tumour-initiating AML cell populations were later identified and described to be of a $CD34^+CD19^-$ or $CD34^+CD10^-$ and $CD34^+CD4^-$ or $CD34^+CD7^-$ phenotype (Cox *et al.*, 2004; Cox *et al.*, 2007).

4.2 Solid tumour cancer stem cells

Using similar approaches involving cell transplantation into immunodeficient mice, CSCs have subsequently been identified in a variety of solid tumours, including breast, brain and liver cancer.

4.2.1 Breast cancer stem cells

Metastatic breast cancer was the first solid tumour in which CSCs were identified and prospectively isolated. The $CD44^{+}CD24^{-/low}Lineage^{-}$ cell population initiated tumours upon transplantation into mice with as few as 100 cells per injection. Importantly, they could be serially passaged and reliably reproduced the heterogeneous phenotype of the original breast cancer. In contrast, unsorted cells from the primary tumour or injection of a large number of alternate phenotypes, such as $CD44^{+}CD24^{+}$ cells, failed to form tumours (Al-Hajj *et al.*, 2003). Furthermore, it was established that increased expression of the detoxifying enzyme aldehyde dehydrogenase (ALDH) identifies the tumorigenic breast stem cell fraction and high ALDH1 activity correlates with poorer prognosis (Ginestier *et al.*, 2007).

4.2.2 Central nervous system cancer stem cells

The discovery of breast CSC was reported in the same year as the identification of tumour-initiating stem cells in the brain. Singh and colleagues identified and prospectively isolated a $CD133^{+}$ population of cells from a range of human brain tumours including medulloblastomas, pilocytic astrocytoma, glioblastoma and anaplastic ependymoma that *in vitro* exhibited stem cell properties and gave rise to heterogeneous cell populations with the same phenotype as the original tumour cells. Upon transplantation of as few as 100 $CD133^{+}$ glioma cells into the frontal lobes of NOD/SCID mice, serially transplantable tumours were initiated that mirrored the original tumour phenotype, whereas no tumours developed after injection of a much larger number of $CD133^{-}$ cells from the same tumour (Singh *et al.*, 2003; Singh *et al.*, 2004).

4.2.3 Liver cancer stem cells

Only very recently have liver CSCs been described. However the mounting evidence is compelling and ever more markers are suggested to describe the population of cells that may be responsible for liver cancer initiation, maintenance and potentially tumour recurrence after HCC resection, as described below.

4.2.3.1 Side population (Hoechst 33342 dye efflux)

The first evidence for the existence of liver CSCs came from Haraguchi and colleagues who performed Hoechst 33342 side population (SP) analyses of various human gastrointestinal cell lines and identified a subpopulation of cells with CSC properties. The SP approach is based on the finding that cells without stem cell characteristics accumulate the fluorescent nucleic acid-binding dye Hoechst 33342, whereas stem cells and CSCs do not as they are capable of effectively effluxing the dye through high activity of adenosine triphosphate (ATP)-binding cassette (ABC) transporters such as the multidrug resistance transporter 1 (MDR1) or breast cancer resistance protein (BCRP, also known as ABCG2). These ABC transporters employ ATP hydrolysis to facilitate substrate export across membranes against steep concentration gradients and thereby protect cells from cytotoxic agents and importantly from chemotherapeutic drugs such as cisplatin and doxorubicin. The authors report that the HCC lines HuH7 and Hep3B contained 0.9% to 1.8% SP cells with CSC properties, respectively, whereas no SP cells could be purified from the less aggressive hepatoma cell line HepG2 (Haraguchi *et al.*, 2006). These results were confirmed shortly after by Chiba and colleagues who identified SP cells in some human liver cell lines, which successfully induced xenograft tumours in NOD/SCID mice upon transplantation of as few as 1000 SP cells, while attempts to produce tumours with 1 x 10^6 non-SP cells failed consistently (Chiba *et al.*, 2006).

4.2.3.2 CD133 (Prominin 1)

Several recent studies have used this glycoprotein initially identified as a marker for CD34+ haematopoietic stem cells and later as a marker of LPCs for the isolation of liver CSCs. Suetsugu *et al.* reported that both the hepatoblastoma cell line HepG2 as well as the human fetal hepatoblast cell line Hc lacked CD133 expression, and that CD133+ cells could only be demonstrated in the human HCC line Huh7. CD133+ cells showed a higher proliferative potential in culture but also a greater ability to initiate tumour growth *in vivo* compared to the CD133- population (Suetsugu *et al.*, 2006). Hepatic cells with a CD133 phenotype have been shown to be more resistant to chemotherapeutic drugs such as doxorubicin and 5-fluorouracil than their CD133- counterparts through preferential activation of the Akt/protein kinase B and Bcl-2 cell survival pathways. Furthermore, resistance of normal stem cells to cyclophosphamide is facilitated by the differentially expressed marker ALDH. Studies on ALDH and CD133+ cells found ALDH expression only in the CD133+ subpopulation and suggested a hierarchical cell organisation with regard to tumorigenicity in the order CD133+ALDH+ > CD133+ALDH- > CD133-ALDH-, which suggests ALDH as an additional marker useful for liver CSC identification (Ma *et al.*, 2007, Ma *et al.*, 2008). In addition, it has been demonstrated that TGFβ signalling can induce CD133 expression in the HCC cell line Huh7 through epigenetic regulation, which results in a significant increase in tumour initiation capacity in these cells compared to CD133- Huh7 cells (You *et al.*, 2010).

4.2.3.3 Epithelial cell adhesion molecule (EpCAM)

Myajima and colleagues identified EpCAM as a biliary and LPC marker, which is expressed in biliary epithelial cells and becomes upregulated in liver upon 2-AAF/PHx and DDC treatment (Okabe *et al.*, 2009). Since EpCAM expression has been reported in many normal epithelial as well as in tumour cells, it is not surprising that it has been suggested as a useful CSC marker. EpCAM+ cells isolated from human HCC tissues were shown to be more tumorigenic and invasive than EpCAM- cells and consistently formed invasive tumours in NOD/SCID mice, even after serial transplantation, whereas the EpCAM- population did not (Yamashita *et al.*, 2009). EpCAM is a direct transcriptional target of Wnt/β-catenin signalling, which has been implicated as a CSC self-renewal pathway (Yamashita *et al.*, 2007). Activation of the Wnt/β-catenin pathway increased the EpCAM+ cell population, whereas knockdown of EpCAM resulted in decreased proliferation, colony formation, migration and drug resistance (Yamashita *et al.*, 2009).

5. Hepatocellular carcinoma

Mortality from chronic liver disease is the most rapidly increasing cause of death in many western nations. The commonest aetiologies contributing to this escalation are chronic viral hepatitis C or B infection, alcoholic and non-alcoholic fatty liver disease. All these conditions can cause fibrosis and, subsequently, cirrhosis and HCC. Much evidence has been gathered demonstrating that HCC can arise from deregulated LPC proliferation and maturation during chronic liver injury in humans and in animal models of liver disease and carcinogenesis.

5.1 HCC: A clinically important end-stage complication of chronic liver disease

End-stage complications of chronic liver disease (cirrhosis and HCC) are the 9th commonest global cause of death and will remain so for at least the next 20 years. Of great concern is the prediction by the World Health Organisation that by 2030, deaths from HCC will for the first time exceed those from non-malignant complications of cirrhosis, such as liver failure and portal hypertension (Mathers *et al.*, 2006).

Most cases of HCC in the western world arise in the setting of established cirrhosis (Bruix & Sherman, 2005; Olsen *et al.*, 2010; Sherman, 2011). The median survival of untreated HCC is in the order of 6-16 months. In view of the poor survival in the absence of therapy, strategies have been implemented to reduce the incidence of HCC through immunisation to prevent chronic HBV infection and screening of high-risk groups (i.e. those with cirrhosis). Despite these approaches, we are still faced with an escalation in the number of cases and requirement for treatment (El-Serag & Mason, 1999; Bruix & Sherman, 2005; Llovet *et al.*, 2005; Mathers & Loncar, 2006; Llovet *et al.*, 2008).

Presently the treatment of choice for HCC is liver resection or orthotopic liver transplantation (OLT), either with or without adjunctive chemotherapy or non-surgical ablative therapy. Liver resection is the treatment of choice for HCC in non-cirrhotic livers, and accounts for 5% of HCC cases in western countries and 40% of cases in non-western countries. Patients with well compensated cirrhosis and who do not have portal hypertension may also be considered for resection, provided that lesions are confined to the liver and enough "functional reserve" of liver is retained to ensure survival of the patient (Bruix & Sherman, 2005; Llovet *et al.*, 2005). Currently, tumour size, number and vascular invasion are still the strongest predictors of survival with up to 70% of subjects surviving five years. Tumour recurrence complicates 70% of cases at five years, reflecting either intrahepatic metastases (true recurrences) or the development of *de novo* tumours (Llovet *et al.*, 2005). Based on comparative genomic hybridisation, DNA fingerprinting using loss of heterozygosity assays, or DNA microarray studies, it is estimated that just over half of recurrences correspond to intrahepatic metastases undetected by the time of resection, whereas less than half are *de novo* HCCs (Chen *et al.*, 2000; Finkelstein *et al.*, 2003; Ng *et al.*, 2003). OLT is indicated in individuals who fulfil the "Milan criteria": patients with a single HCC of up to five centimetres in size or up to three nodules not larger than three centimetres each. Strict adherence to these criteria results in 5-year survival of up to 70% with recurrence rates usually less than 15%(Bismuth *et al.*, 1993; Mazzaferro *et al.*, 1996; Llovet *et al.*, 2005; Mazzaferro *et al.*, 2009).

5.2 Pathogenesis of HCC: The emerging role of LPCs and CSCs

The activation and proliferation of LPCs during chronic liver injury is associated with an inflammatory response that involves activation of resident and recruited inflammatory cells (Fig. 4). These inflammatory cells initiate tissue regeneration by promoting the removal of cellular debris and by stimulating LPCs to proliferate through release of mitogenic growth factors and cytokines (Lowes et al., 2003; Knight et al., 2005). Whilst LPCs play an important role in normal liver repair processes, dysregulation of their proliferation and differentiation has been linked to fibrogenesis and carcinogenesis (Lowes *et al.*, 1999; Clouston *et al.*, 2005; Knight *et al.*, 2008; Ruddell *et al.*, 2009; Tirnitz-Parker *et al.*, 2010). Clear demonstration of a role for LPCs, and possibly CSCs, in HCC development was reported by Shachaf and colleagues (Shachaf *et al.*, 2004). Inactivation of the Myc oncogene was sufficient to induce sustained regression of invasive HCC in a murine model. Tumour cells differentiated into hepatocytes and biliary epithelial cells. This process was associated with rapid loss of expression of the tumour marker α-fetoprotein, increase in expression of liver cell markers CK8 and carcinoembryonic antigen, and in some cells the biliary LPC marker CK19. Many of the "reverted" tumour cells remained dormant as long as Myc remained inactivated; however, Myc reactivation immediately restored their neoplastic features. Using array comparative genomic hybridisation, Shachaf and coworkers confirmed that the dormant liver cells and the restored tumour retained the identical molecular signature and hence were clonally derived from the tumour cells. Thus, tumours have pluripotent capacity to differentiate into normal cellular lineages and tissue structures, while retaining their latent potential to become cancerous

Several other studies have confirmed a LPC phenotype in a substantial number of HCCs. Detailed immunophenotyping revealed that 28–50% of HCCs express markers of LPCs, such as CK7 and CK19. Histologically, these tumours consist of cells that have an intermediate phenotype between LPCs and mature hepatocytes. Furthermore, HCCs that express both hepatocyte and biliary cell markers such as albumin, CK7 and CK19, carry a significantly poorer prognosis and higher recurrence after surgical resection and liver transplantation (Roskams, 2006; Yao & Mishra, 2009). The "precursor-product" relationship between LPCs, CSCs and HCC is further strengthened by the observation that 55% of small dysplastic foci, which represent the earliest premalignant lesions, are comprised of LPCs and intermediate hepatocytes (Weinstein *et al.*, 2001). Finally, inhibition of the LPC response to liver injury using a broad range of pharmacological therapies such as interferon alpha 2b (Lim *et al.*, 2006), COX-II inhibitors (Davies *et al.*, 2006), or tyrosine kinase inhibitors (Knight *et al.*, 2008) is associated with a reduction in the severity of hepatic fibrosis and incidence of HCC. These observations provide more evidence in support of a critical role for LPCs and CSCs in the carcinogenic process. Collectively these studies suggest that anti-inflammatory agents may be useful therapeutically in reducing the incidence of liver cancer among patients with chronic liver pathologies.

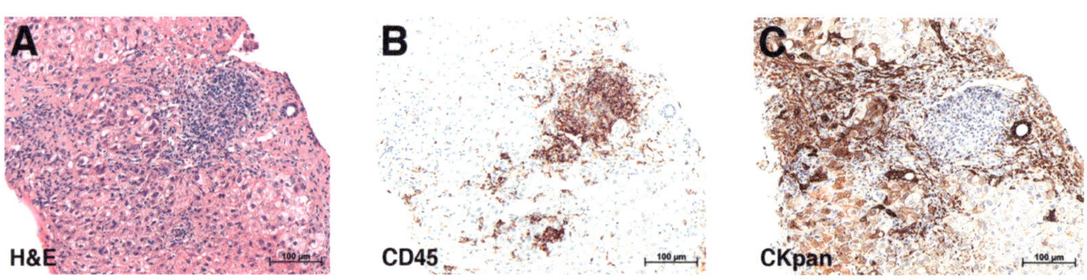

Fig. 4. Co-regulation of inflammatory response and LPC proliferation in hepatitis C patients. Haematoxylin and eosin (H&E) staining of a liver section from a hepatitis C virus-infected patient demonstrates disrupted liver architecture through infiltration and proliferation of small basophilic cells as well as steatotic changes in hepatocytes (A). Staining for the common leukocyte marker CD45 (B) and the biliary LPC marker CKpan (C) suggests co-regulation of the inflammatory response with the Ductular Reaction.

6. Conclusion

The cellular target of transformation leading to HCC is currently undefined. Potential candidates include the hepatocyte and the LPC and they need not be mutually exclusive. However, there is substantial circumstantial as well as some direct evidence implicating LPCs. This view would also be compatible with the increasingly popular theory of the stem cell basis of cancer. In the context of HCC, a variety of animal models, which induce chronic liver injury ultimately produce liver cancers and most of these pathologies display increased proliferation of LPCs. To conform to current views on carcinogenesis i.e. it is a rare event that affects a few cells and there are multiple stages in the process, it is necessary to hypothesise that a minority of LPCs are tumorigenic and that these have incurred the early genetic alterations that have initiated their progression to cancer. The challenge for future strategies to treat liver cancer is to identify these initiated LPCs and to show their direct link to HCC. This should be followed up with studies to elucidate progressive changes at the molecular level, which govern their behaviour and to exploit their vulnerability. Such knowledge will facilitate better diagnosis as well as treatment and prevention of HCC.

7. Acknowledgement

The authors would like to thank Ian Dickson for technical assistance with preparation of the figures.

8. References

Akhurst, B., Croager, E. J., Farley-Roche, C. A., Ong, J. K., Dumble, M. L., Knight, B. & Yeoh, G. C. (2001). A modified choline-deficient, ethionine-supplemented diet protocol effectively induces oval cells in mouse liver. *Hepatology,* 34, 519-522.

Akhurst, B., Matthews, V., Husk, K., Smyth, M. J., Abraham, L. J. & Yeoh, G. C. (2005). Differential lymphotoxin-beta and interferon gamma signaling during mouse liver regeneration induced by chronic and acute injury. *Hepatology,* 41, 327-335.

Al-Hajj, M., Wicha, M. S., Benito-Hernandez, A., Morrison, S. J. & Clarke, M. F. (2003). Prospective identification of tumorigenic breast cancer cells. *Proceedings of the National Academy of Sciences U S A,* 100, 3983-3988.

Alison, M., Golding, M., Lalani, E. N., Nagy, P., Thorgeirsson, S. & Sarraf, C. (1997). Wholesale hepatocytic differentiation in the rat from ductular oval cells, the progeny of biliary stem cells. *Journal of Hepatology,* 26, 343-352.

Arber, N., Zajicek, G. & Ariel, I. (1988). The streaming liver. II. Hepatocyte life history. *Liver,* 8, 80-87.

Bisgaard, H. C., Parmelee, D. C., Dunsford, H. A., Sechi, S. & Thorgeirsson, S. S. (1993). Keratin 14 protein in cultured nonparenchymal rat hepatic epithelial cells: characterization of keratin 14 and keratin 19 as antigens for the commonly used mouse monoclonal antibody OV-6. *Molecular Carcinogenesis,* 7, 60-66.

Bismuth, H., Chiche, L., Adam, R., Castaing, D., Diamond, T. & Dennison, A. (1993). Liver resection versus transplantation for hepatocellular carcinoma in cirrhotic patients. *Annals of Surgery,* 218, 145-151.

Bonnet, D. & Dick, J. E. (1997). Human acute myeloid leukemia is organized as a hierarchy that originates from a primitive hematopoietic cell. *Nature Medicine,* 3, 730-737.

Bralet, M. P., Branchereau, S., Brechot, C. & Ferry, N. (1994). Cell lineage study in the liver using retroviral mediated gene transfer. Evidence against the streaming of hepatocytes in normal liver. *American Journal of Pathology,* 144, 896-905.

Bruix, J. & Sherman, M. (2005). Management of hepatocellular carcinoma. *Hepatology,* 42, 1208-1236.

Bucher, N. L. R. & Malt, R. A. (1971) *Regeneration of the liver and kidney.* Little, Brown and Company, Boston.

Cao, L., Zhou, Y., Zhai, B., Liao, J., Xu, W., Zhang, R., Li, J., Zhang, Y., Chen, L., Qian, H., Wu, M. & Yin, Z. (2011). Sphere-forming cell subpopulations with cancer stem cell properties in human hepatoma cell lines. *BMC Gastroenterology,* 11, 71.

Chen, Y. J., Yeh, S. H., Chen, J. T., Wu, C. C., Hsu, M. T., Tsai, S. F., Chen, P. J. & Lin, C. H. (2000). Chromosomal changes and clonality relationship between primary and recurrent hepatocellular carcinoma. *Gastroenterology,* 119, 431-440.

Chiba, T., Kita, K., Zheng, Y. W., Yokosuka, O., Saisho, H., Iwama, A., Nakauchi, H. & Taniguchi, H. (2006). Side population purified from hepatocellular carcinoma cells harbors cancer stem cell-like properties. *Hepatology,* 44, 240-251.

Clouston, A. D., Powell, E. E., Walsh, M. J., Richardson, M. M., Demetris, A. J. & Jonsson, J. R. (2005). Fibrosis correlates with a ductular reaction in hepatitis C: roles of impaired replication, progenitor cells and steatosis. *Hepatology,* 41, 809-818.

Colombo, F., Baldan, F., Mazzucchelli, S., Martin-Padura, I., Marighetti, P., Cattaneo, A., Foglieni, B., Spreafico, M., Guerneri, S., Baccarin, M., Bertolini, F., Rossi, G., Mazzaferro, V., Cadamuro, M., Maggioni, M., Agnelli, L., Rebulla, P., Prati, D. & Porretti, L. (2011). Evidence of distinct tumour-propagating cell populations with different properties in primary human hepatocellular carcinoma. *PLoS One,* 6, e21369.

Cox, C. V., Evely, R. S., Oakhill, A., Pamphilon, D. H., Goulden, N. J. & Blair, A. (2004). Characterization of acute lymphoblastic leukemia progenitor cells. *Blood,* 104, 2919-2925.

Cox, C. V., Martin, H. M., Kearns, P. R., Virgo, P., Evely, R. S. & Blair, A. (2007). Characterization of a progenitor cell population in childhood T-cell acute lymphoblastic leukemia. *Blood,* 109, 674-682.

Dabeva, M. D. & Shafritz, D. A. (1993). Activation, proliferation, and differentiation of progenitor cells into hepatocytes in the D-galactosamine model of liver regeneration. *American Journal of Pathology,* 143, 1606-1620.

Davies, R. A., Knight, B., Tian, Y. W., Yeoh, G. C. & Olynyk, J. K. (2006). Hepatic oval cell response to the choline-deficient, ethionine supplemented model of murine liver injury is attenuated by the administration of a cyclo-oxygenase 2 inhibitor. *Carcinogenesis,* 27, 1607-1616.

Decker, K. & Keppler, D. (1972). Galactosamine induced liver injury. *Progress in Liver Diseases,* 4, 183-199.

Dunsford, H. A., Karnasuta, C., Hunt, J. M. & Sell, S. (1989). Different lineages of chemically induced hepatocellular carcinoma in rats defined by monoclonal antibodies. *Cancer Research,* 49, 4894-4900.

El-Serag, H. B. & Mason, A. C. (1999). Rising incidence of hepatocellular carcinoma in the United States. *New England Journal of Medicine,* 340, 745-750.

Engelhardt, N. V., Factor, V. M., Medvinsky, A. L., Baranov, V. N., Lazareva, M. N. & Poltoranina, V. S. (1993). Common antigen of oval and biliary epithelial cells (A6) is a differentiation marker of epithelial and erythroid cell lineages in early development of the mouse. *Differentiation,* 55, 19-26.

Evarts, R. P., Nagy, P., Marsden, E. & Thorgeirsson, S. S. (1987). In situ hybridization studies on expression of albumin and alpha-fetoprotein during the early stage of neoplastic transformation in rat liver. *Cancer Research,* 47, 5469-5475.

Falkowski, O., An, H. J., Ianus, I. A., Chiriboga, L., Yee, H., West, A. B. & Theise, N. D. (2003). Regeneration of hepatocyte 'buds' in cirrhosis from intrabiliary stem cells. *Journal of Hepatology,* 39, 357-364.

Farber, E. (1956a). Carcinoma of the liver in rats fed ethionine. *American Medical Association Archives of Pathology,* 62, 445-453.

Farber, E. (1956b). Similarities in the sequence of early histological changes induced in the liver of the rat by ethionine, 2-acetylamino-fluorene, and 3'-methyl-4-dimethylaminoazobenzene. *Cancer Research,* 16, 142-148.

Farber, E. (1967). Ethionine fatty liver. *Advances in Lipid Research,* 5, 119-183.

Fausto, N. & Campbell, J. S. (2003). The role of hepatocytes and oval cells in liver regeneration and repopulation. *Mechanisms of Development,* 120, 117-130.

Fausto, N., Campbell, J. S. & Riehle, K. J. (2006). Liver regeneration. *Hepatology,* 43, S45-53.

Finkelstein, S. D., Marsh, W., Demetris, A. J., Swalsky, P. A., Sasatomi, E., Bonham, A., Subotin, M. & Dvorchik, I. (2003). Microdissection-based allelotyping discriminates de novo tumor from intrahepatic spread in hepatocellular carcinoma. *Hepatology,* 37, 871-879.

Forbes, S. J., Russo, F. P., Rey, V., Burra, P., Rugge, M., Wright, N. A. & Alison, M. R. (2004). A significant proportion of myofibroblasts are of bone marrow origin in human liver fibrosis. *Gastroenterology,* 126, 955-963.

Fujio, K., Evarts, R. P., Hu, Z., Marsden, E. R. & Thorgeirsson, S. S. (1994). Expression of stem cell factor and its receptor, c-kit, during liver regeneration from putative stem cells in adult rat. *Laboratory Investigation,* 70, 511-516.

Fujio, K., Hu, Z., Evarts, R. P., Marsden, E. R., Niu, C. H. & Thorgeirsson, S. S. (1996). Coexpression of stem cell factor and c-kit in embryonic and adult liver. *Experimental Cell Research,* 224, 243-250.

Furuyama, K., Kawaguchi, Y., Akiyama, H., Horiguchi, M., Kodama, S., Kuhara, T., Hosokawa, S., Elbahrawy, A., Soeda, T., Koizumi, M., Masui, T., Kawaguchi, M., Takaori, K., Doi, R., Nishi, E., Kakinoki, R., Deng, J. M., Behringer, R. R., Nakamura, T. & Uemoto, S. (2011). Continuous cell supply from a Sox9-expressing progenitor zone in adult liver, exocrine pancreas and intestine. *Nature Genetics,* 43, 34-41.

Gebhardt, R. (1988). Different proliferative activity in vitro of periportal and perivenous hepatocytes. *Scandinavian Journal of Gastroenterology Supplement,* 151, 8-18.

Ghoshal, A. K. & Farber, E. (1984). The induction of liver cancer by dietary deficiency of choline and methionine without added carcinogens. *Carcinogenesis,* 5, 1367-1370.

Ginestier, C., Hur, M. H., Charafe-Jauffret, E., Monville, F., Dutcher, J., Brown, M., Jacquemier, J., Viens, P., Kleer, C. G., Liu, S., Schott, A., Hayes, D., Birnbaum, D., Wicha, M. S. & Dontu, G. (2007). ALDH1 is a marker of normal and malignant human mammary stem cells and a predictor of poor clinical outcome. *Cell Stem Cell,* 1, 555-567.

Golding, M., Sarraf, C. E., Lalani, E. N., Anilkumar, T. V., Edwards, R. J., Nagy, P., Thorgeirsson, S. S. & Alison, M. R. (1995). Oval cell differentiation into hepatocytes in the acetylaminofluorene-treated regenerating rat liver. *Hepatology,* 22, 1243-1253.

Grisham, J. W. (1962). A morphologic study of deoxyribonucleic acid synthesis and cell proliferation in regenerating rat liver; autoradiography with thymidine-H3. *Cancer Research,* 22, 842-849.

Grisham, J. W. & Porta, E. A. (1964). Origin and Fate of Proliferated Hepatic Ductal Cells in the Rat: Electron Microscopic and Autoradiographic Studies. *Experimental and Molecular Pathology,* 86, 242-261.

Haque, S., Haruna, Y., Saito, K., Nalesnik, M. A., Atillasoy, E., Thung, S. N. & Gerber, M. A. (1996). Identification of bipotential progenitor cells in human liver regeneration. *Laboratory Investigation,* 75, 699-705.

Haraguchi, N., Utsunomiya, T., Inoue, H., Tanaka, F., Mimori, K., Barnard, G. F. & Mori, M. (2006). Characterization of a side population of cancer cells from human gastrointestinal system. *Stem Cells,* 24, 506-513.

Higgins, G. M. & Anderson, R. M. (1931). Experimental pathology of the liver. I. Restoration of the liver of the white rat following partial surgical removal. *Archives of Pathology,* 12, 186-262.

Holic, N., Suzuki, T., Corlu, A., Couchie, D., Chobert, M. N., Guguen-Guillouzo, C. & Laperche, Y. (2000). Differential expression of the rat gamma-glutamyl transpeptidase gene promoters along with differentiation of hepatoblasts into biliary or hepatocytic lineage. *American Journal of Pathology,* 157, 537-548.

Ishii, T., Yasuchika, K., Suemori, H., Nakatsuji, N., Ikai, I. & Uemoto, S. (2010). Alpha-fetoprotein producing cells act as cancer progenitor cells in human cholangiocarcinoma. *Cancer Letters,* 294, 25-34.

Jakubowski, A., Ambrose, C., Parr, M., Lincecum, J. M., Wang, M. Z., Zheng, T. S., Browning, B., Michaelson, J. S., Baetscher, M., Wang, B., Bissell, D. M. & Burkly, L. C. (2005). TWEAK induces liver progenitor cell proliferation. *Journal of Clinical Investigation,* 115, 2330-2340.

Jang, Y. Y., Collector, M. I., Baylin, S. B., Diehl, A. M. & Sharkis, S. J. (2004). Hematopoietic stem cells convert into liver cells within days without fusion. *Nature Cell Biology,* 6, 532-539.

Jensen, C. H., Jauho, E. I., Santoni-Rugiu, E., Holmskov, U., Teisner, B., Tygstrup, N. & Bisgaard, H. C. (2004). Transit-amplifying ductular (oval) cells and their hepatocytic progeny are characterized by a novel and distinctive expression of delta-like protein/preadipocyte factor 1/fetal antigen 1. *American Journal of Pathology,* 164, 1347-1359.

Jiang, Y., Jahagirdar, B. N., Reinhardt, R. L., Schwartz, R. E., Keene, C. D., Ortiz-Gonzalez, X. R., Reyes, M., Lenvik, T., Lund, T., Blackstad, M., Du, J., Aldrich, S., Lisberg, A., Low, W. C., Largaespada, D. A. & Verfaillie, C. M. (2002). Pluripotency of mesenchymal stem cells derived from adult marrow. *Nature,* 418, 41-49.

Kawasaki, S., Makuuchi, M., Ishizone, S., Matsunami, H., Terada, M. & Kawarazaki, H. (1992). Liver regeneration in recipients and donors after transplantation. *Lancet,* 339, 580-581.

Kawasaki, Y., Omori, Y., Li, Q., Nishikawa, Y., Yoshioka, T., Yoshida, M., Ishikawa, K. & Enomoto, K. (2011). Cytoplasmic accumulation of connexin32 expands cancer stem cell population in human HuH7 hepatoma cells by enhancing its self-renewal. *International Journal of Cancer,* 128, 51-62.

Kennedy, S., Rettinger, S., Flye, M. W. & Ponder, K. P. (1995). Experiments in transgenic mice show that hepatocytes are the source for postnatal liver growth and do not stream. *Hepatology,* 22, 160-168.

Kinosita, R. (1937). Studies on the cancerogenic chemical substances. *Transactions of the Japanese Society for Pathology,* 27, 665-727.

Knight, B., Yeoh, G. C., Husk, K. L., Ly, T., Abraham, L. J., Yu, C., Rhim, J. A. & Fausto, N. (2000). Impaired preneoplastic changes and liver tumor formation in tumor necrosis factor receptor type 1 knockout mice. *Journal of Experimental Medicine,* 192, 1809-1818.

Knight, B., Matthews, V. B., Akhurst, B., Croager, E. J., Klinken, E., Abraham, L. J., Olynyk, J. K. & Yeoh, G. (2005). Liver inflammation and cytokine production, but not acute phase protein synthesis, accompany the adult liver progenitor (oval) cell response to chronic liver injury. *Immunology and Cell Biology,* 83, 364-374.

Knight, B. & Yeoh, G. C. (2005). TNF/LTalpha double knockout mice display abnormal inflammatory and regenerative responses to acute and chronic liver injury. *Cell and Tissue Research,* 319, 61-70.

Knight, B., Lim, R., Yeoh, G. C. & Olynyk, J. K. (2007). Interferon-gamma exacerbates liver damage, the hepatic progenitor cell response and fibrosis in a mouse model of chronic liver injury. *Journal of Hepatology,* 47, 826-833.

Knight, B., Tirnitz-Parker, J. E. & Olynyk, J. K. (2008). C-kit inhibition by imatinib mesylate attenuates progenitor cell expansion and inhibits liver tumor formation in mice. *Gastroenterology,* 135, 969-979, 979 e961.

Krause, D. S., Theise, N. D., Collector, M. I., Henegariu, O., Hwang, S., Gardner, R., Neutzel, S. & Sharkis, S. J. (2001). Multi-organ, multi-lineage engraftment by a single bone marrow-derived stem cell. *Cell,* 105, 369-377.

Lagasse, E., Connors, H., Al-Dhalimy, M., Reitsma, M., Dohse, M., Osborne, L., Wang, X., Finegold, M., Weissman, I. L. & Grompe, M. (2000). Purified hematopoietic stem cells can differentiate into hepatocytes in vivo. *Nature Medicine,* 6, 1229-1234.

Lee, K. P., Lee, J. H., Kim, T. S., Kim, T. H., Park, H. D., Byun, J. S., Kim, M. C., Jeong, W. I., Calvisi, D. F., Kim, J. M. & Lim, D. S. (2010). The Hippo-Salvador pathway restrains hepatic oval cell proliferation, liver size, and liver tumorigenesis. *Proceedings of the National Academy of Sciences U S A,* 107, 8248-8253.

Lee, T. K., Castilho, A., Cheung, V. C., Tang, K. H., Ma, S. & Ng, I. O. (2011). CD24(+) liver tumor-initiating cells drive self-renewal and tumor initiation through STAT3-mediated NANOG regulation. *Cell Stem Cell,* 9, 50-63.

Leite, A. R., Correa-Giannella, M. L., Dagli, M. L., Fortes, M. A., Vegas, V. M. & Giannella-Neto, D. (2007). Fibronectin and laminin induce expression of islet cell markers in hepatic oval cells in culture. *Cell and Tissue Research,* 327, 529-537.

Lemire, J. M., Shiojiri, N. & Fausto, N. (1991). Oval cell proliferation and the origin of small hepatocytes in liver injury induced by D-galactosamine. *American Journal of Pathology,* 139, 535-552.

Lenzi, R., Liu, M. H., Tarsetti, F., Slott, P. A., Alpini, G., Zhai, W. R., Paronetto, F., Lenzen, R. & Tavoloni, N. (1992). Histogenesis of bile duct-like cells proliferating during ethionine hepatocarcinogenesis. Evidence for a biliary epithelial nature of oval cells. *Laboratory Investigation,* 66, 390-402.

Lim, R., Knight, B., Patel, K., McHutchison, J. G., Yeoh, G. C. & Olynyk, J. K. (2006). Antiproliferative effects of interferon alpha on hepatic progenitor cells in vitro and in vivo. *Hepatology,* 43, 1074-1083.

Llovet, J. M., Schwartz, M. & Mazzaferro, V. (2005). Resection and liver transplantation for hepatocellular carcinoma. *Seminars in Liver Disease,* 25, 181-200.

Llovet, J. M., Ricci, S., Mazzaferro, V., Hilgard, P., Gane, E., Blanc, J. F., de Oliveira, A. C., Santoro, A., Raoul, J. L., Forner, A., Schwartz, M., Porta, C., Zeuzem, S., Bolondi, L., Greten, T. F., Galle, P. R., Seitz, J. F., Borbath, I., Haussinger, D., Giannaris, T., Shan, M., Moscovici, M., Voliotis, D. & Bruix, J. (2008). Sorafenib in advanced hepatocellular carcinoma. *New England Journal of Medicine,* 359, 378-390.

Locker, J., Reddy, T. V. & Lombardi, B. (1986). DNA methylation and hepatocarcinogenesis in rats fed a choline-devoid diet. *Carcinogenesis,* 7, 1309-1312.

Lombardi, B., Ugazio, G. & Raick, A. N. (1966). Choline-deficiency fatty liver: relation of plasma phospholipids to liver triglycerides. *American Journal of Physiology,* 210, 31-36.

Lombardi, B., Pani, P. & Schlunk, F. F. (1968). Choline-deficiency fatty liver: impaired release of hepatic triglycerides. *Journal of Lipid Research,* 9, 437-446.

Lowes, K. N., Brennan, B. A., Yeoh, G. C. & Olynyk, J. K. (1999). Oval cell numbers in human chronic liver diseases are directly related to disease severity. *American Journal of Pathology,* 154, 537-541.

Lowes, K. N., Croager, E. J., Olynyk, J. K., Abraham, L. J. & Yeoh, G. C. (2003). Oval cell-mediated liver regeneration: Role of cytokines and growth factors. *Journal of Gastroenterology and Hepatology,* 18, 4-12.

Ma, S., Chan, K. W., Hu, L., Lee, T. K., Wo, J. Y., Ng, I. O., Zheng, B. J. & Guan, X. Y. (2007). Identification and characterization of tumorigenic liver cancer stem/progenitor cells. *Gastroenterology,* 132, 2542-2556.

Ma, S., Chan, K. W., Lee, T. K., Tang, K. H., Wo, J. Y., Zheng, B. J. & Guan, X. Y. (2008). Aldehyde dehydrogenase discriminates the CD133 liver cancer stem cell populations. *Molecular Cancer Research,* 6, 1146-1153.

Marquardt, J. U., Raggi, C., Andersen, J. B., Seo, D., Avital, I., Geller, D., Lee, Y. H., Kitade, M., Holczbauer, A., Gillen, M. C., Conner, E. A., Factor, V. M. & Thorgeirsson, S. S. (2011). Human hepatic cancer stem cells are characterized by common stemness traits and diverse oncogenic pathways. *Hepatology,* 54, 1031-1042.

Mathers, C. D. & Loncar, D. (2006). Projections of global mortality and burden of disease from 2002 to 2030. *PLoS Medicine,* 3, e442.

Mazzaferro, V., Regalia, E., Doci, R., Andreola, S., Pulvirenti, A., Bozzetti, F., Montalto, F., Ammatuna, M., Morabito, A. & Gennari, L. (1996). Liver transplantation for the treatment of small hepatocellular carcinomas in patients with cirrhosis. *New England Journal of Medicine,* 334, 693-699.

Mazzaferro, V., Llovet, J. M., Miceli, R., Bhoori, S., Schiavo, M., Mariani, L., Camerini, T., Roayaie, S., Schwartz, M. E., Grazi, G. L., Adam, R., Neuhaus, P., Salizzoni, M., Bruix, J., Forner, A., De Carlis, L., Cillo, U., Burroughs, A. K., Troisi, R., Rossi, M.,

Gerunda, G. E., Lerut, J., Belghiti, J., Boin, I., Gugenheim, J., Rochling, F., Van Hoek, B. & Majno, P. (2009). Predicting survival after liver transplantation in patients with hepatocellular carcinoma beyond the Milan criteria: a retrospective, exploratory analysis. *Lancet Oncology*, 10, 35-43.

Michalopoulos, G. K. & DeFrances, M. C. (1997). Liver regeneration. *Science*, 276, 60-66.

Ng, I. O., Guan, X. Y., Poon, R. T., Fan, S. T. & Lee, J. M. (2003). Determination of the molecular relationship between multiple tumour nodules in hepatocellular carcinoma differentiates multicentric origin from intrahepatic metastasis. *Journal of Pathology*, 199, 345-353.

Ng, Y. K. & Iannaccone, P. M. (1992). Fractal geometry of mosaic pattern demonstrates liver regeneration is a self-similar process. *Developmental Biology*, 151, 419-430.

Okabe, M., Tsukahara, Y., Tanaka, M., Suzuki, K., Saito, S., Kamiya, Y., Tsujimura, T., Nakamura, K. & Miyajima, A. (2009). Potential hepatic stem cells reside in EpCAM+ cells of normal and injured mouse liver. *Development*, 136, 1951-1960.

Oliva, J., French, B. A., Qing, X. & French, S. W. (2010). The identification of stem cells in human liver diseases and hepatocellular carcinoma. *Experimental and Molecular Pathology*, 88, 331-340.

Olsen, S. K., Brown, R. S. & Siegel, A. B. (2010). Hepatocellular carcinoma: review of current treatment with a focus on targeted molecular therapies. *Therapeutic Advances in Gastroenterology*, 3, 55-66.

Omori, N., Omori, M., Evarts, R. P., Teramoto, T., Miller, M. J., Hoang, T. N. & Thorgeirsson, S. S. (1997). Partial cloning of rat CD34 cDNA and expression during stem cell-dependent liver regeneration in the adult rat. *Hepatology*, 26, 720-727.

Overturf, K., al-Dhalimy, M., Ou, C. N., Finegold, M. & Grompe, M. (1997). Serial transplantation reveals the stem-cell-like regenerative potential of adult mouse hepatocytes. *American Journal of Pathology*, 151, 1273-1280.

Paku, S., Schnur, J., Nagy, P. & Thorgeirsson, S. S. (2001). Origin and structural evolution of the early proliferating oval cells in rat liver. *American Journal of Pathology*, 158, 1313-1323.

Paku, S., Nagy, P., Kopper, L. & Thorgeirsson, S. S. (2004). 2-acetylaminofluorene dose-dependent differentiation of rat oval cells into hepatocytes: confocal and electron microscopic studies. *Hepatology*, 39, 1353-1361.

Paradis, V., Youssef, N., Dargere, D., Ba, N., Bonvoust, F., Deschatrette, J. & Bedossa, P. (2001). Replicative senescence in normal liver, chronic hepatitis C, and hepatocellular carcinomas. *Human Pathology*, 32, 327-332.

Petersen, B. E., Goff, J. P., Greenberger, J. S. & Michalopoulos, G. K. (1998a). Hepatic oval cells express the hematopoietic stem cell marker Thy-1 in the rat. *Hepatology*, 27, 433-445.

Petersen, B. E., Zajac, V. F. & Michalopoulos, G. K. (1998b). Hepatic oval cell activation in response to injury following chemically induced periportal or pericentral damage in rats. *Hepatology*, 27, 1030-1038.

Petersen, B. E., Bowen, W. C., Patrene, K. D., Mars, W. M., Sullivan, A. K., Murase, N., Boggs, S. S., Greenberger, J. S. & Goff, J. P. (1999). Bone marrow as a potential source of hepatic oval cells. *Science*, 284, 1168-1170.

Petersen, B. E., Grossbard, B., Hatch, H., Pi, L., Deng, J. & Scott, E. W. (2003). Mouse A6-positive hepatic oval cells also express several hematopoietic stem cell markers. *Hepatology,* 37, 632-640.

Ponder, K. P. (1996). Analysis of liver development, regeneration, and carcinogenesis by genetic marking studies. *Faseb Journal,* 10, 673-682.

Qiu, Q., Hernandez, J. C., Dean, A. M., Rao, P. H. & Darlington, G. J. (2011). CD24-positive cells from normal adult mouse liver are hepatocyte progenitor cells. *Stem Cells and Development,* 20, 2177-2188.

Rabes, H. M. (1976). Kinetics of hepatocellular proliferation after partial resection of the liver. *Progress in Liver Diseases,* 5, 83-99.

Rhim, J. A., Sandgren, E. P., Degen, J. L., Palmiter, R. D. & Brinster, R. L. (1994). Replacement of diseased mouse liver by hepatic cell transplantation. *Science,* 263, 1149-1152.

Rhim, J. A., Sandgren, E. P., Palmiter, R. D. & Brinster, R. L. (1995). Complete reconstitution of mouse liver with xenogeneic hepatocytes. *Proceedings of the National Academy of Sciences U S A,* 92, 4942-4946.

Rogler, L. E. (1997). Selective bipotential differentiation of mouse embryonic hepatoblasts in vitro. *American Journal of Pathology,* 150, 591-602.

Roskams, T. & Desmet, V. (1998). Ductular reaction and its diagnostic significance. *Seminars in Diagnostic Pathology,* 15, 259-269.

Roskams, T., Yang, S. Q., Koteish, A., Durnez, A., DeVos, R., Huang, X., Achten, R., Verslype, C. & Diehl, A. M. (2003a). Oxidative stress and oval cell accumulation in mice and humans with alcoholic and nonalcoholic fatty liver disease. *American Journal of Pathology,* 163, 1301-1311.

Roskams, T. A., Libbrecht, L. & Desmet, V. J. (2003b). Progenitor cells in diseased human liver. *Seminars in Liver Disease,* 23, 385-396.

Roskams, T. (2006). Liver stem cells and their implication in hepatocellular and cholangiocarcinoma. *Oncogene,* 25, 3818-3822.

Rountree, C. B., Mishra, L. & Willenbring, H. (2012). Stem cells in liver diseases and cancer: Recent advances on the path to new therapies. *Hepatology,* 55, 298-306.

Ruddell, R. G., Knight, B., Tirnitz-Parker, J. E., Akhurst, B., Summerville, L., Subramaniam, V. N., Olynyk, J. K. & Ramm, G. A. (2009). Lymphotoxin-beta receptor signaling regulates hepatic stellate cell function and wound healing in a murine model of chronic liver injury. *Hepatology,* 49, 227-239.

Sarraf, C., Lalani, E. N., Golding, M., Anilkumar, T. V., Poulsom, R. & Alison, M. (1994). Cell behavior in the acetylaminofluorene-treated regenerating rat liver. Light and electron microscopic observations. *American Journal of Pathology,* 145, 1114-1126.

Sasaki, K., Kon, J., Mizuguchi, T., Chen, Q., Ooe, H., Oshima, H., Hirata, K. & Mitaka, T. (2008). Proliferation of hepatocyte progenitor cells isolated from adult human livers in serum-free medium. *Cell Transplantation,* 17, 1221-1230.

Schmelzer, E., Zhang, L., Bruce, A., Wauthier, E., Ludlow, J., Yao, H. L., Moss, N., Melhem, A., McClelland, R., Turner, W., Kulik, M., Sherwood, S., Tallheden, T., Cheng, N., Furth, M. E. & Reid, L. M. (2007). Human hepatic stem cells from fetal and postnatal donors. *Journal of Experimental Medicine,* 204, 1973-1987.

Schwartz, R. E., Reyes, M., Koodie, L., Jiang, Y., Blackstad, M., Lund, T., Lenvik, T., Johnson, S., Hu, W. S. & Verfaillie, C. M. (2002). Multipotent adult progenitor cells from bone marrow differentiate into functional hepatocyte-like cells. *Journal of Clinical Investigation,* 109, 1291-1302.

Sell, S. (1978). Distribution of alpha-fetoprotein- and albumin-containing cells in the livers of Fischer rats fed four cycles of N-2-fluorenylacetamide. *Cancer Research,* 38, 3107-3113.

Sell, S. (2001). Heterogeneity and plasticity of hepatocyte lineage cells. *Hepatology,* 33, 738-750.

Shachaf, C. M., Kopelman, A. M., Arvanitis, C., Karlsson, A., Beer, S., Mandl, S., Bachmann, M. H., Borowsky, A. D., Ruebner, B., Cardiff, R. D., Yang, Q., Bishop, J. M., Contag, C. H. & Felsher, D. W. (2004). MYC inactivation uncovers pluripotent differentiation and tumour dormancy in hepatocellular cancer. *Nature,* 431, 1112-1117.

Sherman, M. (2011). Modern approach to hepatocellular carcinoma. *Current Gastroenterology Reports,* 13, 49-55.

Shinozuka, H., Lombardi, B., Sell, S. & Iammarino, R. M. (1978a). Enhancement of DL-ethionine-induced liver carcinogenesis in rats fed a choline-devoid diet. *Journal of the National Cancer Institute,* 61, 813-817.

Shinozuka, H., Lombardi, B., Sell, S. & Iammarino, R. M. (1978b). Early histological and functional alterations of ethionine liver carcinogenesis in rats fed a choline-deficient diet. *Cancer Research,* 38, 1092-1098.

Sidransky, H. & Verney, E. (1969). Influence of ethionine on choline-deficiency fatty liver. *Journal of Nutrition,* 97, 419-430.

Sigal, S. H., Brill, S., Fiorino, A. S. & Reid, L. M. (1992). The liver as a stem cell and lineage system. *American Journal of Physiology,* 263, G139-148.

Simpson, G. E. & Finck, E. S. (1963). The Pattern of Regeneration of Rat Liver after Repeated Partial Hepatectomies. *Journal of Pathology and Bacteriology,* 86, 361-370.

Singh, S. K., Clarke, I. D., Terasaki, M., Bonn, V. E., Hawkins, C., Squire, J. & Dirks, P. B. (2003). Identification of a cancer stem cell in human brain tumours. *Cancer Research* 63, 5821-5828.

Singh, S. K., Hawkins, C., Clarke, I. D., Squire, J. A., Bayani, J., Hide, T., Henkelman, R. M., Cusimano, M. D. & Dirks, P. B. (2004). Identification of human brain tumour initiating cells. *Nature,* 432, 396-401.

Smith, P. G., Tee, L. B. & Yeoh, G. C. (1996). Appearance of oval cells in the liver of rats after long-term exposure to ethanol. *Hepatology,* 23, 145-154.

Solt, D. & Farber, E. (1976). New principle for the analysis of chemical carcinogenesis. *Nature,* 263, 701-703.

Starzl, T. E., Fung, J., Tzakis, A., Todo, S., Demetris, A. J., Marino, I. R., Doyle, H., Zeevi, A., Warty, V., Michaels, M., Kusne, S., Rudert, W. A., & Trucco, M. (1993). Baboon-to-human liver transplantation. *Lancet,* 341, 65-71.

Suetsugu, A., Nagaki, M., Aoki, H., Motohashi, T., Kunisada, T. & Moriwaki, H. (2006). Characterization of CD133+ hepatocellular carcinoma cells as cancer stem/progenitor cells. *Biochemical and Biophysical Research Communications,* 351, 820-824.

Suzuki, A., Sekiya, S., Onishi, M., Oshima, N., Kiyonari, H., Nakauchi, H. & Taniguchi, H. (2008). Flow cytometric isolation and clonal identification of self-renewing bipotent hepatic progenitor cells in adult mouse liver. *Hepatology,* 48, 1964-1978.

Tatematsu, M., Ho, R. H., Kaku, T., Ekem, J. K. & Farber, E. (1984). Studies on the proliferation and fate of oval cells in the liver of rats treated with 2-acetylaminofluorene and partial hepatectomy. *American Journal of Pathology,* 114, 418-430.

Tatematsu, M., Kaku, T., Medline, A. & Farber, E. (1985). Intestinal metaplasia as a common option of oval cells in relation to cholangiofibrosis in liver of rats exposed to 2-acetylaminofluorene. *Laboratory Investigation,* 52, 354-362.

Tee, L. B., Smith, P. G. & Yeoh, G. C. (1992). Expression of alpha, mu and pi class glutathione S-transferases in oval and ductal cells in liver of rats placed on a choline-deficient, ethionine-supplemented diet. *Carcinogenesis,* 13, 1879-1885.

Theise, N. D., Saxena, R., Portmann, B. C., Thung, S. N., Yee, H., Chiriboga, L., Kumar, A. & Crawford, J. M. (1999). The canals of Hering and hepatic stem cells in humans. *Hepatology,* 30, 1425-1433.

Theise, N. D., Badve, S., Saxena, R., Henegariu, O., Sell, S., Crawford, J. M. & Krause, D. S. (2000a). Derivation of hepatocytes from bone marrow cells in mice after radiation-induced myeloablation. *Hepatology,* 31, 235-240.

Theise, N. D., Nimmakayalu, M., Gardner, R., Illei, P. B., Morgan, G., Teperman, L., Henegariu, O. & Krause, D. S. (2000b). Liver from bone marrow in humans. *Hepatology,* 32, 11-16.

Thorgeirsson, S. S. & Grisham, J. W. (2006). Hematopoietic cells as hepatocyte stem cells: a critical review of the evidence. *Hepatology,* 43, 2-8.

Thurman, R. G. & Kauffman, F. C. (1985). Sublobular compartmentation of pharmacologic events (SCOPE): metabolic fluxes in periportal and pericentral regions of the liver lobule. *Hepatology,* 5, 144-151.

Tian, Y. W., Smith, P. G. & Yeoh, G. C. (1997). The oval-shaped cell as a candidate for a liver stem cell in embryonic, neonatal and precancerous liver: identification based on morphology and immunohistochemical staining for albumin and pyruvate kinase isoenzyme expression. *Histochemistry and Cell Biology,* 107, 243-250.

Tirnitz-Parker, J. E., Tonkin, J. N., Knight, B., Olynyk, J. K. & Yeoh, G. C. (2007). Isolation, culture and immortalisation of hepatic oval cells from adult mice fed a choline-deficient, ethionine-supplemented diet. *International Journal of Biochemistry & Cell Biology,* 39, 2226-2239.

Tirnitz-Parker, J. E., Viebahn, C. S., Jakubowski, A., Klopcic, B. R., Olynyk, J. K., Yeoh, G. C. & Knight, B. (2010). Tumor necrosis factor-like weak inducer of apoptosis is a mitogen for liver progenitor cells. *Hepatology,* 52, 291-302.

Tonkin, J. N., Knight, B., Curtis, D., Abraham, L. J. & Yeoh, G. C. (2008). Bone marrow cells play only a very minor role in chronic liver regeneration induced by a choline-deficient, ethionine-supplemented diet. *Stem Cell Research,* 1, 195-204.

Ueberham, E., Aigner, T., Ueberham, U. & Gebhardt, R. (2007). E-cadherin as a reliable cell surface marker for the identification of liver specific stem cells. *Journal of Molecular Histology,* 38, 359-368.

Van Hul, N. K., Abarca-Quinones, J., Sempoux, C., Horsmans, Y. & Leclercq, I. A. (2009). Relation between liver progenitor cell expansion and extracellular matrix deposition in a CDE-induced murine model of chronic liver injury. *Hepatology,* 49, 1625-1635.

Vassilopoulos, G., Wang, P. R. & Russell, D. W. (2003). Transplanted bone marrow regenerates liver by cell fusion. *Nature,* 422, 901-904.

Wang, C., Yang, W., Yan, H. X., Luo, T., Zhang, J., Tang, L., Wu, F. Q., Zhang, H. L., Yu, L. X., Zheng, L. Y., Li, Y. Q., Dong, W., He, Y. Q., Liu, Q., Zou, S. S., Lin, Y., Hu, L., Li, Z., Wu, M. C. & Wang, H. Y. (2012). Hepatitis B virus X (HBx) induces tumorigenicity of hepatic progenitor cells in 3,5-diethoxycarbonyl-1,4-dihydrocollidine-treated HBx transgenic mice. *Hepatology,* 55, 108-120.

Wang, X., Montini, E., Al-Dhalimy, M., Lagasse, E., Finegold, M. & Grompe, M. (2002). Kinetics of liver repopulation after bone marrow transplantation. *American Journal of Pathology,* 161, 565-574.

Wang, X., Foster, M., Al-Dhalimy, M., Lagasse, E., Finegold, M. & Grompe, M. (2003a). The origin and liver repopulating capacity of murine oval cells. *Proceedings of the National Academy of Sciences U S A,* 100 Suppl 1, 11881-11888.

Wang, X., Willenbring, H., Akkari, Y., Torimaru, Y., Foster, M., Al-Dhalimy, M., Lagasse, E., Finegold, M., Olson, S. & Grompe, M. (2003b). Cell fusion is the principal source of bone-marrow-derived hepatocytes. *Nature,* 422, 897-901.

Weinstein, M., Monga, S. P., Liu, Y., Brodie, S. G., Tang, Y., Li, C., Mishra, L. & Deng, C. X. (2001). Smad proteins and hepatocyte growth factor control parallel regulatory pathways that converge on beta1-integrin to promote normal liver development. *Molecular and Cell Biology,* 21, 5122-5131.

Wiemann, S. U., Satyanarayana, A., Tsahuridu, M., Tillmann, H. L., Zender, L., Klempnauer, J., Flemming, P., Franco, S., Blasco, M. A., Manns, M. P. & Rudolph, K. L. (2002). Hepatocyte telomere shortening and senescence are general markers of human liver cirrhosis. *Faseb Journal,* 16, 935-942.

Wilson, J. W. & Leduc, E. H. (1958). Role of cholangioles in restoration of the liver of the mouse after dietary injury. *Journal of Pathology and Bacteriology,* 76, 441-449.

Yamashita, T., Budhu, A., Forgues, M. & Wang, X. W. (2007). Activation of hepatic stem cell marker EpCAM by Wnt-beta-catenin signaling in hepatocellular carcinoma. *Cancer Research,* 67, 10831-10839.

Yamashita, T., Ji, J., Budhu, A., Forgues, M., Yang, W., Wang, H. Y., Jia, H., Ye, Q., Qin, L. X., Wauthier, E., Reid, L. M., Minato, H., Honda, M., Kaneko, S., Tang, Z. Y. & Wang, X. W. (2009). EpCAM-positive hepatocellular carcinoma cells are tumor-initiating cells with stem/progenitor cell features. *Gastroenterology,* 136, 1012-1024.

Yamashita, T., Honda, M., Nio, K., Nakamoto, Y., Yamashita, T., Takamura, H., Tani, T., Zen, Y. & Kaneko, S. (2010). Oncostatin m renders epithelial cell adhesion molecule-positive liver cancer stem cells sensitive to 5-Fluorouracil by inducing hepatocytic differentiation. *Cancer Research,* 70, 4687-4697.

Yanai, H., Nakamura, K., Hijioka, S., Kamei, A., Ikari, T., Ishikawa, Y., Shinozaki, E., Mizunuma, N., Hatake, K. & Miyajima, A. (2010). Dlk-1, a cell surface antigen on foetal hepatic stem/progenitor cells, is expressed in hepatocellular, colon, pancreas and breast carcinomas at a high frequency. *Journal of Biochemistry,* 148, 85-92.

Yang, L., Li, S., Hatch, H., Ahrens, K., Cornelius, J. G., Petersen, B. E. & Peck, A. B. (2002). In vitro trans-differentiation of adult hepatic stem cells into pancreatic endocrine hormone-producing cells. *Proceedings of the National Academy of Sciences U S A,* 99, 8078-8083.

Yao, Z. & Mishra, L. (2009). Cancer stem cells and hepatocellular carcinoma. *Cancer Biology & Therapy,* 8, 1691-1698.

Yokoyama, S., Sells, M. A., Reddy, T. V. & Lombardi, B. (1985). Hepatocarcinogenic and promoting action of a choline-devoid diet in the rat. *Cancer Research,* 45, 2834-2842.

Yoon, S. M., Gerasimidou, D., Kuwahara, R., Hytiroglou, P., Yoo, J. E., Park, Y. N. & Theise, N. D. (2011). Epithelial cell adhesion molecule (EpCAM) marks hepatocytes newly derived from stem/progenitor cells in humans. *Hepatology,* 53, 964-973.

You, H., Ding, W. & Rountree, C. B. (2010). Epigenetic regulation of cancer stem cell marker CD133 by transforming growth factor-beta. *Hepatology,* 51, 1635-1644.

Zajicek, G., Oren, R. & Weinreb, M., Jr. (1985). The streaming liver. *Liver,* 5, 293-300.

Zhang, A., London, R., Schulz, F. M., Giguere-Simmonds, P. W., Delriviere, L., Chandraratana, H., Hardy, K., Zheng, S., Olynyk, J. K. & Yeoh, G. (2010). Human liver progenitor cell lines are readily established from non-tumorous tissue adjacent to hepatocellular carcinoma. *Stem Cells and Development,* 19, 1277-1284.

Zhang, M. & Thorgeirsson, S. S. (1994). Modulation of connexins during differentiation of oval cells into hepatocytes. *Experimental Cell Research,* 213, 37-42.

3

Hepatic Progenitors of the Liver and Extra-Hepatic Tissues

Eva Schmelzer
McGowan Institute for Regenerative Medicine, Department of Surgery, University of Pittsburgh, Pennsylvania, USA

1. Introduction

The liver has a tremendous capacity to regenerate at all developmental stages (for reviews, see (1-3)). Liver cell mass can be restored even after repeated partial hepatectomies as well as after toxic injury. The contribution of stem cells to these processes is still under debate. Adult liver cells have been shown to regenerate liver tissue repeatedly when transplanted serially (4). However, hepatocytes cannot be considered stem cells because they are unipotent (for a glossary of terms, see **Table 1**). This chapter describes various liver progenitors that have been found by different researchers in humans and other mammalian species. Intra and extra-hepatic progenitors are discussed that can give rise to liver lineages. Intra-hepatic progenitors of non-hepatic lineages, such as endothelial or hematopoietic restricted progenitors, are not discussed. Although the focus of this chapter is on progenitors that have been characterized in normal, non-pathological conditions of the liver, oval cells will be described briefly.

Term	Description
Totipotent	Capable to give rise to cells of all three embryonic germ layers (i.e. endoderm, mesoderm and ectoderm) as well as extra-embryonic tissue of the placenta.
Pluripotent	Capable to give rise to cells of all three embryonic germ layers (i.e. endoderm, mesoderm and ectoderm) but not to extra-embryonic tissue. Most commonly used e.g. for embryonic stem cells, which derive from the inner cell mass of the blastocyst.
Multipotent	Capable to give rise to multiple but not all lineages. For example, bone marrow mesenchymal stem cells are considered multipotent.
Bipotent	Able to give rise to two fates. In liver, hepatoblasts are considered bipotential as they can develop into biliary and hepatic lineages.
Unipotent	Able to give rise to only one cell type. Hepatocytes are considered unipotential.
Progenitor	Broad term to describe various types of precursors with different potential.

Term	Description
Stem cell	Cell, which is capable to differentiate into multiple lineages and is also able of self-renewal.
Hepatoblast	Hepatic parenchymal cell of the fetal liver. Defined by its expression of immature protein alpha-fetoprotein and absence of several mature hepatic functions and proteins.
Hepatocyte	Hepatic parenchymal cell of the adult liver. In non-pathological conditions defined by its expression of mature functions and proteins, such as albumin and cytochrome P450 enzymes, and the absence of immature proteins such as alpha-fetoprotein.
Oval cell	Small cells with oval-shaped nuclei that emerge in livers, which have been treated with certain toxins.

Table 1. Common terminology relevant to liver progenitor biology. Further details can be found also in (5, 6).

2. Embryonic liver development

During embryonic development, the liver arises from the definitive endoderm (for reviews on liver development, see (7-9)). The definitive endoderm is an embryonic layer, whereas visceral endoderm is a non-embryonic derived layer, also called extra-embryonic endoderm. The definitive endoderm is one of the three germ layers, which include also ectoderm and mesoderm. The definitive endoderm is initially located beneath the ectoderm and mesoderm. In the mouse, the definitive endoderm layer forms a liver bud between E8.5 and E9.5. This layer will also form the pancreas, lung, stomach, intestine, and thyroid. The cardiac mesoderm and septum transversum mesenchyme release signals, such as fibroblast growth factors (FGF) and bone morphogenetic proteins (BMP), which are necessary to induce liver specification. The septum transversum mesenchyme has been implicated to give rise to stellate cells (also called Ito cells), which are fat and vitamin A-storing and extracellular matrix producing liver cells (10); cells positive for the Lim-homeobox gene (Lhx2) migrate from the septum transversum into the forming liver bud and become desmin and Lhx2 positive stellate cells. Cells in the developing liver bud are termed hepatoblasts and express alpha-fetoprotein (AFP). Hepatoblasts have been described as bipotential progenitors, developing into mature hepatocytes as well as bile duct epithelial cells (cholangiocytes), based on findings from *ex vivo* and *in vitro* studies (11-15). Suppression of transcription factor CCAAT-enhancer-binding protein alpha (CEBPα) has been suggested to induce their specification towards biliary differentiation (16, 17).

3. Human hepatic progenitors in fetal and adult livers

Different hepatic progenitors in human livers have been described. Based on early findings in developmental biology, hepatic stem cells were originally defined as AFP positive hepatoblasts. More recent research, however, reveals that hepatic stem cells are AFP negative and are the precursors to hepatoblasts (12, 18). Furthermore, stem cells of assumed mesendodermal origin capable of multilineage differentiation towards liver- and mesenchymal lineages have been discovered (19). An overview about human hepatic progenitors that have been isolated and characterized is given in **Table 2**.

Publication	Developmental stage of liver tissue	Presumable lineage	Term used by authors	Isolation method	Phenotype	*In vivo* model for repopulation	*In vitro* characteristics
Najimi *et al.* 2007 (20), Khuu *et al.* 2011 (21)	Adult	Mesenchymal	Adult derived human liver stem/ progenitor cell (ADHLSCs)	Culture	Positive: CD90, CD73, CD29, CD44, CD13, HLA-class I. Weak: CD49e, CD49b, CD49f. Negative: CD105, CD133, CD117, CD45, CD34, HLA-DR	$uPA^{+}/^{+}$-SCID with and without 70% hepatectomy	Hepatic functions after induced differentiation
Dan *et al.* 2006 (19)	Fetal	Mesendodermal	Human fetal liver multipotent progenitor cells (hFLMPC)	Culture	Positive: CD34, CD90, CD117, CD326, c-met, SSEA4, CK18, CK19, CD44h, vimentin. Negative: CD133, CD45, AFP, albumin	Rag2$^{-}/^{-}$ $\gamma/^{-}$ retrorsine/ CCl_4	Long-term culture, ~46h PDT, multipotent
Herrera *et al.* 2006 (22)	Adult	Mesendodermal	Human liver stem cells (HLSCs)	Culture	Positive: Albumin, AFP, CD29, CD73, CD44, CD90, vimentin, nestin. Weak: CK8, CK18. Negative: CD34, CD45, CD117, CD133, CK19.	SCID, N-acetyl-p-aminophen	Multipotent, high expansion potential, ~36h PDT
Schmelzer *et al.* 2006, 2007 (12, 18)	Fetal (16-20 weeks of gestation), neonatal, pediatric, adult	Endodermal	Human hepatic stem cells (hHpSC)	MACS, culture	Positive: CD326, CD133, CD56, E-cadherin, CD29, CD44h, claudin3, CK19. Weak: albumin. Negative: AFP.	NOD/SCID	Long-term culture, >150 population doublings, precursors of hepatoblasts
Malhi *et al.* 2002 (23)	Fetal	Endodermal	Human fetal liver progenitor/ stem cells	Culture	Positive: AFP, GGT, CK8, CK19, CD34	SCID CCl_4	Long-term culture

Publication	Develop-mental stage of liver tissue	Presumable lineage	Term used by authors	Isolation method	Phenotype	*In vivo* model for repopula-tion	*In vitro* characteristics
Schmelzer *et al.* 2006, 2007 (12, 18)	Fetal (16-20 weeks of gestation)		Human hepatoblasts	MACS, culture	Positive: AFP. Variable: CD326	NOD/SCID	Can arise from hHpSC colonies in culture

Abbreviations: AFP: alpha-fetoprotein; CCl_4: Carbon tetrachloride; CD: cluster of differentiation; CK: cytokeratin; GGT: γ-glutamyl transpeptidase; HLA: human leukocyte antigen; NOD: non-obese diabetic; SCID: severe-combined immunodeficient; MACS: magnetic activated cells sorting; PDT: population doubling time; uPA: urokinase-type plasminogen activator.

Table 2. Progenitors with hepatic potential isolated from human livers. Details are given in the respective sections.

3.1 Human liver multipotent progenitors

Dan *et al.* isolated liver stem cells co-expressing endodermal and mesenchymal phenotypes from human fetal liver by culture selection on feeder cells (19). These cells could differentiate not only into hepatocytes and bile duct cells, but also into fat, bone, cartilage, and endothelial cells. Because of their multilineage differentiation potential, these cells were termed human fetal liver multipotent progenitor cells (hFLMPC). The *in vivo* percentage of this progenitor was not given, as these cells were isolated by culture selection. Cell surface and intracellular markers included: CD34, CD90, CD117, CD326 (also called epithelial cell adhesion molecule (EpCAM)), c-met, SSEA4, CK18, CK19, CD44h, and vimentin. Cells were negative for albumin, CD133, CD45, and AFP. They could be cultured monoclonal and long-term for up to 100 population doublings. Cells had population doubling times of 46h. Early and late passages demonstrated identical morphology, differentiation potential, and telomere length. Cultured cells formed typical clusters with cells having a high nuclear to cytoplasm ratio. The morphology of these clusters resembled hepatic stem cells colonies described by Schmelzer *et al.* (12). When transplanted into immunotolerant Rag2$^{-/-}$ γ$^{-/-}$ mice (using a modified retrorsine/carbon tetra-chloride model), human-specific albumin in mouse serum and human-specific albumin in sections of the liver could be detected. Liver sections of transplanted mice demonstrated clusters of human hepatocytes. A repopulation of 0.8–1.7% was estimated. The multipotential differentiation potential and resemblance to hepatic stem cell colonies suggests that hFLMPC represent mesendodermal precursors of hepatic stem cells.

Herrera *et al.* isolated a similar population from human adult livers (22) using culture selection. These cells also expressed hepatic and mesenchymal markers. Cell surface and intracellular markers included albumin, AFP, CD29, CD73, CD44, CD90, vimentin and nestin; however, there was a negative expression of CD34, CD45, CD117, CD133, and CK19, and a weakly positive expression of CK8 and CK18. The cells were different from those described by Dan *et al.*, as albumin and AFP expression could be observed and hematopoietic markers CD34 and CD117 were absent. *In vitro*, progenitors differentiated not only into hepatocytes, but also into osteogenic, endothelial, and islet-like, insulin-producing structures. Adipogenic differentiation could not be induced. As these cells were culture-selected, percentages of their *in vivo* occurrence were not established. Cells *in vitro* demonstrated exponential growth rates. When transplanted, human cells could be localized *in vivo* within the liver parenchyma of severe-combined immunodeficient (SCID) mice treated with N-acetyl-p-aminophen.

Mesenchymal progenitors isolated from adult human livers were investigated for their potential to differentiate into hepatocytes (20, 21). Mesenchymal-like cells were obtained by selective culture (not sorting) of total liver cells. FACS analyses of cultured cells revealed a phenotype similar to mesenchymal stem cells with positive expression for CD90, CD73, CD29, CD44, CD13, and HLA-class I, but negative expression for CD105, CD133, CD117, CD45, CD34, and HLA-DR; cells were weakly positive for CD49e and CD49b, and only a minor fraction expressed CD49f. When cells were intrasplenically transplanted into $uPA^{+}/^{+}$-SCID mice, human albumin and AFP positive cells could be observed and human albumin secretion was detected. When transplanted into SCID mice with and without 70% hepatectomy, human albumin gene expression could be measured in mice livers that had undergone hepatectomy, and human albumin positive cells could be detected in mouse liver sections in both models. Potential fusion events were not analyzed. When cells were induced to hepatic lineages *in vitro* (21), hepatic functions were increased compared to non-induced controls, but lower than those of freshly isolated adult liver cells.

3.2 Hepatic stem cells in the human liver

Hepatic stem cells can be isolated from fetal, neonatal, pediatric, and adult human livers with identical characteristics (12, 18), as described by Schmelzer *et al.* Cell surface and intracellular markers include CD326, CD133, CD56, E-cadherin, CD29, Patched (24), claudin 3 (18), CK19, and show weak positivity for albumin. Cells are negative for AFP, CD45, CD34, CD38, CD14, CD90, CD235a, VEGFr, vWF, CD31, CD146, desmin, ASMA, transferrin, connexins, PEPCK, DPP4, CYP450; CD117 is variably expressed. Sonic and Indian Hedgehog signaling pathway components are expressed (24). Stem cells could be selected by MACS sorting as well as under selective culture conditions, which included serum-free medium and culture on plastic. Under these culture conditions, hepatic stem cell colonies formed. These colonies (**Figure 1**) exhibit a typical epithelial morphology of densely packed, small cells with high nucleus-to-cytoplasm ratio. Stem cell colonies are positive for CD326 (**Figure 2**), CD44h, CD56, and weakly express albumin, but are negative for AFP.

Cells were capable of self-renewal, as shown by clonogenic expansion for more than 150 population doublings. 0.5 - 2.5% of all liver cells from all ages were positive for CD326 expression. Hepatic stem cells have a small diameter of about 9 µm. *In vivo*, they are located in the ductal plates in fetal and neonatal livers and in the Canals of Hering in pediatric and adult livers. The Canal of Hering has been previously described as the reservoir of stem cells in postnatal livers (25, 26). Carpentier *et al.* recently studied lineage tracing by using a Cre recombinase Sox9 mouse model and confirmed that ductal plate cells give rise to cholangiocytes, periportal hepatocytes, and adult liver progenitor cells (27). Furuyama *et al.* (28) demonstrated that adult intestinal cells, hepatocytes and pancreatic acinar cells are physiologically supplied from Sox9-expressing progenitors using Cre-based lineage tracing in mice. In CCl_4 mediated liver injury, Sox9-positive progenitors contributed to liver regeneration. Hepatic stem cells have been shown to differentiate into biliary and hepatocytic lineages *in vivo* and *in vitro* (12). Freshly isolated cells or stem cells expanded in culture developed into mature liver tissue expressing human-specific proteins when transplanted into NOD/SCID mice, and lost their expression of stem cell marker CD326, CD133, and CK19. Whether those cells also possess multilineage differentiation potential beyond endodermal fates, i.e. mesodermal or ectodermal, has not yet been investigated. Khan *et al.* transplanted human fetal liver derived CD326^{+} sorted progenitors into patients with liver fibrosis (29). Patients demonstrated improvements in clinical and biochemical parameters and a decrease in mean MELD (model for end-stage liver disease) score at six-month follow-up.

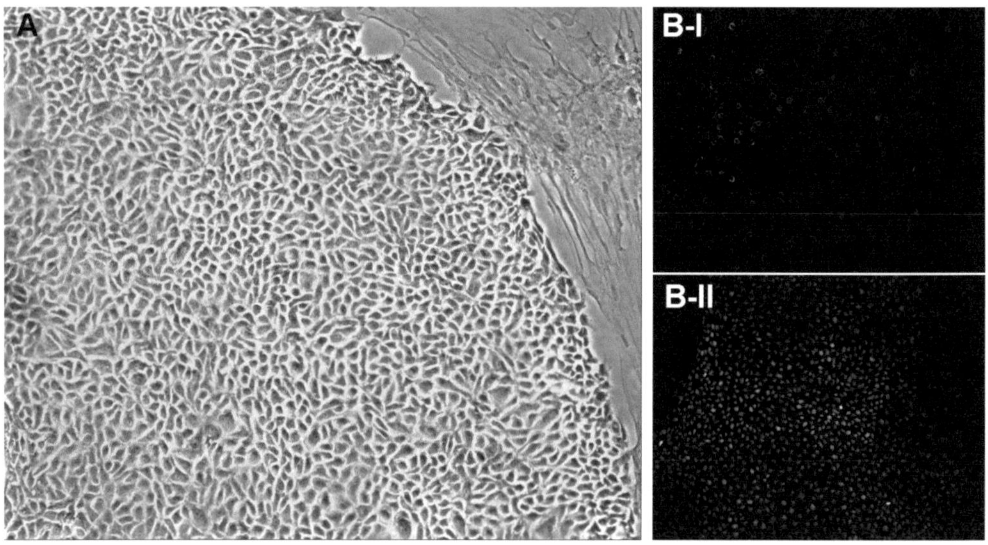

Fig. 1. Human hepatic stem cell colony in culture, established from fetal liver cell suspensions as described in (12); phase contrast microscopy (**A**), and fluorescence microscopy (**B**) of proliferating cells with positive nuclei for incorporated thymidine analog bromodeoxyuridine (**B-I**) and corresponding total nuclei stained with 4',6-diamidino-2-phenylindole (**B-II**).

3.3 Hepatoblasts in human liver

Hepatoblasts are the main parenchymal cell type of the fetal liver and are defined by their expression of AFP. AFP positive cells are rare in normal adult livers, except in livers with severe injury or disease (30-32) (for review, see (33)). Hepatoblasts can give rise to hepatocytes and cholangiocytes, and are therefore also named bipotential progenitors (15). AFP-negative hepatic stem cells are the precursors to hepatoblasts that can mature into AFP-positive hepatoblasts (12). Human fetal hepatoblasts could be cultured long-term and clonally, and contributed to liver parenchyma when transplanted into SCID mice (23). Hepatoblasts express biliary and hepatocyte markers such as CK19, CK14, gamma glutamyl transpeptidase, glucose-6-phosphatase, glycogen, albumin, AFP, E-cadherin (34), α-1 microglobulin, HepPar1, glutamate dehydrogenase, and dipeptidyl peptidase IV (15, 18).

Human hepatoblasts do not express the mesenchymal or hematopoietic markers CD90, vimentin, and CD34 (34). In mice, hepatoblasts express the surface marker Dlk-1 (35-37), which was subsequently demonstrated to be expressed by human fetal hepatoblasts as well (34). Mouse fetal liver cells sorted for Dlk-1 can be cultured long-term; transplantation of Dlk-1 positive cells into the spleen gives rise to hepatocytes in the liver. Several signaling pathways and transcription factors contribute towards differentiation into either cell type. In mice, Notch signaling controls differentiation towards biliary epithelium by upregulation of HNF1β but downregulation of HNF1α, HNF4, and C/EBPα (38), and, in turn, suppression of C/EBPα expression in periportal hepatoblasts is suggested to induce biliary epithelial differentiation by increasing HNF6 and HNF1β expression (17).

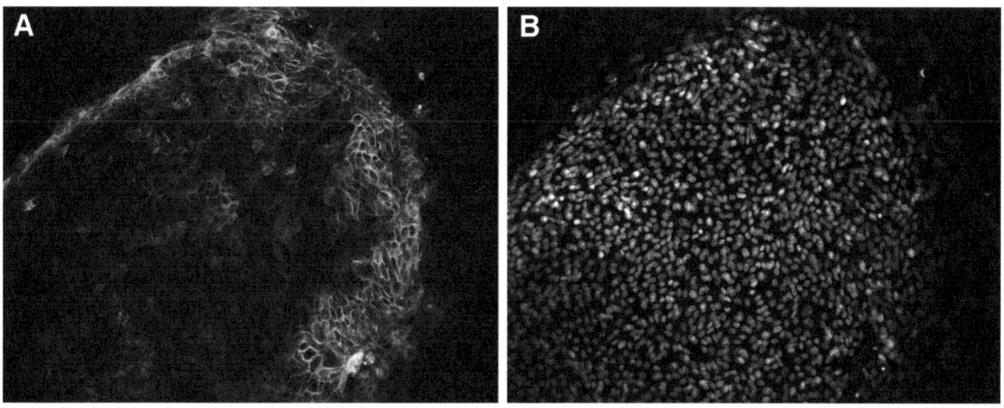

Fig. 2. Human hepatic stem cell colonies established as described in (12) are positive for CD326 (**A**). Fluorescence microscopy for the transmembrane glycoprotein CD326 (also named epithelial cell adhesion molecule (EpCAM)) in (**A**), and corresponding nuclei stained with 4',6-diamidino-2-phenylindole in (**B**).

4. Murine and rat hepatic progenitors in the fetal and adult liver

Various surface markers have been applied to identify hepatic stem or progenitor populations in rodents.

Germain *et al.* described the bipotential capacity of fetal rat liver cells to differentiate into hepatic and biliary cells *in vitro* (39), as did Kubota *et al.* using clonal cultures (40). Small hepatocytes were detected in non-parenchymal fractions of adult rat liver cells (41-44). These small hepatocytes produced colonies that expressed hepatic and biliary markers. A similar type of colony could be obtained when adult liver cell clusters were placed into culture (45). Suzuki *et al.* sorted progenitor populations from fetal mice and rat livers with a phenotype of c-met$^+$, CD49f$^+$, CD117$^-$, CD45$^-$, and TER119$^-$ (46-50). Sorted cells developed into albumin and glycogen positive cells when transplanted into retrorsine-treated adult rats that had undergone two-third partial hepatectomy. Cells negative for c-met or positive for CD45 could not repopulate recipient livers. These progenitors could be also cultured clonally. Feng *et al.* (51) demonstrated that these cells could also undergo pancreatic differentiation in culture as well as *in vivo* when transplanted into alloxan-induced diabetic mice. Similarly, Nierhoff *et al.* (35) demonstrated that fetal mouse liver epithelial cells positive for AFP or E-cadherin did not express hematopoietic stem cell markers CD34, CD117, Ter119, or CD45, but were positive for progenitor markers Sca-1 and Pan-cytokeratin. Both E-cadherin positive sorted as well as unsorted fetal liver cell fractions from wild type mice gave rise to liver parenchyma when transplanted into retrorsine treated DPPIV^{-}/$^{-}$ mice.

As described for human hepatoblasts above, mouse fetal liver hepatoblasts have been shown to express the surface marker Dlk-1 (35-37). Dlk-1 positive sorted mouse fetal liver cells can be cultured long-term and, when transplanted into the spleen, give rise to hepatocytes in the liver. Dabeva *et al.* (52) described the re-population potential of wild type fetal rat liver cells when transplanted into DPPIV^{-}/$^{-}$ rat models. These models included knockouts that had undergone two-third partial hepatectomy and were either treated with retrorsine or not. In rats treated with retrorsine, which blocked proliferation of endogenous hepatocytes, mainly bipotential, transplanted progenitors were observed expressing AFP, albumin, and CK19. In non-treated rats, transplanted cells expressed mainly either hepatocytic or biliary markers.

The positive expression of aldehyde dehydrogenase (ALDH) has been used as a feature to select progenitors from adult mouse liver (53). ALDH^{+} cells were shown to have stem cell characteristics and to express markers of human hepatic stem cells such as CD326, CK19, CD133, and Sox9.

Various hepatic progenitor cell lines have been developed from normal, genetically modified, or toxin treated rodents (54-62). Several of these lines were described as bipotential *in vitro* or when transplanted *in vivo*.

5. Oval cells

Oval cells were first described in rodents, emerging when the liver is exposed to certain toxins (for review, see (63)) (64). Termed "oval cells" because of their oval shaped nucleus, these small cells have a diameter of less than 10 μm. They are located near the portal triads and expand in the livers of animals exposed to oncogenic insults. The term "oval cells" frequently refers to liver stem cells or progenitors. However, oval cells can be distinguished from normal hepatic progenitors phenotypically and in their growth regulatory requirements (65). Several protocols have been shown to lead to the emergence of oval cells: administration of 2-acetylamino fluorine or dipin in combination with partial hepatectomy; administration of carbon tetrachloride, 3-methyl-diaminobenzidine, galactosamine, furane, or 3,5-diethoxycarbonyl-1,4-dihydrocollidine; etluonine addition to a choline-deficient diet; or transgenic albumin-urokinase-type plasminogen-activator mice.

Oval cells were described as positive for several surface and intracellular markers (including hematopoietic and mesenchymal markers not found on normal epithelial hepatic stem cells) such as CD34 (66), CD117 (67), AFP, CK14, CK19 (68), GGT, OC.2, OV-6, and CD90 (69). CD90, however, was subsequently demonstrated to be expressed not by oval cells but by myofibroblasts (70).

Some primary liver tumors are suggested to emerge from oval cells (71).

6. Hepatic progenitors found in various mammalian species

Few data have been published on hepatic progenitors from species other than human or rodent. In general, pigs are used as an animal model closely resembling human physiology and metabolic functions. This makes the pig model more favorable than the rodent model. However, this model is scarcely used due to obvious constraints in keeping animals. Kano *et al.* (72, 73) investigated hepatic progenitors isolated by culture selection from non-parenchymal liver cell suspensions of six-seven months old pigs. Cell clusters in culture were positive for the hepatic markers AFP, albumin, transferrin, CK18, CK7, and c-met, but did they not express biliary markers such as gamma-glutamyltransferase, CK19, and CK14, although they were positive for oval cell marker OV6. Duct-like structures emerged from clusters expressing biliary epithelial markers. Clonal cell growth could be established (74). Comparable cells could be obtained (75) by isolating small liver cells from pigs that had undergone partial hepatectomy. In addition to the hepatic markers albumin and AFP, these cells also expressed biliary marker CK19 and were positive for OV6. In culture, cells were positive for stem-cell factor, CD117, CD90, AFP, CK19, and OV6. Fetal porcine liver cells were used to establish colonies of pluripotent progenitors (76, 77).

7. Extra-hepatic sources of potential liver progenitors

Several extra-hepatic sources have been described to harbor progenitors able to differentiate into hepatic lineages *in vitro* and *in vivo*. It is widely debated whether cells of extra-hepatic origin are able to differentiate into hepatic cell types or if they fuse with the recipient's liver cells when transplanted. Tissue sources include bone marrow, adipose tissue, umbilical cord, and peripheral blood. Hepatic differentiation potentials of embryonic stem cells (ESC), placenta derived stem cells, or induced pluripotent stem cells (iPS cells) are not discussed here; further literature can be found in reviews (78-82).

Bone marrow cells or bone marrow derived hematopoietic stem cells have been suggested to be able to trans-differentiate into hepatic lineages. Petersen *et al.* performed initial experiments with cross-strain and cross-sex bone marrow and liver transplantations in rats (83). When male bone marrow was transplanted into female recipients and liver damage was induced, Y-chromosome positive cells could be detected in the female livers. Also, when male dipeptidyl peptidase (DPPIV) positive bone marrow was transplanted into female DPPIV negative recipients and liver damage was induced, DPPIV positive cells could be detected in the female livers. A further approach included transplantations of major histocompatibility complex class II L21-6 isozyme negative whole livers into positive enzyme expressing rats; after induction of liver damage, positive enzyme expressing cells could be detected. Alison *et al.* (84) investigated human female livers from patients who had received male bone marrow transplants. Y-chromosome positive cells that co-expressed CK8 were detected in the female livers. About 0.5 - 2% of all livers cells were Y-chromosome positive. Theise *et al.* described further *in vivo* experiments on the possible contribution of bone marrow cells towards hepatic lineages in mice (85) and humans (86). Whole bone marrow cells or $CD34^{+}lin^{-}$ sorted cells from male mice were transplanted into female recipients; up to 2.2% (bone marrow) or about 0.7% ($CD34^{+}lin^{-}$) Y-chromosome positive cells could be detected within the female livers. In human patients who had undergone cross-sex bone marrow transplantation, Y-chromosome positive cells could be observed in female livers. 4 - 43% of cholangiocytes and 4 - 38% of hepatocytes were positive for Y-chromosome. Lagasse *et al.* (87) intravenously injected adult wild type bone marrow cells in FAH-/- mice, an animal model of tyrosinemia type I. The mice were rescued and biochemical functions were regained. Only purified hematopoietic stem cells gave rise to donor-derived hematopoietic and hepatic regeneration from total bone marrow cells. However, subsequently published studies revealed that the majority of those liver cells, which were assumed to be donor derived differentiated bone marrow cells, are instead rather the product of donor cells fusing with host liver cells (88, 89). Other studies demonstrated bone marrow cells contributed nothing or very little to liver lineages *in vivo* (90-92). Jang *et al.* (93) and Harris *et al.* (94) could show, however, that a minor percentage (up to 0.1%) of bone marrow cells can contribute to liver cells *in vivo* without fusion. Most evidence to date indicates that only a minority of the observed trans-differentiation events is actually due to differentiation of bone marrow cells into liver lineages and the majority of observed trans-differentiated cells are indeed fusion events.

Similar to the findings of the above described *in vivo* studies, *in vitro* studies of the hepatic differentiation potential of hematopoietic stem cells produced contradicting findings (95-99). Overall, results from *in vitro* studies suggest that bone marrow hematopoietic stem cells can differentiate only barely, if at all, into hepatic lineages.

Mesenchymal stem cells (MSCs), which have similar characteristics, have been isolated from various tissue sources; MSCs from sources such as bone marrow (100-105), skin (106), umbilical cord (107, 108) and adipose tissues (109-115) have been analyzed for the potential to differentiate towards hepatic lineages *in vitro* and *in vivo*. MSC markers from various tissues show similar surface marker expression profiles, described first as classical MSC markers by Pittenger *et al* (116), which were CD29, CD44, CD71, CD90, CD106, CD120a, and CD124. Culture selected clonal bone marrow derived MSCs expressed mesenchymal cell-specific markers (e.g. CD13, CD29, CD44, and CD90), and were negative for hematopoietic markers such as CD3, CD14, CD34, and CD45 (100). When transplanted in SCID mice, non-fused human cells could be detected in the liver. Adipose tissue derived stem cells were described to differentiate into hepatic lineages (109-115). Adipose tissue derived MSCs were characterized to potentially express CD9, CD13, CD29, CD44, CD49d, CD54, CD73, CD90, CD105, CD146, CD166, osteopontin and osteonectin, and to be negative for hematopoietic and endothelial markers such as CD45, CD34 and CD31. Marker expressions and hepatic potential are further summarized in current reviews (117, 118). In general, most *in vivo* transplantation studies using MSCs did not exclude donor cell fusion with host cells. Only one study (Aurich *et al.* (112)) demonstrated the integration of non-fused human adipose MSCs in the livers of mice that had undergone combined toxin induced liver damage and hepatectomy.

Lee *et al.* (119) transplanted green fluorescent protein mouse gallbladder epithelial cells into non-fluorescent SCID mice that had undergone retrorsine treatment and either partial hepatectomy before transplantation or carbon tetrachloride treatment following transplantation. Within one to four months after transplantation, green fluorescent protein positive cells could be detected within the recipient mice. These cells expressed mostly biliary markers, but cells positive for hepatic markers could be detected as well.

Zhao *et al.* isolated hematopoietic stem cells from peripheral blood (120) and demonstrated their *in vitro* multilineage differentiation potential; treatment of cultures with HGF induced cells to acquire a round or oval-like flattened morphology. Most of the cells were positive for intracellular albumin and AFP expression; some cells demonstrated CK7 expression. Sun *et al.* (121) showed that human umbilical blood cells integrated into livers of rat chimeras, and these cells were positive for human hematopoietic, biliary, and hepatic proteins. Crema *et al.* isolated $CD133^{+}$ cord blood cells (122). Transplantation into liver-damaged SCID mice resulted in clusters of human-derived cells expressing human leucocyte antigen-class I, HepPar1, and OV6 antigens. Within these clusters, human albumin, AFP, and CK19 could be detected. Human umbilical blood cells demonstrated *in vitro* hepatocyte-like differentiation and expression of hepatic proteins when transplanted in rodents with induced liver damage (107, 123, 124).

Conclusively, it appears that extra-hepatic progenitors integrate into the liver only to a very minor percentage and only when severe liver damage is induced. The majority of these events appear to be due to fusion and not differentiation. The observed improvements of liver functions by mesenchymal cells could be attributed to their secretion of growth factors and cytokines and immunosuppressive properties (111, 125-127).

8. Conclusion

Although there is still some debate about the detailed characteristics that identify hepatic progenitors, much progress has been achieved during recent years in defining, isolating, characterizing, and transplanting various types of progenitors. This is especially the case for hepatic progenitors isolated from human livers. Hepatic progenitors represent a population with potential advantages over total liver cell suspensions or hepatocytes for cell transplantation in patients (29, 128), for review see (129). Because of their high proliferation and differentiation potential a major advantage for transplantation of stem cells over total liver cell suspensions would be the requirement for less cell numbers to inject, which would decrease the risks associated with transplanting high cell numbers. In addition, because of their proliferation and differentiation potential, progenitors could be used in applications such as extracorporeal liver support systems (130, 131), and may be used as an alternative cell source in pharmacological screening models. Cultures of progenitors also provide an easy *in vitro* tool to study principles of developmental biology.

9. References

[1] Michalopoulos GK. Liver regeneration: alternative epithelial pathways. Int J Biochem Cell Biol. Feb;43(2):173-9.

[2] Michalopoulos GK. Liver regeneration after partial hepatectomy: critical analysis of mechanistic dilemmas. Am J Pathol. Jan;176(1):2-13.

[3] Michalopoulos GK. Liver regeneration. J Cell Physiol. 2007 Nov;213(2):286-300.

[4] Overturf K, al-Dhalimy M, Ou CN, Finegold M, Grompe M. Serial transplantation reveals the stem-cell-like regenerative potential of adult mouse hepatocytes. Am J Pathol. 1997;151(5):1273-80.

[5] Schmelzer E, McClelland RE, Melhem A, Zhang L, Yao H, Wauthier E, et al. Hepatic stem cell and the liver's maturational lineages: Implications for liver biology, gene expression, and cell therapies. In: Potten C, Clarke R, Wilson J, Renehan A, editors. Tissue stem cells. New York: Taylor and Francis; 2006. p. 161-214.

[6] Dan YY, Yeoh GC. Liver stem cells: a scientific and clinical perspective. J Gastroenterol Hepatol. 2008 May;23(5):687-98.
[7] Zaret K. Early liver differentiation: genetic potentiation and multilevel growth control. Curr Opin Genet Dev. 1998;8(5):526-31.
[8] Zaret KS. Hepatocyte differentiation: from the endoderm and beyond. Curr Opin Genet Dev. 2001;11(5):568-74.
[9] Zaret KS. Regulatory phases of early liver development: paradigms of organogenesis. Nat Rev Genet. 2002 Jul;3(7):499-512.
[10] Kolterud A, Wandzioch E, Carlsson L. Lhx2 is expressed in the septum transversum mesenchyme that becomes an integral part of the liver and the formation of these cells is independent of functional Lhx2. Gene Expr Patterns. 2004 Sep;4(5):521-8.
[11] Shiojiri N. The origin of intrahepatic bile duct cells in the mouse. J Embryol Exp Morphol. 1984 Feb;79:25-39.
[12] Schmelzer E, Zhang L, Bruce A, Wauthier E, Ludlow J, Yao H, et al. Human Hepatic Stem Cells from Fetal and Postnatal Donors. J Exp Med. 2007;204(8):1973-87.
[13] Shiojiri N, Koike T. Differentiation of biliary epithelial cells from the mouse hepatic endodermal cells cultured in vitro. Tohoku J Exp Med. 1997 Jan;181(1):1-8.
[14] Shiojiri N, Mizuno T. Differentiation of functional hepatocytes and biliary epithelial cells from immature hepatocytes of the fetal mouse in vitro. Anat Embryol (Berl). 1993 Mar;187(3):221-9.
[15] Haruna Y, Saito K, Spaulding S, Nalesnik MA, Gerber MA. Identification of bipotential progenitor cells in human liver development. Hepatology. 1996 Mar;23(3):476-81.
[16] Shiojiri N, Takeshita K, Yamasaki H, Iwata T. Suppression of C/EBP alpha expression in biliary cell differentiation from hepatoblasts during mouse liver development. J Hepatol. 2004 Nov;41(5):790-8.
[17] Yamasaki H, Sada A, Iwata T, Niwa T, Tomizawa M, Xanthopoulos KG, et al. Suppression of C/EBPalpha expression in periportal hepatoblasts may stimulate biliary cell differentiation through increased Hnf6 and Hnf1b expression. Development. 2006 Nov;133(21):4233-43.
[18] Schmelzer E, Wauthier E, Reid LM. The Phenotypes of Pluripotent Human Hepatic Progenitors. Stem Cells. 2006;24(8):1852-8. Epub 2006 Apr 20.
[19] Dan YY, Riehle KJ, Lazaro C, Teoh N, Haque J, Campbell JS, et al. Isolation of multipotent progenitor cells from human fetal liver capable of differentiating into liver and mesenchymal lineages. Proc Natl Acad Sci U S A. 2006;103(26):9912-7. Epub 2006 Jun 16.
[20] Najimi M, Khuu DN, Lysy PA, Jazouli N, Abarca J, Sempoux C, et al. Adult-derived human liver mesenchymal-like cells as a potential progenitor reservoir of hepatocytes? Cell Transplant. 2007;16(7):717-28.
[21] Khuu DN, Scheers I, Ehnert S, Jazouli N, Nyabi O, Buc-Calderon P, et al. In vitro differentiated adult human liver progenitor cells display mature hepatic metabolic functions: a potential tool for in vitro pharmacotoxicological testing. Cell Transplant. 2011;20(2):287-302.
[22] Herrera MB, Bruno S, Buttiglieri S, Tetta C, Gatti S, Deregibus MC, et al. Isolation and characterization of a stem cell population from adult human liver. Stem Cells. 2006 Dec;24(12):2840-50.

[23] Malhi H, Irani AN, Gagandeep S, Gupta S. Isolation of human progenitor liver epithelial cells with extensive replication capacity and differentiation into mature hepatocytes. J Cell Sci. 2002 Jul 1;115(Pt 13):2679-88.
[24] Sicklick JK, Li YX, Melhem A, Schmelzer E, Zdanowicz M, Huang J, et al. Hedgehog signaling maintains resident hepatic progenitors throughout life. Am J Physiol Gastrointest Liver Physiol. 2006;290(5):G859-70. Epub 2005 Dec 1.
[25] Theise ND, Saxena R, Portmann BC, Thung SN, Yee H, Chiriboga L, et al. The canals of Hering and hepatic stem cells in humans. Hepatology. 1999;30(6):1425-33.
[26] Roskams TA, Theise ND, Balabaud C, Bhagat G, Bhathal PS, Bioulac-Sage P, et al. Nomenclature of the finer branches of the biliary tree: canals, ductules, and ductular reactions in human livers. Hepatology. 2004;39(6):1739-45.
[27] Carpentier R, Suner RE, van Hul N, Kopp JL, Beaudry JB, Cordi S, et al. Embryonic ductal plate cells give rise to cholangiocytes, periportal hepatocytes, and adult liver progenitor cells. Gastroenterology. Oct;141(4):1432-8, 8 e1-4.
[28] Furuyama K, Kawaguchi Y, Akiyama H, Horiguchi M, Kodama S, Kuhara T, et al. Continuous cell supply from a Sox9-expressing progenitor zone in adult liver, exocrine pancreas and intestine. Nat Genet. Jan;43(1):34-41.
[29] Khan AA, Shaik MV, Parveen N, Rajendraprasad A, Aleem MA, Habeeb MA, et al. Human fetal liver-derived stem cell transplantation as supportive modality in the management of end-stage decompensated liver cirrhosis. Cell Transplant. 2010;19(4):409-18.
[30] Sakamoto S, Yachi A, Anzai T, Wada T. AFP-producing cells in hepatitis and in liver cirrhosis. Ann N Y Acad Sci. 1975;259:253-8.
[31] Abelev GI. Alpha-fetoprotein in ontogenesis and its association with malignant tumors. Adv Cancer Res. 1971;14:295-358.
[32] Zhang L, Theise N, Chua M, Reid LM. The stem cell niche of human livers: symmetry between development and regeneration. Hepatology. 2008 Nov;48(5):1598-607.
[33] Abelev GI, Eraiser TL. Cellular aspects of alpha-fetoprotein reexpression in tumors. Semin Cancer Biol. 1999 Apr;9(2):95-107.
[34] Terrace JD, Currie IS, Hay DC, Masson NM, Anderson RA, Forbes SJ, et al. Progenitor cell characterization and location in the developing human liver. Stem Cells Dev. 2007 Oct;16(5):771-8.
[35] Nierhoff D, Ogawa A, Oertel M, Chen YQ, Shafritz DA. Purification and characterization of mouse fetal liver epithelial cells with high in vivo repopulation capacity. Hepatology. 2005;42(1):130-9.
[36] Tanimizu N, Nishikawa M, Saito H, Tsujimura T, Miyajima A. Isolation of hepatoblasts based on the expression of Dlk/Pref-1. J Cell Sci. 2003;116(Pt 9):1775-86.
[37] Tanimizu N, Saito H, Mostov K, Miyajima A. Long-term culture of hepatic progenitors derived from mouse Dlk+ hepatoblasts. J Cell Sci. 2004 Dec 15;117(Pt 26):6425-34.
[38] Tanimizu N, Miyajima A, Tanimizu N, Nishikawa M, Saito H, Tsujimura T, et al. Notch signaling controls hepatoblast differentiation by altering the expression of liver-enriched transcription factors. Isolation of hepatoblasts based on the expression of Dlk/Pref-1. J Cell Sci. 2004;117(Pt 15):3165-74.
[39] Germain L, Blouin MJ, Marceau N. Biliary epithelial and hepatocytic cell lineage relationships in embryonic rat liver as determined by the differential expression of

cytokeratins, alpha-fetoprotein, albumin, and cell surface-exposed components. Cancer Res. 1988;48(17):4909-18.
[40] Kubota H, Reid LM. Clonogenic hepatoblasts, common precursors for hepatocytic and biliary lineages, are lacking classical major histocompatibility complex class I antigen. Proc Natl Acad Sci U S A. 2000;97(22):12132-7.
[41] Tateno C, Yoshizato K. Growth and differentiation in culture of clonogenic hepatocytes that express both phenotypes of hepatocytes and biliary epithelial cells. Am J Pathol. 1996;149(5):1593-605.
[42] Mitaka T, Kojima T, Mizuguchi T, Mochizuki Y. Growth and maturation of small hepatocytes isolated from adult rat liver. Biochem Biophys Res Commun. 1995;214(2):310-7.
[43] Mitaka T, Mikami M, Sattler GL, Pitot HC, Mochizuki Y. Small cell colonies appear in the primary culture of adult rat hepatocytes in the presence of nicotinamide and epidermal growth factor. Hepatology. 1992;16(2):440-7.
[44] Mitaka T, Sato F, Mizuguchi T, Yokono T, Mochizuki Y. Reconstruction of hepatic organoid by rat small hepatocytes and hepatic nonparenchymal cells. Hepatology. 1999;29(1):111-25.
[45] Tateno C, Yoshizato K. Long-term cultivation of adult rat hepatocytes that undergo multiple cell divisions and express normal parenchymal phenotypes. Am J Pathol. 1996;148(2):383-92.
[46] Suzuki A, Nakauchi H, Taniguchi H. In vitro production of functionally mature hepatocytes from prospectively isolated hepatic stem cells. Cell Transplant. 2003;12(5):469-73.
[47] Suzuki A, Zheng YW, Fukao K, Nakauchi H, Taniguchi H. Liver repopulation by c-Met-positive stem/progenitor cells isolated from the developing rat liver. Hepatogastroenterology. 2004 Mar-Apr;51(56):423-6.
[48] Suzuki A, Zheng Y, Kondo R, Kusakabe M, Takada Y, Fukao K, et al. Flow-cytometric separation and enrichment of hepatic progenitor cells in the developing mouse liver. Hepatology. 2000;32(6):1230-9.
[49] Suzuki A, Taniguchi H, Zheng YW, Takada Y, Fukunaga K, Seino K, et al. Clonal colony formation of hepatic stem/progenitor cells enhanced by embryonic fibroblast conditioning medium. Transplant Proc. 2000;32(7):2328-30.
[50] Suzuki A, Zheng Yw YW, Kaneko S, Onodera M, Fukao K, Nakauchi H, et al. Clonal identification and characterization of self-renewing pluripotent stem cells in the developing liver. J Cell Biol. 2002;156(1):173-84. Epub 2002 Jan 07.
[51] Feng RQ, Du LY, Guo ZQ. In vitro cultivation and differentiation of fetal liver stem cells from mice. Cell Res. 2005;15(5):401-5.
[52] Dabeva MD, Petkov PM, Sandhu J, Oren R, Laconi E, Hurston E, et al. Proliferation and differentiation of fetal liver epithelial progenitor cells after transplantation into adult rat liver. Am J Pathol. 2000;156(6):2017-31.
[53] Dolle L, Best J, Empsen C, Mei J, Van Rossen E, Roelandt P, et al. Successful isolation of liver progenitor cells by aldehyde dehydrogenase activity from naive mice. Hepatology. 2012;55(2):540-52.

[54] Strick-Marchand H, Weiss MC. Inducible differentiation and morphogenesis of bipotential liver cell lines from wild-type mouse embryos. Hepatology. 2002;36(4 Pt 1):794-804.
[55] Strick-Marchand H, Morosan S, Charneau P, Kremsdorf D, Weiss MC. Bipotential mouse embryonic liver stem cell lines contribute to liver regeneration and differentiate as bile ducts and hepatocytes. Proc Natl Acad Sci U S A. 2004;101(22):8360-5. Epub 2004 May 20.
[56] Rogler LE. Selective bipotential differentiation of mouse embryonic hepatoblasts in vitro. Am J Pathol. 1997;150(2):591-602.
[57] Richards WG, Yoder BK, Isfort RJ, Detilleux PG, Foster C, Neilsen N, et al. Isolation and characterization of liver epithelial cell lines from wild-type and mutant TgN737Rpw mice. Am J Pathol. 1997;150(4):1189-97.
[58] Ott M, Rajvanshi P, Sokhi RP, Alpini G, Aragona E, Dabeva M, et al. Differentiation-specific regulation of transgene expression in a diploid epithelial cell line derived from the normal F344 rat liver. J Pathol. 1999;187(3):365-73.
[59] Grisham JW. Cell types in long-term propagable cultures of rat liver. Ann N Y Acad Sci. 1980;349:128-37.
[60] Grisham JW, Coleman WB, Smith GJ. Isolation, culture, and transplantation of rat hepatocytic precursor (stem-like) cells. Proc Soc Exp Biol Med. 1993;204(3):270-9.
[61] Tsao MS, Smith JD, Nelson KG, Grisham JW. A diploid epithelial cell line from normal adult rat liver with phenotypic properties of 'oval' cells. Exp Cell Res. 1984;154(1):38-52.
[62] Lee LW, Tsao MS, Grisham JW, Smith GJ. Emergence of neoplastic transformants spontaneously or after exposure to N-methyl-N'-nitro-N-nitrosoguanidine in populations of rat liver epithelial cells cultured under selective and nonselective conditions. Am J Pathol. 1989;135(1):63-71.
[63] Newsome PN, Hussain MA, Theise ND. Hepatic oval cells: helping redefine a paradigm in stem cell biology. Curr Top Dev Biol. 2004;61:1-28.
[64] Farber E. Similarities in the sequence of early histological changes induced in the liver of the rat by ethionine, 2-acetylamino-fluorene, and 3'-methyl-4-dimethylaminoazobenzene. Cancer Res. 1956;16(2):142-8.
[65] Oh SH, Hatch HM, Petersen BE. Hepatic oval 'stem' cell in liver regeneration. Semin Cell Dev Biol. 2002 Dec;13(6):405-9.
[66] Omori N, Omori M, Evarts RP, Teramoto T, Miller MJ, Hoang TN, et al. Partial cloning of rat CD34 cDNA and expression during stem cell-dependent liver regeneration in the adult rat. Hepatology. 1997;26(3):720-7.
[67] Fujio K, Evarts RP, Hu Z, Marsden ER, Thorgeirsson SS. Expression of stem cell factor and its receptor, c-kit, during liver regeneration from putative stem cells in adult rat. Lab Invest. 1994;70(4):511-6.
[68] Bisgaard HC, Nagy P, Ton PT, Hu Z, Thorgeirsson SS. Modulation of keratin 14 and alpha-fetoprotein expression during hepatic oval cell proliferation and liver regeneration. J Cell Physiol. 1994 Jun;159(3):475-84.
[69] Petersen BE, Goff JP, Greenberger JS, Michalopoulos GK. Hepatic oval cells express the hematopoietic stem cell marker Thy-1 in the rat. Hepatology. 1998;27(2):433-45.

[70] Dezso K, Jelnes P, Laszlo V, Baghy K, Bodor C, Paku S, et al. Thy-1 is expressed in hepatic myofibroblasts and not oval cells in stem cell-mediated liver regeneration. Am J Pathol. 2007 Nov;171(5):1529-37.

[71] Faris RA, Monfils BA, Dunsford HA, Hixson DC. Antigenic relationship between oval cells and a subpopulation of hepatic foci, nodules, and carcinomas induced by the "resistant hepatocyte" model system. Cancer Res. 1991;51(4):1308-17.

[72] Kano J, Noguchi M, Kodama M, Tokiwa T. The in vitro differentiating capacity of nonparenchymal epithelial cells derived from adult porcine livers. Am J Pathol. 2000;156(6):2033-43.

[73] Kano J, Tokiwa T, Zhou X, Kodama M. Colonial growth and differentiation of epithelial cells derived from abattoir adult porcine livers. J Gastroenterol Hepatol. 1998 Sep;13 Suppl:S62-9.

[74] Tokiwa T, Yamazaki T, Ono M, Enosawa S, Tsukiyama T. Cloning and characterization of liver progenitor cells from the scattered cell clusters in primary culture of porcine livers. Cell Transplant. 2008;17(1-2):179-86.

[75] He Z, Feng M. Activation, isolation, identification and culture of hepatic stem cells from porcine liver tissues. Cell Prolif. 2011 Dec;44(6):558-66.

[76] Talbot NC, Pursel VG, Rexroad CE, Jr., Caperna TJ, Powell AM, Stone RT. Colony isolation and secondary culture of fetal porcine hepatocytes on STO feeder cells. In Vitro Cell Dev Biol Anim. 1994;30A(12):851-8.

[77] Talbot NC, Rexroad CE, Jr., Powell AM, Pursel VG, Caperna TJ, Ogg SL, et al. A continuous culture of pluripotent fetal hepatocytes derived from the 8-day epiblast of the pig. In Vitro Cell Dev Biol Anim. 1994;30A(12):843-50.

[78] Chun YS, Chaudhari P, Jang YY. Applications of patient-specific induced pluripotent stem cells; focused on disease modeling, drug screening and therapeutic potentials for liver disease. Int J Biol Sci.6(7):796-805.

[79] Asgari S, Pournasr B, Salekdeh GH, Ghodsizadeh A, Ott M, Baharvand H. Induced pluripotent stem cells: a new era for hepatology. J Hepatol. Oct;53(4):738-51.

[80] Chen X, Zeng F. Directed hepatic differentiation from embryonic stem cells. Protein Cell. Mar;2(3):180-8.

[81] Hannoun Z, Filippi C, Sullivan G, Hay DC, Iredale JP. Hepatic endoderm differentiation from human embryonic stem cells. Curr Stem Cell Res Ther. Sep;5(3):233-44.

[82] Miki T, Marongiu F, Ellis EC, Dorko K, Mitamura K, Ranade A, et al. Production of hepatocyte-like cells from human amnion. Methods Mol Biol. 2009;481:155-68.

[83] Petersen BE, Bowen WC, Patrene KD, Mars WM, Sullivan AK, Murase N, et al. Bone marrow as a potential source of hepatic oval cells. Science. 1999;284(5417):1168-70.

[84] Alison MR, Poulsom R, Jeffery R, Dhillon AP, Quaglia A, Jacob J, et al. Hepatocytes from non-hepatic adult stem cells. Nature. 2000;406(6793):257.

[85] Theise ND, Badve S, Saxena R, Henegariu O, Sell S, Crawford JM, et al. Derivation of hepatocytes from bone marrow cells in mice after radiation-induced myeloablation. Hepatology. 2000;31(1):235-40.

[86] Theise ND, Nimmakayalu M, Gardner R, Illei PB, Morgan G, Teperman L, et al. Liver from bone marrow in humans. Hepatology. 2000;32(1):11-6.

[87] Lagasse E, Connors H, Al-Dhalimy M, Reitsma M, Dohse M, Osborne L, et al. Purified hematopoietic stem cells can differentiate into hepatocytes in vivo. Nat Med. 2000;6(11):1229-34.

[88] Wang X, Willenbring H, Akkari Y, Torimaru Y, Foster M, Al-Dhalimy M, et al. Cell fusion is the principal source of bone-marrow-derived hepatocytes. Nature. 2003 Apr 24;422(6934):897-901.

[89] Vig P, Russo FP, Edwards RJ, Tadrous PJ, Wright NA, Thomas HC, et al. The sources of parenchymal regeneration after chronic hepatocellular liver injury in mice. Hepatology. 2006;43(2):316-24.

[90] Cantz T, Sharma AD, Jochheim-Richter A, Arseniev L, Klein C, Manns MP, et al. Reevaluation of bone marrow-derived cells as a source for hepatocyte regeneration. Cell Transplant. 2004;13(6):659-66.

[91] Menthena A, Deb N, Oertel M, Grozdanov PN, Sandhu J, Shah S, et al. Bone marrow progenitors are not the source of expanding oval cells in injured liver. Stem Cells. 2004;22(6):1049-61.

[92] Popp FC, Slowik P, Eggenhofer E, Renner P, Lang SA, Stoeltzing O, et al. No contribution of multipotent mesenchymal stromal cells to liver regeneration in a rat model of prolonged hepatic injury. Stem Cells. 2007 Mar;25(3):639-45.

[93] Jang YY, Collector MI, Baylin SB, Diehl AM, Sharkis SJ. Hematopoietic stem cells convert into liver cells within days without fusion. Nat Cell Biol. 2004;6(6):532-9. Epub 2004 May 09.

[94] Harris RG, Herzog EL, Bruscia EM, Grove JE, Van Arnam JS, Krause DS. Lack of a fusion requirement for development of bone marrow-derived epithelia. Science. 2004 Jul 2;305(5680):90-3.

[95] Lian G, Wang C, Teng C, Zhang C, Du L, Zhong Q, et al. Failure of hepatocyte marker-expressing hematopoietic progenitor cells to efficiently convert into hepatocytes in vitro. Exp Hematol. 2006;34(3):348-58.

[96] Yamada Y, Nishimoto E, Mitsuya H, Yonemura Y. In vitro transdifferentiation of adult bone marrow Sca-1+ cKit- cells cocultured with fetal liver cells into hepatic-like cells without fusion. Exp Hematol. 2006;34(1):97-106.

[97] Avital I, Inderbitzin D, Aoki T, Tyan DB, Cohen AH, Ferraresso C, et al. Isolation, characterization, and transplantation of bone marrow-derived hepatocyte stem cells. Biochem Biophys Res Commun. 2001;288(1):156-64.

[98] Crosby HA, Kelly DA, Strain AJ. Human hepatic stem-like cells isolated using c-kit or CD34 can differentiate into biliary epithelium. Gastroenterology. 2001;120(2):534-44.

[99] Miyazaki M, Akiyama I, Sakaguchi M, Nakashima E, Okada M, Kataoka K, et al. Improved conditions to induce hepatocytes from rat bone marrow cells in culture. Biochem Biophys Res Commun. 2002;298(1):24-30.

[100] Tao XR, Li WL, Su J, Jin CX, Wang XM, Li JX, et al. Clonal mesenchymal stem cells derived from human bone marrow can differentiate into hepatocyte-like cells in injured livers of SCID mice. J Cell Biochem. 2009 Oct 15;108(3):693-704.

[101] Ayatollahi M, Soleimani M, Tabei SZ, Kabir Salmani M. Hepatogenic differentiation of mesenchymal stem cells induced by insulin like growth factor-I. World J Stem Cells. Dec 26;3(12):113-21.

[102] Ghaedi M, Soleimani M, Shabani I, Duan Y, Lotfi AS. Hepatic differentiation from human mesenchymal stem cells on a novel nanofiber scaffold. Cell Mol Biol Lett. Mar;17(1):89-106.
[103] Hwang S, Hong HN, Kim HS, Park SR, Won YJ, Choi ST, et al. Hepatogenic Differentiation of Mesenchymal Stem Cells in a Rat Model of Thioacetamide-induced Liver Cirrhosis. Cell Biol Int. 2012;36(3):279-88.
[104] Lin N, Lin J, Bo L, Weidong P, Chen S, Xu R. Differentiation of bone marrow-derived mesenchymal stem cells into hepatocyte-like cells in an alginate scaffold. Cell Prolif. Oct;43(5):427-34.
[105] Pournasr B, Mohamadnejad M, Bagheri M, Aghdami N, Shahsavani M, Malekzadeh R, et al. In vitro differentiation of human bone marrow mesenchymal stem cells into hepatocyte-like cells. Arch Iran Med. Jul;14(4):244-9.
[106] De Kock J, Vanhaecke T, Biernaskie J, Rogiers V, Snykers S. Characterization and hepatic differentiation of skin-derived precursors from adult foreskin by sequential exposure to hepatogenic cytokines and growth factors reflecting liver development. Toxicol In Vitro. 2009 Dec;23(8):1522-7.
[107] Kakinuma S, Tanaka Y, Chinzei R, Watanabe M, Shimizu-Saito K, Hara Y, et al. Human umbilical cord blood as a source of transplantable hepatic progenitor cells. Stem Cells. 2003;21(2):217-27.
[108] Campard D, Lysy PA, Najimi M, Sokal EM. Native umbilical cord matrix stem cells express hepatic markers and differentiate into hepatocyte-like cells. Gastroenterology. 2008 Mar;134(3):833-48.
[109] Seo MJ, Suh SY, Bae YC, Jung JS. Differentiation of human adipose stromal cells into hepatic lineage in vitro and in vivo. Biochem Biophys Res Commun. 2005 Mar 4;328(1):258-64.
[110] Banas A, Teratani T, Yamamoto Y, Tokuhara M, Takeshita F, Osaki M, et al. Rapid hepatic fate specification of adipose-derived stem cells and their therapeutic potential for liver failure. J Gastroenterol Hepatol. 2009 Jan;24(1):70-7.
[111] Banas A, Teratani T, Yamamoto Y, Tokuhara M, Takeshita F, Osaki M, et al. IFATS collection: in vivo therapeutic potential of human adipose tissue mesenchymal stem cells after transplantation into mice with liver injury. Stem Cells. 2008 Oct;26(10):2705-12.
[112] Aurich H, Sgodda M, Kaltwasser P, Vetter M, Weise A, Liehr T, et al. Hepatocyte differentiation of mesenchymal stem cells from human adipose tissue in vitro promotes hepatic integration in vivo. Gut. 2009 Apr;58(4):570-81.
[113] Talens-Visconti R, Bonora A, Jover R, Mirabet V, Carbonell F, Castell JV, et al. Hepatogenic differentiation of human mesenchymal stem cells from adipose tissue in comparison with bone marrow mesenchymal stem cells. World J Gastroenterol. 2006 Sep 28;12(36):5834-45.
[114] Banas A, Teratani T, Yamamoto Y, Tokuhara M, Takeshita F, Quinn G, et al. Adipose tissue-derived mesenchymal stem cells as a source of human hepatocytes. Hepatology. 2007 Jul;46(1):219-28.
[115] Okura H, Komoda H, Saga A, Kakuta-Yamamoto A, Hamada Y, Fumimoto Y, et al. Properties of hepatocyte-like cell clusters from human adipose tissue-derived mesenchymal stem cells. Tissue Eng Part C Methods. Aug;16(4):761-70.

[116] Pittenger MF, Mackay AM, Beck SC, Jaiswal RK, Douglas R, Mosca JD, et al. Multilineage potential of adult human mesenchymal stem cells. Science. 1999 Apr 2;284(5411):143-7.
[117] Baer PC. Adipose-derived stem cells and their potential to differentiate into the epithelial lineage. Stem Cells Dev. Oct;20(10):1805-16.
[118] Al Battah F, De Kock J, Vanhaecke T, Rogiers V, Goette M. Current status of human adipose-derived stem cells: differentiation into hepatocyte-like cells. ScientificWorldJournal.11:1568-81.
[119] Lee SP, Savard CE, Kuver R. Gallbladder epithelial cells that engraft in mouse liver can differentiate into hepatocyte-like cells. Am J Pathol. 2009 Mar;174(3):842-53.
[120] Zhao Y, Glesne D, Huberman E. A human peripheral blood monocyte-derived subset acts as pluripotent stem cells. Proc Natl Acad Sci U S A. 2003;100(5):2426-31.
[121] Sun Y, Xiao D, Zhang RS, Cui GH, Wang XH, Chen XG. Formation of human hepatocyte-like cells with different cellular phenotypes by human umbilical cord blood-derived cells in the human-rat chimeras. Biochem Biophys Res Commun. 2007 Jun 15;357(4):1160-5.
[122] Crema A, Ledda M, De Carlo F, Fioretti D, Rinaldi M, Marchese R, et al. Cord blood CD133 cells define an OV6-positive population that can be differentiated in vitro into engraftable bipotent hepatic progenitors. Stem Cells Dev. Nov;20(11):2009-21.
[123] Moon YJ, Lee MW, Yoon HH, Yang MS, Jang IK, Lee JE, et al. Hepatic differentiation of cord blood-derived multipotent progenitor cells (MPCs) in vitro. Cell Biol Int. 2008 Oct;32(10):1293-301.
[124] Moon YJ, Yoon HH, Lee MW, Jang IK, Lee DH, Lee JH, et al. Multipotent progenitor cells derived from human umbilical cord blood can differentiate into hepatocyte-like cells in a liver injury rat model. Transplant Proc. 2009 Dec;41(10):4357-60.
[125] Puissant B, Barreau C, Bourin P, Clavel C, Corre J, Bousquet C, et al. Immunomodulatory effect of human adipose tissue-derived adult stem cells: comparison with bone marrow mesenchymal stem cells. Br J Haematol. 2005 Apr;129(1):118-29.
[126] Isoda K, Kojima M, Takeda M, Higashiyama S, Kawase M, Yagi K. Maintenance of hepatocyte functions by coculture with bone marrow stromal cells. J Biosci Bioeng. 2004;97(5):343-6.
[127] Parekkadan B, van Poll D, Suganuma K, Carter EA, Berthiaume F, Tilles AW, et al. Mesenchymal stem cell-derived molecules reverse fulminant hepatic failure. PLoS One. 2007;2(9):e941.
[128] Habibullah CM, Syed IH, Qamar A, Taher-Uz Z. Human fetal hepatocyte transplantation in patients with fulminant hepatic failure. Transplantation. 1994 Oct 27;58(8):951-2.
[129] Dhawan A, Puppi J, Hughes RD, Mitry RR. Human hepatocyte transplantation: current experience and future challenges. Nat Rev Gastroenterol Hepatol. May;7(5):288-98.
[130] Ring A, Gerlach J, Peters G, Pazin BJ, Minervini CF, Turner ME, et al. Hepatic Maturation of Human Fetal Hepatocytes in Four-Compartment Three-Dimensional Perfusion Culture. Tissue Eng Part C Methods. 2010;16(5):835-45.

[131] Schmelzer E, Triolo F, Turner ME, Thompson RL, Zeilinger K, Reid LM, et al. Three-Dimensional Perfusion Bioreactor Culture Supports Differentiation of Human Fetal Liver Cells. Tissue Eng Part A. 2010;16(6):2007-16.

4

Possible Roles of Nuclear Lipids in Liver Regeneration

M. Viola-Magni[1] and P.B. Gahan[2]
[1]*Perugia University, Enrico Puccinelli Foundation*
[2]*King's College London*
[1]*Italy*
[2]*UK*

1. Introduction

Although no lipids were considered to be present inside the nuclear membrane (Berg, 1951), their presence in chromatin was first demonstrated cytochemically by Chayen et al (1957) in *Vicia faba* root apices and liver nuclei. Sphingomyelin was further demonstrated biochemically to represent some 7% of isolated calf thymus nucleohistone preparations (Chayen and Gahan, 1958), the presence of sphingomyelin being confirmed by X-ray diffraction studies (Wilkins M. H. F, personal communication). Nevertheless, the lipids and carbohydrate present in the nuclei were considered to be minor components, most of them being due to contamination during chromatin separation (Tata et al.1972). In contrast, some biochemical measurements showed the presence of neutral lipids (Song and Rebel 1987) and phospholipids in nuclei and chromosomes from a large variety of tissues (Chayen et al 1959, a, b, Gahan 1965a). The criticism linked to possible contamination cannot be applied to the cytochemical evidence that showed the presence of chromatin-associated phospholipid material in a broad range of tissues (Idelman 1957, 1958a,b, Chayen et al. 1959a,b, La Cour et al. 1958, Gahan 1965a,b, Cave and Gahan 1971, Gahan et al. 1974, Gahan et al. 1987, Viola-Magni et al. 1985a). In a combined autoradiographic and biochemical analysis, it was shown that H^3 -ethalomine incorporated into *Vicia faba* root nuclei was localised at the level of chromatin and nucleoli rather than at the level of the nuclear membrane. Hepatocyte nuclei treated with Triton and hypotonic solutions liberate chromatin that contains 10% of the total nuclear lipids. The composition of fatty acids demonstrated an enrichment of palmitic acid and a reduction in arachidonic acid (Albi et al. 1994) thus supporting the idea that these lipids cannot be derived from the nuclear membrane. In addition, the chromatographic separation of phospholipids has demonstrated an enrichment of both sphingomyelin and phosphatidylserine with respect to the nuclear membrane composition (Albi et al. 1994). The data were also confirmed by studying the turnover of phospholipids at the level of the microsomes, nuclear membrane and chromatin from hepatocytes (Viola-Magni et al. 1986). In rats injected with radioactive phosphorus, the peak of incorporation was observed after 6 h in microsomes and nuclear membranes, but only after 9h in the chromatin. This confirmed a lack of possible chromatin contamination.

A clear demonstration was obtained by labelling the fatty acids of the nuclear membrane by radio-iodination. Hepatocyte nuclei were separated and then radio-iodinated; the chromatin extracted from them was unlabelled, whereas all the label present in the nuclei was recovered from the nuclear membrane fraction. Radio-iodination of isolated chromatin showed the presence of label thus confirming the presence of lipids. (Albi et al.1994).

The presence of nuclear phospholipids was also demonstrated in a large variety of tissues including tumour cells (Splanger et al. 1975, Upreti et al. 1983).

Cocco et al. (1988) demonstrated the presence of phosphoinositides which may act as nuclear signals through the generation of DAG (diacylglycerol) due to specific phospholipase activity (D' Santos et al. 1998, Cocco et al. 2001, Martelli et al. 2001, Irvine 2003). The presence of neutral lipids was demonstrated by Song and Rebel (1987) and of cholesterol (CHO) by Albi and Viola-Magni (2002). The presence in chromatin of the enzymes sphingomyelinase (SMase), sphingomyelin synthase (SMsynthase), phospholipases of phosphatidylcholine (PC) and phosphatidylinositol (PI) and sphingomyelin reverse synthase associated demonstrated the existence of a metabolic cycle for such phospholipids. (Albi and Viola-Magni 2004, Albi 2011).

There is evidence for the presence of a phospholipid-calcium-dependent protein kinase C (PKC) in nuclei together with the enzymes involved with phospholipid turnover (Alessenko and Burlakova (2002). Protein kinase C interacts with the nuclear phosphoinositol and sphingomyelin cycle products. This fact implies the possibility that signal transduction events could also occur at the nuclear level during the induction of cell proliferation, differentiation and apoptosis.

In this review, it is intended to consider the composition of the lipids present in chromatin, the enzymes associated with the metabolism of these lipids, their possible roles in normal hepatocytes, the cell cycle and regenerating liver.

2. Composition of the chromatin associated phospholipids in normal v regenerating liver

After 70% hepatectomy, the liver first regenerates the hepatocytes prior to regenerating the other cell types. The first peak of new hepatocytes is observed after 24 h in 30 day-old rats with a second peak occurring after 36 h. The other cell types, including Kupffer cells and endothelial cells, blood vessels and others, proliferate on the third day (Bresnick 1971).

The synthesis of phospholipids was studied after partial hepatectomy (Viola-Magni et al. 1985 b) in the both hepatocyte nuclei and chromatin. The whole nuclei showed an increase in phospholipid synthesis after six hours reaching a peak at 12 h, after which, a constant level was maintained until 48 h. The synthesis of phospholipids in the chromatin increased at 12h to reach a peak at 18h, which level remained until 24 h. This was followed by a peak at 30 h, a timing that marks the end of the first proliferation peak and the start of the second wave of mitosis (Viola-Magni et al, 1985b). It is to be noted that DNA synthesis starts at 12h after partial hepatectomy to reach a maximum at about 24h (Viola-Magni 1985b). This shows that the initiation of both DNA and phospholipid synthesis are occurring at about the same time.

Since the second peak of DNA synthesis starts at 30 h, the end of the DNA synthesis of the first peak and the initiation of the second happens at the same time as the consequencial peak of lipids observed is the algebraic summation of the two events.

The single phospholipids behave differently during hepatocyte regeneration after partial hepatectomy.

Chromatin phospholipids represent about 10% of the total nuclear lipids (Viola-Magni et al, 1985a) of which sphingomeylin represents some 35% of all nuclear phospholipids (Albi et al, 1994). During liver regeneration, a decrease is observed in the amount of SM at the beginning of S-phase (Albi and Viola-Magni 1997a) followed by an increase at the end of S-phase. The approximate remaining amounts of phospholipids in the nucleus are phosphatidylethanolamine (PE) 10%, PI 19%, phosphatidylserine (PS) 22% and PC 14% (Albi et al. 1994).

3. Roles of individual phospholipids

3.1 SM behaviour

It has been hypothesized that SM may have a role in stabilising the DNA molecule. The decrease of SM at the start of the S phase may be associated with the unwinding of the DNA helix and the increase of SM at the end of S-phase may be linked to the rewinding of the DNA helix. A similar behaviour of SM was also observed in other models by different authors (Stillman 1996, Alessenko and Chatterjee 1995).

3.2 PS behaviour

PS is also one the PLs present in a higher amount in chromatin with respect to the level seen in the nuclear membrane. PS increases when DNA synthesis starts during liver regeneration. A possible role for PS in the chromatin may be the stimulation of DNA polymerase as has been shown *in vitro* by Manzoli et al. (1981).

3.3 PC behaviour

This PL is mostly present in the nuclear membrane with only a small amount in the chromatin. The chromatin PC has a different composition to that of the microsomal fraction in that it contains many unsaturated forms of the monoenic fraction with respect to the microsomal PC that was enriched with tetraene and exaene fractions (Albi et al. 1994).

The chromatin PC does not present a particular modification during liver regeneration except that DAG, a product of PC, increases at 12 h hand 30 h in parallel with the initiation of the two waves of proliferating activity when DNA synthesis starts (Viola-Magni et al.1985b).

3.4 PE behaviour

Although PE represents 22% of the total PL present in chromatin, its behaviour is similar to that of SM. However, no precise indication as to its role(s) in liver regeneration has been observed.

3.5 PI behaviour

PI represents 19% of the PLs present in chromatin (Viola-Magni et al. 1985a, Albi et al. 1994). No specific variations in the amounts of PI have been described during liver regeneration although it may have a role through its degradation enzymes by producing DAG (Albi et al. 2003a).

4. Phospholipid-associated enzymes

4.1 Sphingomyelinase

Sphingolmyelinase was first demonstrated in chromatin by Albi and Viola Magni (1997a). This enzyme is well known as a lysosomal enzyme in the acid form and as a cytoplasmic enzyme in the neutral form (Slife et al. 1989). It is present in many tissues e.g. hepatocytes, the nervous system and various cell cultures.

The hydrolysis of SM by SMAse results in the production of ceramide that has many physiological functions. This reaction is stimulated by many factors including interferon, (Kim et al. 1991), interleukin1 (Ballou et al. 1992), 1-25 OH vitamin D (Okasaki et al. 1989, 1990) and TNF (Dressler et al. 1992, Jayadev et al. 1994). Ceramide can be further hydrolysed to sphingosine that inhibits the protein kinase C present in the hepatocyte nuclei.

The enzyme was evaluated in both the nuclear membrane and chromatin fractions isolated from the hepatocytes. The enzyme activity reached a maximum at pH 7.6 in the nuclear membrane fraction and at pH 8.4 in the chromatin fraction. The reactions versus protein content show linear reactions for each enzyme up to 400 mg protein. In contrast, the reactions versus time showed that the nuclear membrane enzyme rose linearly from zero time whilst the chromatin enzyme remained low until 90 minutes when it rose sharply to reach its maximum value. The Km of the nuclear membrane enzyme is 3.9×10^{-4} M and that of the chromatin enzyme is 2.4×10^{-5} M implying that the nuclear membrane enzyme is more similar to that of the plasma membrane. In contrast, the chromatin enzyme appears to be similar to that present in the microsomal fraction.. The specific activity is 9.12 nmoles/10 minutes for the nuclear membrane SMase and 1.39 nmoles /90 minutes for the chromatin SMase (Albi and Viola-Magni 1997a, Table 1).

Generally, the production of ceramides results in a block at G0/G1 in the cell cycle (Riboni et al. 1992; Gomez-Munoz et al. 1995). However, the increased ceramide levels at 12 h after partial hepatectomy coincide with the start of DNA synthesis. Given the differences between the SMases present in the nuclear membrane and chromatin fractions, it is possible to hypothesis that the two enzymes are different and that the enzyme present in the chromatin may play a different role to that of the nuclear membrane and so may not necessarily result in a G0/G1 block.

4.2 Sphingomyelin synthase

The synthesis of sphingomyelin may be obtained through two pathways:

The first involves the reaction between CDP- choline and N-acylsphingosine (Scribney and Kennedy 1958) whilst the second consists of phosphocholine transfer from lecithin to ceramide. This reaction is catalysed by the enzyme phosphatidylcholine:ceramide phosphocholine transferase or sphingomyelin synthase (Diringer et al.1972). SM synthase was found, initially in the microsomes of kidney, lung, liver, spleen and heart (Ullman et al. 1974). Its subcellular localisation is in the Golgi apparatus (Jeckel et al. 1990; Futeman et al.

1990) from which it can be transported to the plasma membrane by vesicular flow (Koval and Pagano 1991;.van Meer and Burger 1992).

The evaluation of SM synthase activity in the nuclear membrane and chromatin fractions showed this enzyme to have different characteristics for each fraction. In the nuclear membrane the optimum pH was 7.6 whereas in the chromatin it was pH 8.4.

The Km was 1.68 x 10^{-4} and 3.59 x10^{-5} for the nuclear membrane and chromatin fractions, respectively. SM synthase activity was 770 pmol/mg protein/min (Vmax 1.1nmol/mg protein/min) in the nuclear membrane and 288 pmol/mg protein/min (Vmax 297 pmol/mg protein/min) in the chromatin. These characteristics exclude a possible contamination by the cytoplasmic structures since the specific activity is higher both in nuclear membrane and in the chromatin with respect to that found in the whole homogenate (Albi and Viola-Magni 1999a, Table 1).

The presence of this enzyme, together with SMase, can help to explain the possible variations in chromatin SM content as observed during liver regeneration.

4.3 Reverse sphingomyelin-synthase

This enzyme utilises SM as a source of phosphorylcholine and is one of the mechanisms involved in PC synthesis. Other mechanisms for the biosynthesis of PC are the Kennedy pathway (Kent 1990), phosphatidylethanolamine methylation (Stetten 1941), lyso-PC-acylation (Marinetti et al. 1958) and base- exchange from phophatidylserine (Bijerve 1971).

It is difficult to suppose that PC will be synthesized in the cytoplasm and transferred to the nuclei since the PC modifications observed occur in a very short time. The base- exchange component was demonstrated in the nucleus (Albi and Viola-Magni 1997b). The presence of reverse SM-synthase may favour a more rapid exchange of PC by using DAG and phosphorylcholine derived from SM.

The presence of this enzyme was demonstrated both in the nuclear membrane and in the isolated chromatin (Albi et al. 2003a). The activity found in the whole homogenate was 0.93 pmol/mg protein/min, in the cytosol 2.61 pmol/mg protein/min and in the nuclear membrane 0.87 pmol/mg protein/min. A higher level of activity was observed in the chromatin at 37.09 pmol/mg protein/min. The optimum pH was 8.4 as for the other chromatin enzymes probably because the maximum solubilisation of chromatin observed at this pH may favour enzyme activity expression.

The reaction was linear with respect to both time and protein concentration. The activity was 9.5 pmol/mg protein/min when DAG was added and increased to 50 pmol/mg protein/min in the presence of SM. Equally, the Km values were 3.56 x 10^{-5} M for exogenous SM and 1.12 x 10^{-4} for exogenous DAG so obeying the Michaelis-Menten kinetics (Table 1). It is not clear at the moment if the SM-synthase and reverse -SM-synthase are the same enzyme or are two different enzymes. The ratio DAG/ceramide depends upon their activities and, therefore, it is necessary to take into account eventual differences (Table 1).

The activity of SM-synthase was measured in the various sub-cellular fractions i.e. whole homogenate, cytosol, nuclear membrane and chromatin fractions. The ratio between SM-synthase /reverse SM-synthase was also determined for these fractions. The higher ratio value was observed in the nuclear membrane fraction of 885.05 indicating that the synthesis of PC may be due to an alternative enzymatic reaction. The SM-synthase activity in the chromatin fraction was only 7.49 higher with respect to the reverse SM-synthase with a consequently lower value for the DAG/ceramide ratio.

4.4 Phosphorylcholine-dependent phospholipase C

This enzyme hydrolyses PC to produce phosphorylcholine that may be used for SM synthesis and DAG that may control many cellular functions (Exton 1990).

Phosphorylcholine-dependent phospholipase C has been determined in hepatocytes and especially in the nuclear membrane and chromatin fractions in which two different isoforms were demonstrated (Baldassarre et al. 1997). In fact the PC present in these two fractions differs in content and turnover (Viola-Magni et al. 1985b, 1986). Since other enzymes such as SMase and SM-synthase were demonstrated, the presence of additional enzymes may help to understand the nuclear DAG/ceramide ratio and how it may be involved in regulating different cellular functions such cell duplication, differentiation and apoptosis. Therefore, the hepatocyte nuclei were separated and the chromatin and nuclear membrane fractions extracted for the determination of the presence and activity of phosphorylcholine-dependent phospholipase C. The enzyme activity in the nuclear membrane was 1.76 nmol/mgprotein/min (Vmax 3.01 nmol/mg protein/min) whilst that in the chromatin fraction was 8.4 times lower (Vmax 0.22 nmol/mg protein/min). The phosphorylcholine-dependent phospholipase C had a pH optimum of 7.6 in the nuclear membrane and 8.4 in the chromatin; its activity was linear during the first 45 min of incubation in the range from 100 to 400mg protein. The enzyme activity followed regular Michaelis-Menten kinetics in both preparations the Km values being 2.46 x 10^{-4} M for the nuclear membrane fraction and 7.8 3 x10^{-5} M for the chromatin fraction (Albi and Viola-Magni. 1999b, Table 1).

This enzyme is Ca^{++} independent and, therefore, may stimulate protein kinase C present in the nuclei since there is no variation in the Ca^{++} concentration that may interfere with its activity (Buchner 1995). The existence of nuclear PKC forms has been shown in the liver (Rogue et al. 1990) and their function may be in maintaining DNA structure or favouring DNA synthesis and repair through the action on laminin B which is localised at the sites of DNA replication (Moir et al. 1994).

4.5 Phosphatidylinositol-dependent phospholipase C

The amount and turnover of phosphatidylinositol in the chromatin fraction were different with respect to those of the nuclear membrane fraction (Viola-Magni et al. 1986). This could be due to a different enzyme activity such as that of phosphatidylinositol-dependent phospholipase C since various enzyme isoforms exist that may be activated by different stimuli (Martelli et al. 2000, 2001, Santi et al. 2001).

The activity of phosphatidylinositol-dependent-phospholipase C was determined in both the nuclear membrane and the chromatin fractions. Two peaks of pH were present in the nuclear membrane fraction, a first peak appearing at pH 7.6 followed by a second peak at pH 8.4-8 (Albi et al. 2003b).

In contrast, the chromatin fraction showed only a small peak at pH 7.6 with a sharper peak at pH 8.6.

This behaviour indicates the presence of at least two different isoforms that are quantitatively different between the two fractions. The presence of two isoforms, beta1 in the chromatin and gamma1 in both the nuclear membrane and chromatin fractions, were demonstrated using specific antibodies coupled with electron microscopy (Neri et al. 1997). However, the delta1 isoform that is present in the cytoplasm was absent from the nuclei.

The PI content in the nuclear membrane fraction was 15.2 μg/mg protein and 1.05 μg/mg protein in the chromatin fraction i.e. fifteen times less. The enzymatic activity evaluated under optimal conditions was 121.43 pmol/mg protein/min in the nuclear membrane fraction and 369.05 pmol/mg protein/min in the chromatin fraction i.e. more than three times higher in the chromatin with respect to nuclear membrane. The Km was 5.77×10^{-5}M for the chromatin associated PI-PLC and 3.89×10^{-3}M for this enzyme associated with the nuclear membrane fraction having a Vmax of 3.3 nmol/mg protein/min and 0.034 nmol/mg protein/min, respectively (Table 1). These results indicate a greater substrate affinity of the chromatin-associated enzyme. It has been demonstrated that this enzyme has a role in cell proliferation (Sun et al. 1997).

Nuclear membranes				Chromatin		
Enzymes	pH	K_m	Sp. activity	pH	K_m	Sp. activity
SMase	7.6	3.9×10^{-4}M	9.12 moles/10 min	8.4	2.4×10^{-5}M	1.39 nmoles/90 m
SM synthase	7.6	1.68×10^{-4}M	770 pmol/mg protein/min	8.4	3.59×10^{-5}M	288 pmol/mg protein/min
Reverse SM synthase	7.6	-------------	0.87 pmol/mg protein/min	8.4	* 3.56×10^{-5}M ** 1.12×10^{-4}M	37.09 pmol/mg protein/min
PC-PLC	7.6	2.46×10^{-4}M	1.76nmol/mg protein/min	8.4	7.83×10^{-5}M	21 pmol/mg protein/min
PI-PLI	7.6,8.4-8.8	3.89×10^{-3}M	121.43 pmol/mg protein/min	7.6 - 8.6	5.77×10^{-5}M	369.05 pmol/mg protein/min

*: V/SM substrate conc.
**: V/DAG substrate conc.

Table 1. Characteristic differences of nuclear membrane and chromatin PLs enzymes

5. The roles of phospholipid-associated enzymes in normal hepatocytes

The role of the phospholipid-associated enzymes present in the chromatin seems to be related to the control of a number of cell events through the balance between the levels of ceramide and DAG in the nucleus (Albi and Viola-Magni 2003c, Albi et al. 2008). When the ceramide increases, the SM synthase is stimulated to produce DAG. When there is an increase in DAG, reverse-SM-synthase is activated together with SMase to yield an increased production of ceramide in order to reach an equilibrium. It is possible that the increase in ceramide may favour the production of sphingosine that can act as a pro-apoptotic stimulus. (Tsugane et al.1999).

This enzyme system may also be involved with gene expression by controlling the transfer of RNA to the cytoplasm. After enzymatic digestion with DNase and RNase, it was possible to isolate a complex containing a small amount of DNA, RNA, proteins and PLs. The RNA is RNase insensitive and behaves as double-stranded RNA. There are only two PLs invloved, namely, SM and PC (Albi et al. 1996; Micheli et al.1998).

The enzymes SMase, SM synthase and reverse SM-synthase are present. If the complex is treated with SMase, the undigested RNA becomes RNase sensitive. Therefore, the presence of SM appears to protect the RNA from digestion. SMase aids the digestion by causing a decrease of SM that returns to the normal value through the activation of SM-synthase that exploits the PC derived from phosphorylcholine. The amount of PC may be restored through the activation of the enzyme reverse SM-synthase (Micheli et al. 1998).

The products of PL metabolism may act also as internal signals by activating other nuclear proteins such as PKC or favouring the synthesis of polymerases through the presence of PS (Albi et al. 1991).

6. The role of cholesterol

Albi and Viola-Magni (2002) have demonstrated the presence of CHO in hepatocyte chromatin. Previous researchers have attributed many functions to CHO metabolism (Luskey 1988) and an increase in its concentration has been demonstrated in both cancer and proliferating cells (Rao 1986). In liver regeneration, the amount of chromatin CHO changes during the first 24 h. Two fractions were demonstrated in the chromatin fraction, one of which is a free fraction and the other that is only extractable after SMase or proteinase K digestion (Albi and Viola-Magni, 2002).

After partial hepatectomy, the bound CHO increased reaching a peak after six hours whereas the free CHO reached a peak only after 18 h. This may be explained as being due to the increased SMase activity and to the block of SM synthesis which favours the transformation of the bound CHO fraction to the free fraction.

At 24 h, SM synthase activity increased and the ratio between bound and free CHO returned to the normal value seen in the non-dividing hepatocytes.

It was demonstrated that inhibition of the melavonate-CHO pathway with nitrogen bisphosphate arrested the cells in S-phase with a reduction in the expression of cdk2 and cdk4, whereas the expression of cdk21 increased (Reszka et al. 2001). The role of CHO in liver regeneration may be linked to the stimulation of the activities of cyclin-dependent kinases. In order to clarify this point the analysis of cyclin behaviour, especially that of A and B1, during liver regeneration may be of interest.

7. The cell cycle

7.1 The behaviour of cyclin A and B1

Cyclin A is present in normal liver and in sham operated rats. During liver regeneration, the behaviour of cyclin A and B1 expression was analysed by Splewak &.Thorgeirsson (1997). They showed an increased amount of both cyclin A and B1 between 12 h and 22 h post hepatectomy when the hepatocytes entered the G1-S phase transition as shown by the tenfold rise in the cyclin mRNA level, its level remaining high between 24 and 48 h and returning to the normal value only after 72 h.

During S-phase, cyclin A is associated with p32cdk2 kinase, whereas during the transition G2-M it forms a complex with p34cdc2 (Pagano et al. 1992).

The mRNA levels of cyclin B1 and p34cdc2 behave in a similar manner remaining low for the S-phase period and increasing only after 20 h to reach a peak at 26 h. This is followed by a decline at 34 h with a new peak forming between 38 h and 44 h. The two peaks correspond to the G2-M phases of the first wave and second waves of the hepatocyte cell cycle.

The great majority of p34cdc2 is linked to cyclin B1 whereas only 25% is linked to cyclin A (Loyer et al. 1994). The presence of the phosphorylated form of cyclin A may represent an inactivation of this cyclin during the G2-M transition.

7.2 Cell cycle regulation by chromatin-associated phospholipids (Table 2)

The increase in chromatin-bound CHO during the first six hours activates the kinases with PS favouring DNA polymerase alpha synthesis and PIP. At 12 h post-hepatectomy a transient increase of DAG is due to the hydrolysis of PI followed the activation of the chromatin enzyme PI-PLC.

The increase of DAG favours the translocation of PKC into the nucleus (Divecha et al. 1991). The synthesis of cyclin A increases as it complexes with p32cdk2 kinase. When the S-phase starts, a more consistent DAG peak is evident due to the hydrolysis of PC by the enzyme PC-PLC. At the same time, there is a decrease in SM due to the activation of SMase and consequently an increase in the free CHO present in the nucleus.

hours	Events
6	increase of bound CHO, activation of PS, synthesis of DNA polymerase
12	increase of DAG due to PI PLI, and then to PC-PLI nuclear translocation of protein kinase C Decrease of SM due to SMase, increase of free CHO which stimulates Cyclin A complex
18	DNA synthesis starts, cyclin A complex activity increases, increase of SM due to the activity of SM synthase, decrease of free CHO fraction
22	decrease of cyclin A activity, increase of cyclin B complex
24	DNA synthesis peak
26	cyclin B peak, cell transition from G2 to M
34	decrease of cyclin B
36	second peak of DNA synthesis
38-44	second peak of cyclin B and end of second S-phase

Table 2. Molecular events in relation to the time after hepatectomy

However, the complexing of cyclin A with p32cdk2 kinase increased up to 22h i.e. the end of S-phase. After 18h, SM synthase activity increased with the increase in SM reducing the CHO free fraction and hence the activity of cyclinA.

Cyclin B complexing with p34cdc2 increased its activity so favouring the transition of G1-M phase. The phosphoinositides also decrease during S-phase leading to an inhibition of DNA synthesis through an increase in the activity of dephosphorylating enzymes (York and Majerus 1994). The small fraction of cyclin A complexed with p32 cdc2 is inactivated by phosphorylation, the cells progressing to G2-M (Splewak & Thorgeirsson 1997). The complex B-p34cdc2 also decreases followed by a second peak between 38-44 h post-hepatectomy corresponding to the second mitotic peak.

8. Conclusions

Hepatic regeneration provides a good model for studying the mechanisms controlling cell proliferation and the ways in which they might be modified. This last aspect is very important in liver diseases and transplantation.

It is well known that cell duplication is delayed after hepatectomy for 12 h in the rat and this is not due to the operation effect as shown by the sham operated rats in which this may only be justified for the first four hours.

Much attention has been paid to the possible factors that may be activated during this period. Three hypotheses were made: the original one supposed that there was a single humoral factor, the second concerned the activation of a pathway involving many components and the more recent one, the activation of multiple pathways; the latter is the one most accepted today (Fausto 2006).

The cytokine pathway is activated in the first phase of liver regeneration which stimulates quiescent hepatocytes, growth factors then override a restriction point in G1 the entrance into cycle being associated with Rb phosphorylation, increased expression of the Rb family member p 107 together with cyclins D, E and A that form cdk4/cyclin D and cdk2/ cyclin E complexes (Menjo et al. 1998, Albrect et al. 1998).

The events preceding the entrance into the cell cycle were intensively studied , but less attention was dedicated to events inside the nuclei that favour the initiation of S-phase and G2/M transition. The presence of lipids in chromatin represents a component that appears essential for the two events. The activation of phospholipases causes a transient increase of DAG due to the hydrolysis of PLI followed by a more consistent peak due to the hydrolysis of PC. At the same time, the decrease in SM, due to an increase of activity of SMase, causes an increase in the free cholesterol fraction thus favouring DNA duplication.

The cyclin A complex is activated in parallel and the cells progress to the S phase.

When the S-phase is near completion, SM synthase increases the SM fraction and the cholesterol free fraction decreases. At the same time, cyclin B1 complex is activated thus favouring the cells' transit from G2 to M. Therefore, it is clear that all external stimuli that may favour liver regeneration can act only through the modification of the chromatin components of which lipids seem to have an important role. They may be also independent from external stimuli and may function as an internal balance. In fact, the increase in DAG due to PC hydrolysis stimulates SMase with the liberation of ceramide that may stimulate reverse sphingomyelin synthase to form new PC so favouring the liberation of DAG. On the other hand, the increase in SM reduces the ceramide present.

This internal clock appears to control cell activity favouring proliferation, differentiation when the cells remain in Go or apoptosis when the ceramide present is transformed into sphingosine.

It is clear from this that the role of phospholipids must be considered for a role in cell duplication regulation favouring the regeneration process in liver.

9. References

Albi, E., Viola Magni, M.P., Lazzarini, R. & Gahan, P.B. (1991). *Chromatin phospholipid changes during rat liver development.* Cell Biochem. Funct. 9, 119-123.

Albi, E., Mersel, M., Leray, C., Tomassoni, M.L. & Viola Magni, M.P. (1994). *Rat liver chromatin phospholipids.* Lipids 29, 715-719.

Albi, E., Micheli, M. & Viola Magni, M.P. (1996). *Phospholipids and nuclear RNA.* Cell Biol. Intern. 20, 407-412.

Albi, E. &Viola Magni, M.P. (1997a). *Chromatin neutral spingomyelinase and its role in hepatic regeneration.* Biochim. Biophys. Res. Commun. 236, 29-33.

Albi, E. & Viola Magni, M.P. (1997b). *Choline base exchange activity in rat hepatocyte nuclei and nuclear membrane.* Cell Biol. Intern. 21, 217-221.

Albi, E. & Viola Magni, M.P. (1999a). *Sphingomyelin-synthase in rat liver nuclear membrane and chromatin.* FEBS Lett. 460, 369-372.

Albi, E. & Viola Magni, M.P. (1999b). *Phosphatidylcholine-dependent phospholipase C in rat liver chromatin.* Biochem. Biophys. Res. Commun. 265, 640-643.

Albi, E. & Viola Magni, M.P. (2002). *The presence and the role of chromatin cholesterol in rat liver regeneration.* J. Hepatol. 36, 395-400.

Albi, E., Pieroni, S., Viola Magni, M.P. & Sartori, C. (2003a). *Chromatin sphingomyelin changes in cell proliferation and/or apoptosis induced by ciprofibrate.* J. Cell. Physiol. 196, 354-361

Albi, E., Rossi, G., Maraldi, N.M., Viola Magni, M.P., Cataldi, S., Solimando, L.& Zini, N. (2003b). *Involvement of nuclear phosphatidylinositoldependent phospholipases C in cell cycle progression during rat liver regeneration.* J. Cell. Physiol. 197, 181-188.

Albi, E. & Viola-Magni M.P. (2003c). *The metabolism of nuclear phospholipids in cell function and regulation.* Recent Res.Devel.Biophys.Biochem., 3, 45-63

Albi, E. & Viola-Magni, M.P. (2004). *The role of intranuclear lipids.* Biology of the Cell 96, 657-667

Albi E., Cataldi S. & Rossi G. (2008). *The nuclear ceramide/diacylglicerol balance depends on the physiological state of thyroid cells and changes during UV-C radiation induced apoptosis.* Arch. Biochem. Biophys. 478, 52-58

Albi, E. (2011) *Role of intranuclear lipids in health and disease* Clin. Lipidol. 6,59-69

Albrecht, J.H., Poon R.Y., Ahonen C.L., Rieland B.M., Deng C. & Crary C.S. (1998). *Involvement of p21 and p27 in the regulation of CDK activity and cell cycle progression in the regenerating liver.* Oncogene 16, 2141-2150

Alessenko, A. & Burlakova, E.B. (2002). *Functional role of phospholipids in nuclear events.* Bioelectrochemistry 58, 13-21.

Alessenko, A.& Chatterjee, S. (1995). *Neutral sphingomyelinase: localization in rat liver nuclei and involvement in regeneration/proliferation.* Mol. Cell. Biochem. 143, 169-174.

Baldassare, J.J., Jarpe, M.B., Alferes, L. & Raben, D.M. (1997). *Nuclear translocation of RhoA mediates the mitogen-induced activation of phospholipasem D involved in nuclear envelope signal transduction.* J. Biol. Chem. 272, 4911-4914.

Ballou, L. R., Chao, C. P., Holness, M. A., Barker, S. C., & Raghow, R. (1992). *Interleukin-1-mediated PGE2\production and sphingomyelin metabolism. Evidence for the regulation of cycloossigenase gene expression by sphingosine and ceramide.* J. Biol. Chem. 267, 20044-20050.

Bjerve, K.S. (1971). *The Ca(2+) stimulated incorporation of choline into microsomal lecithin subspecies in vitro.* FEBS Lett. 17, 14-16.

Berg, N.O. (1951). *A histological study of masked lipids; stainability, distribution and functional variations.* Acta Pathol. Microbiol Scand Suppl 90:1-192.

Bresnick, E, (1971). *Regenerating liver an experimental model for study of growth.* Methods Cancer Res., 6,347-397

Buchner, K. (1995). *Protein kinase C in the transduction of signals toward and within the cell nucleus.* Eur. J. Biochem. 288, 211-221.

Cave, C.F. & Gahan, P.B. (1971). *A cytochemical and autoradiographic investigation of nucleolar phospholipids.* Caryologia 23, 303-312.

Chayen, J.& Gahan, P.B. (1958). *Lipid components in nucleohistone.* Biochem. J. 69, 49P.

Chayen, J., La Cour, L.F. & Gahan, P.B. (1957). *Uptake of benzopyrene by chromosomal phospholipids.* Nature 180, 652-65

Chayen, J., Gahan, P.B. & La Cour, L.F. (1959a). *The masked lipids of nuclei. Quart.* J.Microscop. Sci. 100, 3

Chayen, J., Gahan,P.B. & La Cour, L.F. (1959b). *The nature of a chromosomal phospholipid* Quart. J. Microscop Sc.100,325

Cocco, L., Martelli, A.M., Gilmour, R.S., Ognibene, A., Manzoli, F.A. & Irvine, R.F. (1988). *Rapid changes in phospholipid metabolism in the nuclei of Swiss 3T3 cells induced by treatment of the cells with insulinlike growth factor I.* Biochem. Biophys. Res. Commun. 154, 1266-1272.

Cocco, L., Martelli, A.M., Gilmour, R.S., Rhee, S.G. & Manzoli, F.A. (2001). *Nuclear phospholipase C and signalling.* Biochim. Biophys. Acta 1530, 1-14.

Diringer, H., Marggraf, W.D., Koch, M.A. & Anderer, F.A. (1972). *Evidence for a new biosynthetic patway of sphingomyelin in SV 40 transformed mouse cells.* Biochem. Biophys. Res. Commun. 47, 1345^1351.

Divecha, N., Banfic, H.& Irvine, R.F. (1991). *The polyphosphoinositide cycle exists in the nuclei of Swiss 3T3 cells under the control of a receptor (for IGF-I) in the plasma membrane, and stimulation of the cycle increases nuclear diacylglycerol and apparently induces translocation of proteinkinase C to the nucleus.* EMBO J. 10, 3207-3214.

Dressler, K. A., Mathias, S., and Kolesnick, R. N. (1992). *Tumor necrosis factor-alpha activates the sphingomyelin signal transduction patway in a cell-free system* Science 255, 1715-1718.

D'Santos, C.S., Clarke, J.H. & Divecha, N. (1998). *Phospholipid signalling in the nucleus. Een DAG uit het leven van de inositide signalering in de nucleus.* Biochim. Biophys. Acta. 1436, 202-232.

Exton, J.H. (1990). Signaling through phosphatidylcholine breackdown. Biol. Chem. 265, 1-4.

Fausto, N., Campbel J.S. & Riehte, K. J (2006). *Liver regeneration.* Hepathology, 43, 545-563 [8]

Futerman, A.H., Stieger, B., Hubbard, A.L. & Pagano, R. (1990). *Spingomyelin synthesis in rat liver occurs predominantly at the cis and medial cisternae of the Golgi apparatus.* J. Biol. Chem. 265, 8650-8657.

Gahan P.B. (1965a). *Histochemical evidence for the presence of lipids on the chromosomes of animal cells.* Exp. Cell Res.39,136-144.

Gahan, P.B. (1965b). *The possible presence of aldehydes and carbohydrates in chromosomes and interphase nuclei.* Histochemie 5, 289-296.

Gahan, P.B., Bartlett, R., Cleland, L., & Olsen, K., (1974). *Cytochemical evidence for the presence of phospholipids on human chromosomes.* Histochem. J. 6, 219-222.

Gahan, P.B., Viola Magni, M.P. & Cave, C.F. (1987). *Chromatin and nucleolar phospholipids.* Basic Appl. Histochem. 31, 343-353.

Gomez-Munoz, A., Waggoner, D.W., O'Brien, L. & Brindley, D.N., (1995). *Interaction of ceramides, sphingosine, and sphingosine 1-phosphate in regulating DNA synthesis and phospholipase D activity.* J.Biol.Chem.270,26318-26325.

Kent, C. (1990). *Regulation of phosphatidylcholine biosynthesis Prog.* Lipid Res. 29, 87-105.

Kim, M. Y., Linardic, C., Obeid, L. & Hannun, Y. (1991). *Identification of sphingomyelin turnover as an effector mechanism for the action of tumor necrosis factor alpha and gamma-interferon. Specific role in cell differentiation.* Biol. Chem. 266, 484-489. 25

Koval, M. and Pagano, R.E. (1991). *Intracellular transport and metabolism of shingomyelin.* Biochim. Biophys. Acta 1082, 113^125.

Idelman, S. (1957). *Existence d'un complexe lipides-nucléoprotéines à groupements sulfhydridés au niveau du chromosome.* Comptes Rend. Acad. Sci. 244, 1827-1829.

Idelman, S. (1958a). *Localisation du complexe lipiedes-protéines å sulfhydrylés au sein du chromosome.* Comptes Rend. Acad. Sci. 246, 1098-1101.

Idelman, S. (1958b). *Démasquage des lipids du chromosome géant des glandes salivaires de Choronome par difestion enzymatique des proteins.* Comptes Rend. Acad. Sci. 246, 3282-3286.

Irvine, R.F. (2003). *Nuclear lipid signalling.* Nat. Rev. Mol. Cell Biol. 4, 349-360

Jayadev, S., Linardic, C. M. & Hannun, Y. A. (1994). *Identificationof arachidonic acid as a mediator of sphingomyelin hydrolysis in response to tumor necrosis factor alpha.* J. Biol. Chem.. 269, 5757-5763.

Jeckel, D., Karrenbauer, A., Birk, R. & Schmidt, R.R. (1990). *Sphingomyelin is synthesized in the cis Golgi.* FEBS Lett. 261, 155^157.

La Cour, L.F., Chayen, J. & Gahan, P.B. (1958). *Evidence for lipid material in chromosomes.* Exp. Cell Res. 14, 469-474.

Loyer ,P., Gialse,D., Carlou, S., Baffet, G., Meyer, L. & Guguen-Guillouzo, C. (1994). *Expression and activation of cdks (1and 2) and cyclins in the cell cycle progression during liver regeneration.* J.Biol:Chem. 269, 2491-2500

Luskey, K.L. (1988). *Regulation of cholesterol synthesis:mechanism for controlof HMG-CoA reductase.Recent Prog.* Horm. Res., 44, 35-51.

Manzoli, F.A., Capitani, S., Mazzotti, G., Barnabei, O. & Maraldi, N.M. (1981). *Role of chromatin phsopholipids on template availability and ultrastructure of isolated nuclei.* Adv. Enzyme Regul. 20, 247-262.

Marinetti, G.V., Erbland, J., Witter J.F., Petix, J. & Stoltz, E. (1958). *Metabolic patways of lysolecithin in a soluble rat-liver system.* Biochim.Biophys. Acta 30, 223-226

Martelli, A.M., Bortul, R., Tabellini, G., Aluigim, M., Peruzzi, D., Bareggi, R., Narducci, P. & Cocco, L. (2001). *Re-examination of the mechanisms regulating nuclear inositol lipid metabolism.* FEBS Lett. 505, 1-6.

Martelli, A. M., Billi, A.M., Manzoli, L., Faenza, I., Aluigi, M., Falconi, M., De Poi, A., Gilmour, R.S. & Cocco, I. (2000). Insulin *selectively stimulates nuclear phosphoinositide-specific phospholipase C (PI-PLC) beta1 activity through a miogeno activated protein (MAP) kinase dependent serine phosphorylation.* FEBS Lett 486, 230-236

Menjo, M., Ikeda K. & Nakanishi, M. (1998). *Regulation of G1 cyclin-dependent kinases in liver regeneration.* J. Gastroenterol Hepattol 13,100-105

Micheli, M., Albi, E., Leray, C. & Viola Magni, M.P. (1998). *Nuclear sphingomyelin protects RNA from RNase action.* FEBS Lett. 431, 443-447.

Moir, D.R., Mountag-Lowy, M. & Goldman, R. D. (1994). *Dynamic properties of nuclear laminins:laminin B is associated with sites of DNA replication.* J. Cell Biol. 125, 1201-1212.

Neri, L.M., Ricci, D., Carini, C., Marchisio, M., Capitani, S. & Bertagnolo, V. (1997). *Changes of nuclear PI-PLC gamma 1 during rat liver re generation.* Cell Signal.9,353-362

Okazaki, T., Bell, R. M. & Hannun, Y. A. (1989). *Spingomyelin turnover induced by vitamin D3 in HL-60 cells. Role in cell differentiation.* J. Biol. Chem. 22.

Okazaki, T., Bielawska, A., Bell, R.M. & Hannun, Y.A. (1990). *Role of ceramide as a lipid mediator of 1-alpha-25 dihydroxyvitamin D3-induced HL-60 cell differentiation.* J. Biol. Chem. 265, 15823-15831. 5441.,

Pagano, M., Pepperfcok, R., Verde, F., Ansorge, W. & Draetta, G. (1992). *Cyclin A is required at two points in the human cycle.* EMBO J. 11,961-971.

Rao, K.N. (1986). *Regulatory aspects of cholesterol metabolismin cells with dfifferent degrees of replication.* Toxicol.pathol. 14, 430-437.

Reszka, A.A., Halasy-Nagy, J. & Rodan, G.A. (2001). *Nitrogenbisphosphonate block retinoblastoma phosphorylation and cell growth by inhibiting the cholesterol biosynthetic pathway in a keratinocyte model for esophageal irritation.* Mol. Pharmacol. 59, 193-202.

Riboni, L., Bassi, R., Sonnino, S. & Tettamanti, G. (1992). *Formation of free sphingosine and ceramide from exogenous ganglioside GM1 by cerebellar granule cells in culture.* FEBS Letters 300, 188-192

Rogue, P., Labourdette, G., Masmoudi, A., Yoshida, Y., Huang, F. L., Huang, K. P., Zwiller, J., Vincendom, G., & Malviya, A. N. (1990). *Rat liver nuclei protein kinase C is the isozyme type II.* J. Biol. Chem. 265, 4161-4165.

Santi, P., Zini, N., Santi, S., Riccio, M., Guiliani Piccarei, G., De Pol, A. & Maraldi N.M. (2001). *Increased activity and nuclear localizationof inositol lipid signal transduction enzymes in rat hepatoma cells.* Int.J.Oncol.18,165-174

Scribney, M. & Kennedy, E.P. (1958). *The enzymatic synthesis of sphingomyelin.* J. Biol. Chem. 233, 1315-1322.

Song, M. & Rebel, G. (1987). *Rat liver nuclear lipids. Composition and biosynthesis.* Basic Appl. Histochem. 31, 377-387.

Slife, C.W., Wang, E., Hunter, R., Wang, S., Burgess, C., Liotta, D.C. & Merill, A.M. (1989). *Free sphingosine formation from endogenous substrates by a liver plasma membrane system with a divalent cation dependence and a neutral pH optimum.* J.Bio. Chem. 264, 10371-10377

Spangler, M., Coetzee, M.L., Katlyl, S.L., Morris, H.P. & Ove, P. (1975). *Some biochemical characteristics of rat liver and Morris hepatoma nuclei and nuclear membranes.* Cancer Res. 35, 3145-3145.

Splewak Rinaudo, J.A. &.Thorgeirsson S.S. (1997). *Detection of a tyrosine -phosphorilated form of Cyclin A during liver regeneration.* Cell Growth and Differentiation 8) 301-309

Stetten, D. (1941). *Biological relationship of choline, ethanolamine and related compounds.* J.Biol..Chem. 138,437-438

Stillman, B., (1996). *Cell cycle control of DNA replication.* Science 274, 1659-1661

Sun, B., Murray N.R. & Field A.P. (1997). *A role for nuclear phosphatidylinositolspecific phospholipase C in the G2/M transition.* J:Bio.Chem:272,26313-26317

Tata, J.R., Hamilton, M.J. & Cole, R.D. (1972). *Membrane phospholipids associated with nuclei and chromatin: melting profile, template activity and stability of chromatin.* J. Mol. Biol. 67, 231-346

Tsugane, K., Tamiya-Koizumi, K., Nagino, M., Nimura, Y. & Yoshida, S. (1999). *A possible role of nuclear ceramide and sphingosine in hepatocyte apoptosis in rat liver.* J. Hepatol. 31, 8-17.

Ullman, M.D. & Radin, N.S. (1974). *The enzymatic formation of spingomyelin from ceramide and lecithin in the mouse liver.* J. Biol. Chem. 249, 1506^ 1512.

Upreti, G.C., DeAutmno, R.J. & Wood, R. (1983). *Membrane lipids of hepatic tissue. II.* Phospholipids from subcellular fractions of liver and hepatoma

van Meer, G. & Burger, K.N. (1992). *Sphingolipid trafficking-sortedout?* Trends Cell Biol. 11,332-337

Viola Magni, M.P., Gahan, P.B. & Pacy, J. (1985a). *Phospholipid in plant and animal chromatin.* Cell Biochem. Funct. 3, 71-78.

Viola Magni, M.P., Gahan, P.B., Albi, E., Iapoce, R. & Gentilucci, P.F. (1985b). *Chromatin phospholipids and DNA synthesis in hepatic cells.* Bas. Appl. Histochem. 29, 253-259.

Viola Magni, M.P., Gahan, P.B., Albi, E., Iapoce, R. & Gentilucci, P.F. (1986). *Phospholipids in chromatin: incorporation of in different subcellular fraction of hepatocytes.* Cell Biochem. Funct. 4, 283-288.

York, J.D. & Majerus, P.W. (1994). *Nuclear phosphatidylinositols decrease during S-phase of the cell cycle in HeLa cells.* J. Biol. Chem. 269, 7847-7850

5

Matrix Restructuring During Liver Regeneration is Regulated by Glycosylation of the Matrix Glycoprotein Vitronectin

Haruko Ogawa[1], Kotone Sano[2],
Naomi Sobukawa[1] and Kimie Asanuma-Date[1]
[1]*Graduate School of Advanced Sciences and Humanities, and Glycoscience Institute, Ochanomizu University,*
[2]*Faculty of World Heritage, Department of Liberal Arts, Cyber University, Tokyo, Japan*

1. Introduction

There are three major approaches for regenerative medicine. The most innovative approach among them is: induction of target cells from various stem cells such as induced pluripotent stem cells (iPS cells) or embryonic stem cells (ES cells) and implantation of them to regenerate the organ. The second approach is: in vitro tissue regeneration that involves preparation of artificial tissue by combining human cells with scaffolding biomaterials and growth factors. The third is: promotion of self-regeneration through controlling the repair activity of each tissue, which most organisms do naturally, is a more fundamental approach, but it will also be important in cell therapy to regulate tissue organization after induction of differentiation.

Because tissue homeostasis depends on spatially and temporally controlled expression of multifunctional adhesive glycoproteins and receptors, many studies have examined the changes of expression of extracellular matrix (ECM) molecules during tissue remodeling, inflammation and invasion by cancer cells (DeClerck, Y.A., et al. 2004; Seiffert, D. 1997; Kato, S., et al. 1992; Hughes, R.C. 1997) on the one hand. On the other hand, there is increasing evidence that glycosylations post-translationally modulate various biological phenomena by altering the activity and specificity or the stability of glycoproteins through the biosignaling functions of oligosaccharides (Varki, A. 1993; Varki, A., et al., 2009). During tissue remodeling, the glycosylated ECM molecules are different from those of normal tissue owing to the changes in the expression of many proteins that are responsible for glycan synthesis (Dalziel, M., et al. 1999). However, the glycan modulation of most glycoproteins that are involved in tissue remodeling has remained unknown.

When the three big lobes of a liver are excised, the remaining liver recovers its former mass and function within about two weeks in humans or 7 to 10 days in rats (Diehl, A.M. and Rai, R.M. 1996). ECM degradation occurs in the early stage of this process, followed by biosynthesis of the matrix, cell proliferation, and cell differentiation. During this process, many glycosyl transferases (Bauer, C.H., et al. 1976; Serafini-Cessi, F. 1977; Okamoto, Y., et al. 1978; Ip, C. 1979; Oda-Tamai, S., et al. 1985; Miyoshi, E., et al. 1995) and total glycoconjugates in the liver have been reported to change (Okamoto, Y. and Akamatsu, N. 1977; Kato, S. and Akamatsu, N. 1984; Kato, S. and Akamatsu, N. 1985; Ishii, I., et al. 1985). However, the nature of the links between such glycans changes, and the process of tissue remodeling has remained unclear. We consider it important to identify which molecules play important roles in the tissue remodeling during liver regeneration, and we will discuss the glycan modulation of one extracellular molecule, vitronectin, in this chapter. Vitronectin is a multifunctional adhesive glycoprotein that plays a central role in tissue remodeling by connecting pericellular tissue lysis with cell adhesion and motility.

We found that the glycans of vitronectin drastically change during liver regeneration after partial hepatectomy. In our studies to determine the glycan structures during the initial stage of the liver regeneration after partial hepatectomy of rats, we found that alterations in glycosylation, especially decreased sialylation of vitronectin, modulate the biological activities of vitronectin during tissue-remodeling processes by multiple steps (Uchibori-Iwaki, H., et al. 2000; Sano, K., et al. 2010). Liver regeneration is a normal repair process, while fibrosis and cirrhosis are considered to be excessive and abnormal repair processes that often give rise to cancer. In this context, elucidating how alterations of glycans occur and understanding how glycans modulate the glycans on vitronectin is useful in order to develop a strategy to regulate matrix remodeling in regeneration and deposition in liver cirrhosis. Therefore, we aimed to elucidate glycan modulation during liver regeneration after partial hepatectomy. We focused on the changes in vitronectin during liver regeneration, especially the changes of the glycan moiety, which plays a crucial role in controlling survival of hepatic stellate cells.

1.1 Structure and function of vitronectin

Vitronectin is a multifunctional adhesive glycoprotein that originates mainly in hepatocytes and circulates in the blood at high concentrations (0.2 mg/ml in human and no more than 0.1 mg/ml in rats). Vitronectin was first isolated from human serum by Holmes in 1967 as an 'α-1 protein' (Holmes, R. 1967), and has been referred to as 'serum spreading factor', 'epibolin', or 'S-protein'. It induces cell growth in vitro and is known as a major cell-adhesive component in cell culture mediums (Hayman, E.G., et al. 1983). Vitronectin is present as an ECM component in the liver, as well as various other organs, including skeletal muscle, kidney, and brain (Seiffert, D. 1997; Seiffert, D., et al. 1991). Most vitronectin in normal plasma is present as an inactive monomer form that does not bind to various ligands in the plasma (Gebb, C., et al. 1986; Izumi, M., et al. 1989). *In vivo*, vitronectin is activated in the presence of certain ligands such as heparin, type-1 plasminogen activator inhibitor (PAI-1), thrombin-AT-III, membrane attack complex of complements, and through a partial conformational change and multimerization process (Preissner, K.T. and Muller-Berghaus, G. 1987). Tissue vitronectin is considered to be present as an active multimeric form. Conformation-dependent binding of vitronectin was also observed for sulfatide (Gal(3-SO_4)β1-1ceramide), cholesterol 3-sulfate, and various phospholipids including phosphatidylserine, but not gangliosides, while vitronectin bound to cholesterol 3-sulfate regardless of its conformational state (Yoneda, A., et al. 1998). The binding of vitronectin to the membrane lipids and β-endorphin-binding activities were found to be attributable to hemopexin domain 2 and hemopexin domain 1, as well as type I collagen and heparin (Yoneda, A., et al. 1998).

The ligand-binding sites on the domain structure of vitronectins are shown in Fig. 1. Vitronectin regulates the proteolytic degradation of matrix and fibrinolysis through binding with urokinase-type plasminogen activator, PAI-1, and urokinase receptor (Seiffert, D. 1997; Schvartz, I., et al. 1999; Preissner, K.T. 1991; Pretzlaff, R.K., et al. 2000). Besides this function, vitronectin plays a key role in cell adhesion and cellular motility during tissue remodeling through binding to major ECM receptors, integrins such as αvβ1, αvβ3, αvβ5, αvβ6, αvβ8, and other ECM components like collagen and proteoglycans (Schvartz, I., et al. 1999). In plasma, vitronectin has been shown to regulate coagulation and thrombolytic and complement systems (Preissner, K.T. 1991). By providing a link between plasmin-regulated matrix proteolysis and integrin-mediated cell migration, activated vitronectin is considered to play a central role in matrix remodeling.

Fig. 1 also shows the glycosylation structures of vitronectins and ligand-binding sites. *N*-glycosylation sites of vitronectins are well conserved among mammals, and vitronectin contains almost fully sialylated complex-type *N*-linked glycans at these sites. Most vitronectins, except normal human, additionally contain various amounts of *O*-linked glycans, which differ by species: bovine, rabbit, and especially chicken vitronectins possess more *O*-glycans than *N*-glycans. Human vitronectin does not contain *O*-glycans in the normal state, while rat vitronectin contain several *O*-glycans in the connecting region (Sano, K., et al. 2010).

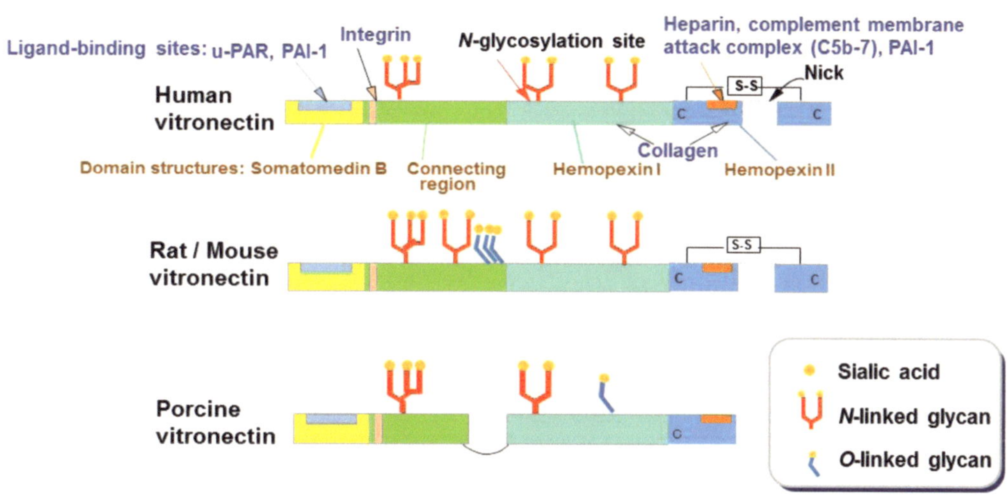

Fig. 1. Glycosylation and domain structure models of vitronectins. Human, rat/mouse, and porcine vitronectin contain three, four, and two *N*-glycosylation sites, respectively (Ogawa, H., Yoneda, A., et al. 1995) (Yoneda, A., Ogawa, H., et al. 1993) (Sano, K., Miyamoto, Y., et al. 2010). The numbers of *O*-glycosylation sites differ by animal species. The major ligand-binding sites that have been identified are indicated at the top of the figure (Yoneda, A., Ogawa, H., et al. 1998) (Preissner, K.T. 1991) (Preissner, K.T. and Seiffert, D. 1998).

2. Modulation of vitronectin activities by glycosylation

De-*N*-glycosylation of plasma vitronectin by enzyme treatment significantly attenuated the cholesterol sulfate-binding activity while it increased the collagen-binding activity. De-*O*-glycosylation or desialylation of vitronectin contributed to the stability for proteolysis (Uchibori, H., Ogawa, H., et al. 1992). These findings suggest that glycosylations modulate the ligand binding activities and the half-life of vitronectin *in vivo*. We prepared various recombinant domains of human vitronectin and mutants with certain domains deleted and expressed them separately in *E. coli* as fusion proteins. Using these recombinants, sulfatide-, phosphatidylserine-, cholesterol 3-sulfate-, type I collagen-, heparin-, and beta-endorphin-binding activities were found to be attributable to hemopexin domains 2 and 1. The possibility was suggested that the presence of a somatomedin B domain and/or connecting region flanking hemopexin domain 1 inactivated its heparin binding. Further, it was indicated that some of the ligand binding activities were modulated by glycosylation of plasma vitronectin, which enables modulation of its biological activities by a change in glycosylation accompanying the physiological or pathological state of the liver.

2.1 Changes in vitronectin during liver regeneration

2.1.1 Changes in collagen-binding activity of plasma vitronectin during early stage of liver regeneration

In this study, we used the liver regeneration of rats induced by two-thirds partial hepatectomy as a model system to study whether and how vitronectin plays a role in tissue remodeling after hepatectomy and how glycosylations are involved in the physiological processes. Vitronectin was purified from plasma at different times during the liver regeneration process and analyzed by SDS-PAGE (Uchibori-Iwaki, H., et al. 2000). As shown in Fig. 2A, each vitronectin showed one band on SDS-PAGE, and 24 h after partial hepatectomy (PH-VN) plasma vitronectin had shifted to a low migration position compared to vitronectins purified from plasma of non-operated (NO-VN) or partially hepatectomized rats and sham-operated rats (SH-VN), suggesting that the molecular mass of vitronectin had shrunk to 65 kDa at 24 h after partial hepatectomy from the 68-69 kDa of other vitronectins. At 24 h after operation, the yield of PH-VN had decreased to 1/3 that of sham-operated rats, and it was restored by 240 h, when liver regeneration was completed. At this time point, the amino acid composition did not change significantly, and the composition divergence (Black, J.A. and Harkins, R.N. 1977) of PH-VN was 0.040 when taking that of SH-VN as 0.0. All three vitronectins had the same N-terminal sequence, indicating that the three vitronectins had high homology among the primary sequence (Uchibori-Iwaki, H., et al. 2000).

As shown in Fig. 2B, the purified vitronectins bound to type I collagen by ELISA in a concentration-dependent manner, and PH-VN was found to exhibit much greater binding to collagen, about 3 times higher than that of SH-VN and NO-VN (Uchibori-Iwaki, H., et al. 2000). The enhanced binding of PH-VN to immobilized collagen shown by ELISA was supported by surface plasmon resonance (SPR), as shown in Fig. 2C. The relative affinity per monomer of PH-VN is remarkably high compared with those of NO- and SH-VN, especially at the lower concentrations (Sano, K., et al. 2007).

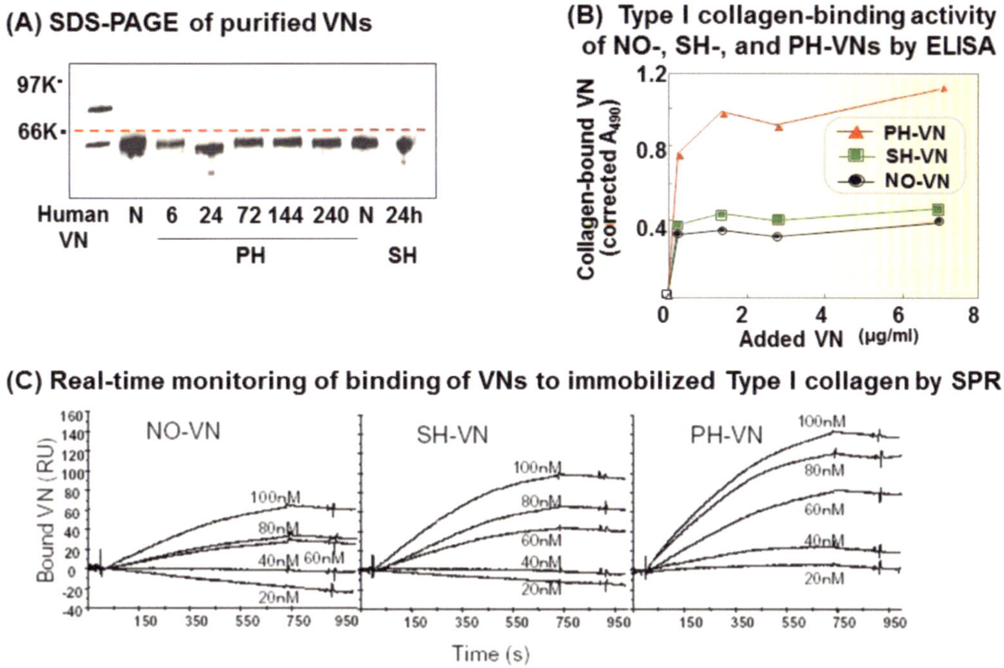

Fig. 2. Changes in electrophoretic mobility and collagen-binding of rat vitronectin at 24 h after partial hepatectomy. (A) SDS-PAGE of vitronectins from non-operated rats (N), partially hepatectomized (PH) rats at 6-240h after operation, and sham-operated (SH) rats at 24 h after operation. (B) Type I collagen (1 µg/100 µL) was coated onto wells of microtiter plates. After blocking with 5% BSA, various concentrations of purified vitronectins were added to each well. The bound vitronectin was measured using HRP-conjugated rabbit anti-human vitronectin IgGs and ELISA. The absorbance of collagen-bound vitronectin was corrected for the antibody reactivity of each vitronectin. (C) Collagen was immobilized on a CM5 sensor chip, and each vitronectin in PBS was injected onto the sensor chip at a flow rate of 20 µL/min at 20°C. The change of resonance units (RU) was corrected by subtracting the value on the BSA-immobilized reference cell.

2.1.2 Changes in glycosylation and carbohydrate concentration of vitronectin during early stage of liver regeneration

As shown in Fig. 3, the carbohydrate analyses of the three vitronectins indicated that total carbohydrate contents of PH-VN and SH-VN decreased to one-third and one-half of that of NO-VN, respectively, and that a remarkable decrease in sialic acids and amounts of glycans occurred due to partial hepatectomy. The lectin reactivity of the three vitronectins indicated that these vitronectins contain complex-type *N*-linked oligosaccharides. The reactivity toward *Phaseolus vulgaris* lectin L_4 (L-PHA) varied remarkably among vitronectins, and PH-VN showed marked reactivity with L-PHA, but SH- and NO-VNs reacted only slightly, suggesting that tri- or tetraantennary lactosamine-branched structures multiplied dramatically after partial hepatectomy. The specificity of PVL toward clustered sialyl residues (Ueda, H., Kojima, K., et al. 1999) (Ueda, H., Matsumoto, H., et al. 2002), the remarkably decreased reactivity of PH-VN with *Psathyrella velutina* lectin (PVL), together with the decrease in the reactivities with concanavalin A (Con A) and *Lens culinaris* lectin (LCA), indicate that the sialylated *N*-glycans markedly decreased after partial hepatectomy, which agrees well with the decreased amounts of carbohydrates including sialyl residues of the PH-VN. The changes in branching glycans would be attributable to the activity of several glycosyltransferases, which have been reported to increase (Miyoshi, E. , et al. 1995; Okamoto, Y., et al. 1978), while the decreased *N*-glycosylation of vitronectin at 24 h after partial hepatectomy could be attributed to the attenuation of the oligosaccharide transferase activity in microsomes (Oda-Tamai, S., et al. 1985).

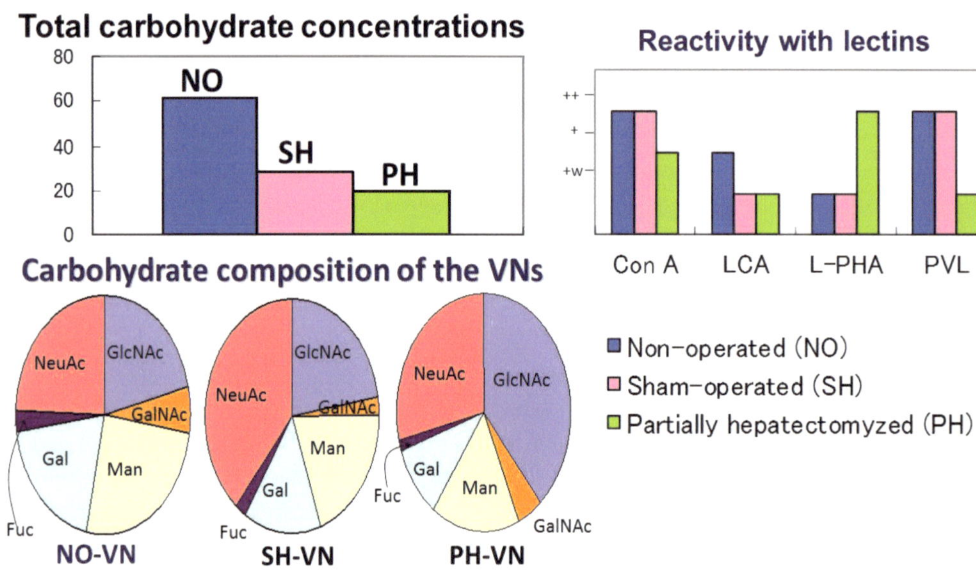

Fig. 3. The carbohydrate concentration, composition, and reactivity with lectins of vitronectins (VN) from non-operated (NO), sham-operated (SH), and partially hepatectomized (PH) rats.

2.1.3 Mechanism of enhanced collagen binding by change of vitronectin glycosylation

To study the enhancement of the mechanism for collagen binding, NO-VN was deglycosylated by sequential exoglycosidase treatments and collagen binding activity was analyzed by ELISA. As shown in Fig. 4A, collagen-binding of vitronectin gradually increased with step-wise trimming of glycans. Deglycosylated vitronectin (NG) showed collagen-binding activity three times higher than that of control vitronectin, suggesting that the enhancement of collagen binding of PH-VN is due to the changes in glycosylation (Sano, K., et al. 2007).

The deglycosylated NO-VNs were analyzed for multimer formation by ultracentrifugation, and the multimer sizes were calculated from the weight average molecular weight of vitronectin (Fig. 4B). The multimer sizes were gradually increased by step-wise deglycosylation, accompanied with an increase of the amounts of multimer vitronectins, which were cross-linked by disulfide-bonds, as measured by the intensity ratio of bands in SDS-PAGE before and after reduction, as shown in Fig. 4C. The enhanced collagen-binding activity of PH-VN was attributable to a multivalent effect that was due to the increase in the sizes and amounts of multimer vitronectins. The increase in multimer vitronectins in active form in various ligand-binding activities will accelerate the matrix incorporation of PH-VNs.

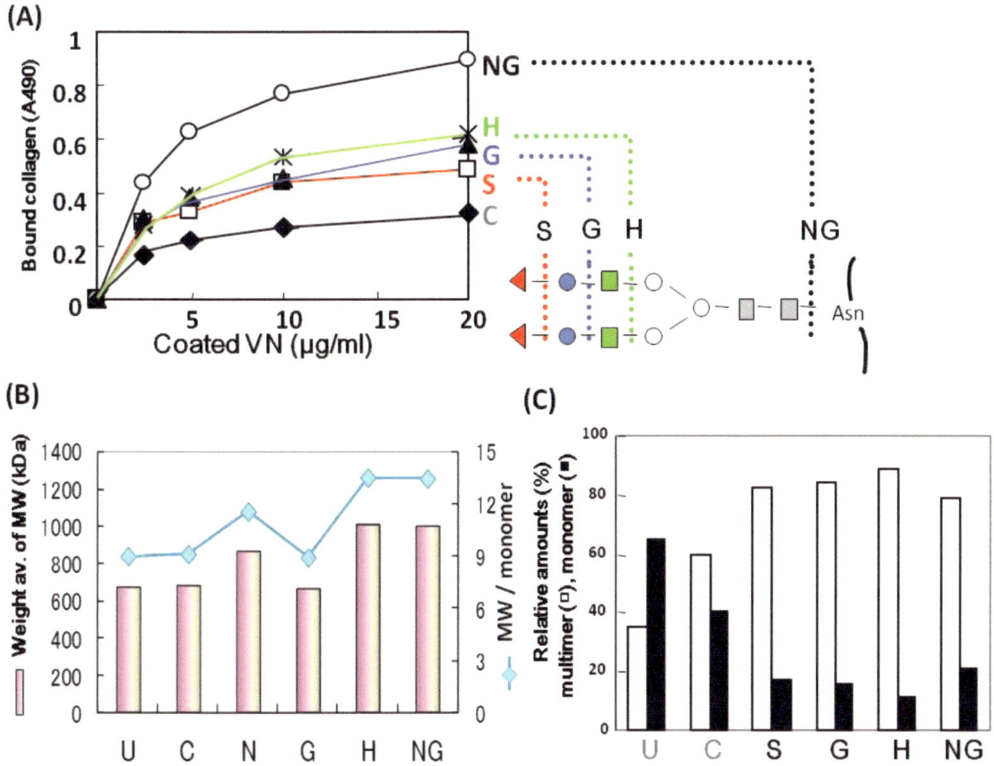

Fig. 4. The collagen binding activities (A), molecular weight and multimerization (B), and relative amounts of multimer (white bar) to monomer (black bar) of glycan-trimmed human vitronectin. The typical complex-type glycan structures of mammalian vitronectin was sequentially trimmed by sialidase (S), β-galactosidase (G), β-hexosaminidase (H), and *N*-glycosidase F (NG). U: untreated vitronectin; C: control vitronectin incubated without enzyme.

2.1.4 Effects of glycosylation of vitronectin on hepatic stellate cell spreading

Hepatic stellate cells are fibrotic cells that are induced during hepatic inflammation and are the major source of the newly synthesized ECM during hepatic fibrosis, whereas the survival or apoptosis of hepatic stellate cells is critical for the development or resolution, respectively, of liver fibrosis in chronic liver diseases (Benyon, R.C. and Arthur, M.J. 2001). In the normal liver, hepatic stellate cells have a low proliferation rate and produce trace amounts of ECM. As liver fibrosis progresses, hepatic stellate cells proliferate, but during the resolution of fibrosis there is extensive apoptosis that coincides with degradation of the liver scar (Benyon, R.C. and Arthur, M.J. 2001). It was reported that the activation of hepatic stellate cells increased the expression of integrin αvβ3, which is the major receptor of vitronectin on the cell surface, and promotes their proliferation and survival (Kato, S. and Akamatsu, N. 1985). We determined the structure and changes of rat vitronectin glycans during liver regeneration, and observed the relationship between the survival signaling of hepatic stellate cells and glycosylation of vitronectin.

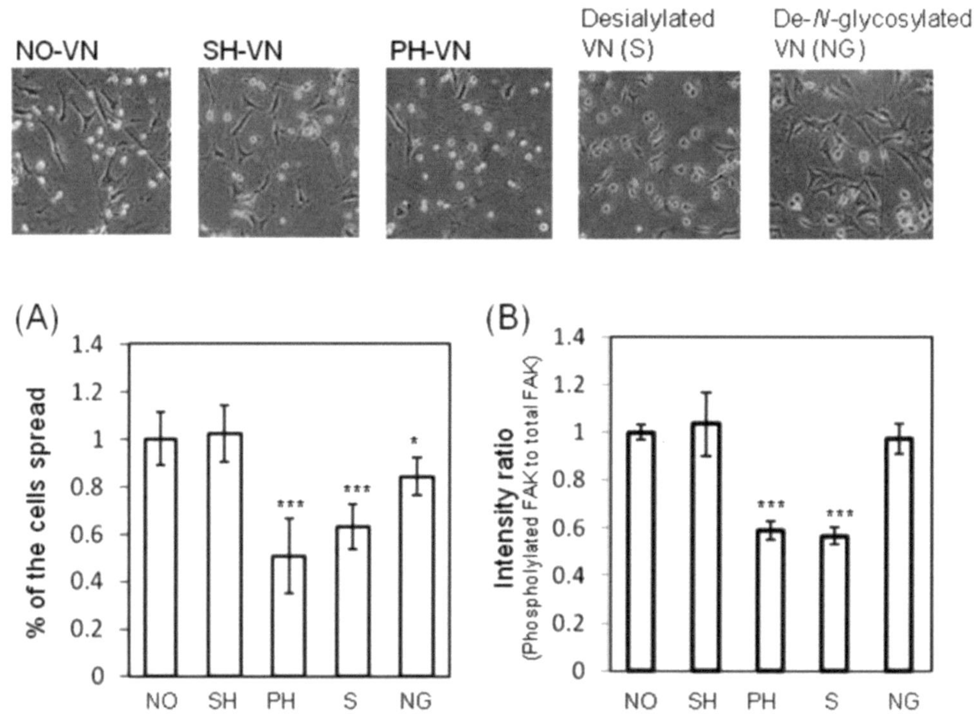

Fig. 5. Spreading and FAK-phosphorylation of HSCs on vitronectins. HSCs were plated on substrates coated with 10 μg/mL of vitronectins purified from NO, SH, or PH rats or desialylated (S) or de-*N*-glycosylated (NG) vitronectin. After 90-min incubation at 37°C in 5% CO_2, the % of the cells spread and FAK-phosphorylated were assessed by taking those on NO-VN as 1. Photomicrographs at ×40 magnification are shown on the upper panel. The data were analyzed by Student's *t*-test. The data represent the means S.D. (n=6); ***, p<0.001; *, p<0.05 compared with that on NO-VN.

In this study, vitronectins were purified from rat plasma at 24 hours after partial hepatectomy, sham-operation, or non-operated and designated as PH-, SH-, and NO-VN, respectively. The effect of PH-VN on HSC spreading was decreased to 1/2 of that of NO- or SH-VN (Fig. 5A). HSC spreading was also decreased on neuraminidase-treated vitronectin compared with untreated vitronectin, whereas it was decreased less on de-*N*-glycosylated NO-VN. These results indicate the importance of glycosylation, particularly sialylation, of vitronectin in HSC spreading. The effect of de-*N*-glycosylation was small compared with that of desialylation, because many sialic acid residues still remained on the *O*-glycans after

N-glycosidase F (PNGase F) treatment. In addition, PNGase F converts asparagine to aspartate, which may reduce the effect of the decrease of the negative charge of sialic acids. Because a clear difference between de-*N*-glycosylated and non-treated samples in cell spreading was still observed, suggesting the contribution of fibronectin-glycans to some extent, it cannot be concluded from this result that *O*-glycans contribute more to the decreased HSC spreading activity of PH-VN than *N*-glycans do.

To address the effects of glycosylation of vitronectin on integrin-mediated signaling, the focal adhesion kinase (FAK) of HSCs was compared among NO-, SH-, and PH-VN. As shown in Fig. 5B, the amount of phosphorylated FAK on PH-VN was decreased in proportion to cell spreading. These results suggest that the change in vitronectin glycosylation due to partial hepatectomy is able to regulate activation of the integrin-mediated signaling pathway. In addition, the effect of phosphorylated FAK on neuraminidase-treated NO-VN was decreased to an extent similar to that on PH-VN.

2.1.5 Site-specific glycosylations of rat vitronectin

In the early stage of liver regeneration, the synthesis of total DNA increased while the synthesis of total glycoproteins decreased within 48 h after partial hepatectomy (Okamoto, Y. and Akamatsu, N. 1977). The contradictory decrease of total glycoprotein synthesis in regenerating rat liver is due to the attenuation of oligosaccharide transferase activity in microsomes (Oda-Tamai, S., et al. 1985). Alterations in the glycan structure of total hepatic glycoproteins have been also suggested during liver regeneration (Kato, S. and Akamatsu, N. 1985; Ishii, I., et al. 1985). However, the changes in the glycans of a particular glycoprotein have not been well characterized. For these reasons, we investigated the site-specific glycosylations and changes after partial hepatectomy in rat plasma vitronectin.

Liquid chromatography/mass spectrometry analysis (LC/MS^n) of Glu-C glycopeptides determined the site-specific glycosylation of each vitronectin. Four potential sites, Asn^{86}, Asn^{96}, Asn^{167}, and Asn^{240} were revealed to be *N*-glycosylated, while the peptides of residues Thr^{110}-Thr^{124} were *O*-glycosylated. The most frequent *N*-glycan structures site-specifically found for each site are shown in Fig. 6. At Asn^{86}, Asn^{96}, and Asn^{240}, biantennary complex-type trisialoglycans with or without core fucosylation and with different amounts of *O*-acetylated NeuNAc were deduced (Fig. 6), whereas biantennary hybrid-type *N*-glycans were found to be the major structures at Asn^{167}. In the Thr^{110}–Thr^{124} region, the highly sialylated glycans were detected in the negative ion mode spectrum, and analysis in positive ion mode revealed that a Hex-HexNAc unit was located in the inner region of the glycans. From the results of lectin reactivity (*18*), it was inferred that one to three sialylated core-1 type molecules were attached.

These *O*-glycans contained disialic acid, which was chemically confirmed by fluorometric C7/C9 analyses and mild acid hydrolysate-fluorometric anion-exchange chromatography (Yasukawa, Z., Sato, C., et al. 2005). PH-VN had less disialyl *O*-glycans and complex-type N-glycans, but more core-fucosylated *N*-glycans than NO-VN (Sano, K., et al. 2010).

At the same time, alterations in the glycosylation of fibronectin (FN) after PH were different from those of vitronectin. The carbohydrate concentration of PH-FN decreased to 66% of that of NO- and SH-FNs. LC/MS^n revealed that eight kinds of complex-type *N*-glycan structures were present in NO-, SH-, and PH-FNs, and that bi- and trisialobiantennary glycans were the major structures (Sano, K., et al. 2008). Hybrid-type *N*-glycans and disialyl *O*-glycans were not detected. These results indicate that the alterations in the glycosylations of fibronectin and vitronectin were significantly different in the early stage of liver regeneration and demonstrate that these glycoproteins play different biological roles in the promotion of tissue remodeling processes.

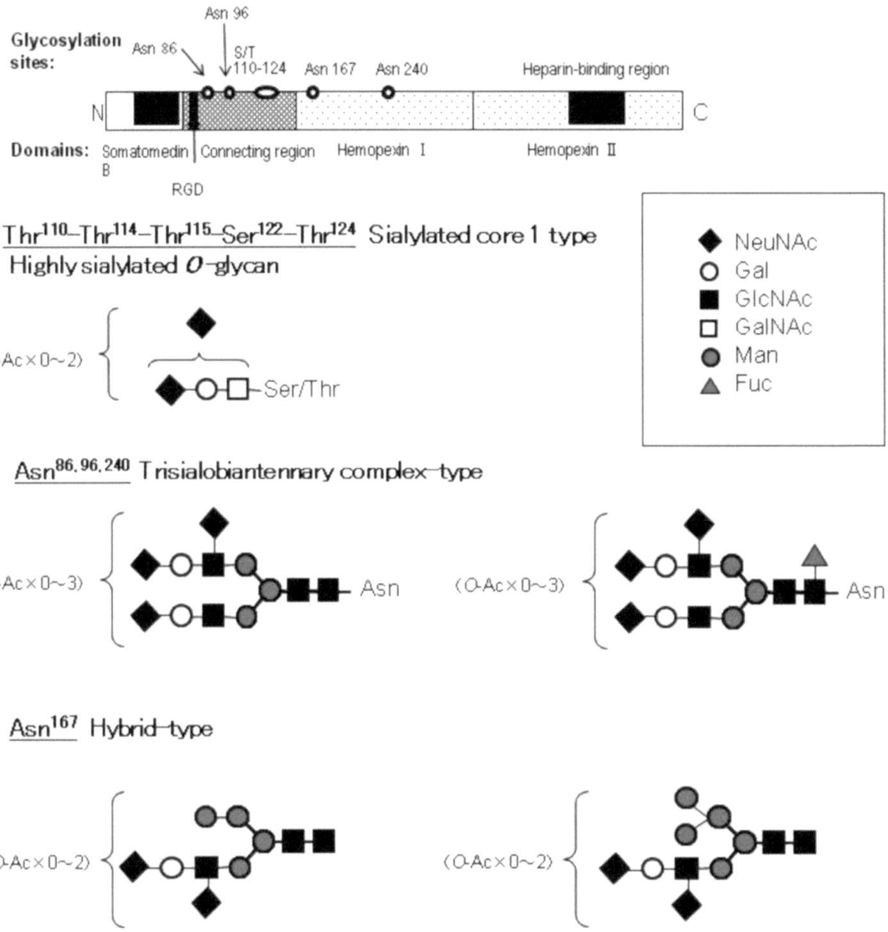

Fig. 6. Site-specific glycosylation of rat plasma vitronectin. The four glycosylation sites on rat plasma vitronectin and the glycan structures at each glycosylation site were determined by glycopeptide analyses using LC/MS^n. The most frequent glycan structures that were present at each site are presented.

2.1.6 Change in isoelectric points and oligosialylation of vitronectin

Highly sialylated *O*-glycans, which have a diNeuAc structure and were markedly decreased in PH-VN, affect the isoelectric points of vitronectins. Immunostaining of vitronectins after two-dimensional PAGE showed that each vitronectin has two components, pI 4.0 and 5.7 in NO-VN, that both shifted to higher pI, pI 4.3 and 6.0 in SH-VN, and further to pI 4.6 and 6.0 in PH-VN (Fig. 7). The pI of NO-VN was converted to one basic point, pI 6, after neuraminidase treatment, and only the more acidic component of pI 4.1 reacted with mAb S2–566, which specifically recognizes the Neu5Acα2,8Neu5Acα2,3Gal structure (Yasukawa, Z., et al. 2006).

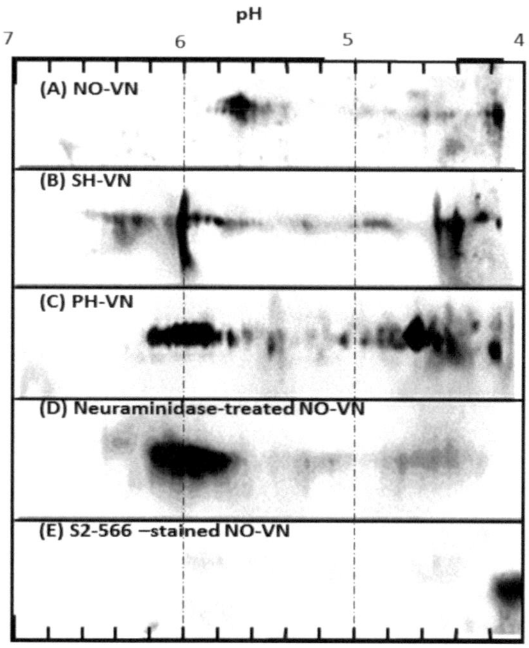

Fig. 7. Two-dimensional PAGE and western blotting of vitronectins. The first electrophoresis was isoelectric focusing. The second electrophoresis was SDS-PAGE under reducing conditions on 7.5% polyacrylamide gel, followed by blotting onto a polyvinylidene difluoride membrane and immunostaining using sheep anti-vitronectin IgGs (A–D) or the anti-oligosialic acid monoclonal antibody S2–566 (E), and HRP-conjugated secondary antibodies. Membranes were developed with ECL-Plus.

The results were also supported by the immunodetection of each vitronectin after 2D-PAGE where both hyper- and hyposilalylated molecules were present and the hypersialylation in PH-VN was markedly attenuated (Fig. 3). These analyses showed that the oligosialic acid on the *O*-glycan significantly affected the pI of the acidic partially hepatectomized vitronectin fraction. In addition to the decrease of oligosialyl epitopes in PH-VN, undersialylation of both *N*- and *O*-glycans was found in the basic PH-VN fraction.

The presence of disialic acid structures in some glycoproteins was previously described (Finne, J., et al. 1977), and neural cell adhesion molecules have been very thoroughly studied (Finne, J., Krusius, T., et al. 1977). Changes in expression of the oligosialic epitope on serum glycoproteins under inflammatory conditions were also reported (Yasukawa, Z., et al. 2005) (Yasukawa, Z., et al. 2007). The fact that the amounts of disialic acid structures in PH-VN were decreased compared with those in NO-VN indicates that the inflammation caused by partial hepatectomy reduces disialylation on vitronectin. The results indicate that the decreased sialylation plays a key role in regulating the function of vitronectin in liver regeneration.

In contrast to PH-VN, the glycosylation and biological activities of vitronectin in cirrhotic plasma were differentially changed. Yamada's group reported that vitronectin, which is active in collagen-binding in plasma, increased and correlated with certain fibrous markers (Yamada, S., et al. 1997, Yamada, S., et al. 1996), that the concentration of vitronectin in liver tissue was significantly increased in chronic liver disease compared with that in normal controls, and that vitronectin was colocalized at fibrous sites (Kobayashi, J., et al. 1994). Several reports supported the observation; therefore, the immunoreactivity of vitronectin in liver can be considered a marker of chronic/mature fibrosis. The vitronectins in untreated plasma exist mainly in native inactive form and exhibit low collagen binding. Urea treatment of cirrhotic and normal plasma revealed that the ratio of active to inactive vitronectin in cirrhotic plasma increased to more than twice that in normal plasma, promoting the incorporation of vitronectin from plasma into the matrix proceeds in spite of the fact that the vitronectin concentration in cirrhotic plasma was only 70% of that in normal plasma (Suzuki, R., et al. 2001). It is important to elucidate the changes in the glycosylation and biological activities of vitronectin in cirrhotic plasma and compare them with those of liver regeneration.

2.2 Summary

The present study attempted to determine how alterations of glycans modulate the biological activities of vitronectin during the initial stage of liver regeneration. Plasma vitronectin was purified from partially hepatectomized and sham-operated rats at 24 hours after operation and from non-operated rats. We found that the glycosylation of vitronectin changed and decreased markedly after surgery. The multimer sizes of PH- and SH-VNs significantly increased compared with NO-VN, and the change was accompanied by an increase in collagen binding. It was indicated that these changes were due to the changes in glycosylation of vitronectin, especially decreased sialylation, which increased the size and amount of the multimers to enhance the collagen-binding activity by a multivalent effect. Adhesion and spreading of rat hepatic stellate cells on PH-VN was decreased to 1/2 of that on NO- or SH-VN. Similarly, desialylated NO-VN decreased the spreading of rat hepatic stellate cells to 1/2 of that of control vitronectin, indicating the importance of sialylation of vitronectin for activation of rat hepatic stellate cells. LC/MS^n of vitronectin glycopeptides determined the site-specific glycosylation and the presence of highly sialylated *O*-glycans, which dramatically decreased after partial hepatectomy. Understanding the functional modulation of glycans on vitronectin may contribute to development of a strategy to regulate liver regeneration and matrix fibrosis in liver cirrhosis.

3. New method to detect changes in sialylation

As described in the previous sections, in the initial stage of the liver regeneration, alterations in sialylation of vitronectin modulate the important biological activities of vitronectin during tissue-remodeling processes by multiple steps. Like vitronectin, the sialylation of other glycoproteins changes under pathological conditions as well as during developmental stages, and altered sialylation often has significant implications in the physiological role of glycoproteins (Varki, A., Commings, R.D., Esko, J. D., Freeze, H. H.,Stanley, P., H., Bertozzi, C.R., Hart, G.W., Etzler, M.E. 2009)(Rutishauser, U. 1998). The aim of this section is to describe a fundamental method using chromatofocusing that enables detection of the changes in sialylation and separation of the matrix glycoproteins according to their pI in small amounts. The strategy is technically feasible and is applicable to various glycoproteins for a number of biological systems.

3.1.1 Principle of chromatofocusing

Chromatofocusing was originally a method of purification by separating proteins according to their pI. The pI of a protein is the pH at which the protein has zero surface charge. If a buffer, initially adjusted to the first pH, is run through an ion exchange column (see Fig. 8A) and followed by another buffer of a second pH, a pH gradient is formed in the column (Fig. 8B) (Sluyterman, L.A.A. E., O 1978; Sluyterman, L.A. and Wijdenes, J. 1978). If this pH gradient is used to elute proteins bound to the ion exchanger, the proteins in the column are eluted in the order of their pI. If the pH of the mobile phase around a protein is higher than the protein's pI, the protein has a positive charge. Therefore, it is dissociated from the anion exchange column and eluted from the column (Fig. 8A-C). If the pH of the mobile phase is lower, the protein has a negative charge and remains in the column (Shan, L. and Anderson, D.J. 2001). In this study, we applied chromatofocusing to detect the changes in sialylation of glycoproteins during liver degeneration.

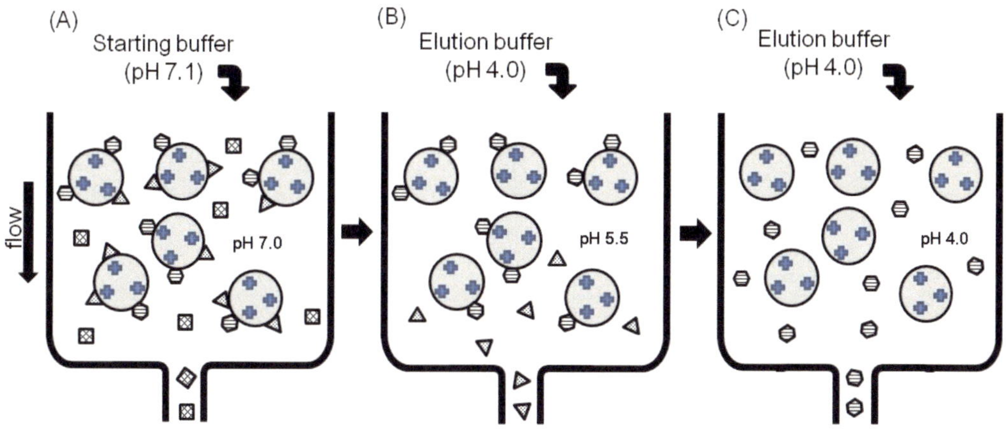

Fig. 8. Proteins having different pI are separated from each other as they pass through the anion exchanger column. A sample containing proteins that are ▩ (pI = 7.0), △ (pI = 5.5), and ⬣ (pI = 4.0) was applied to the column. (A) When the first buffer (pH 7.0) was poured into the column, the pH in the column is 7.0; therefore, rectangle proteins (pI >7.0) are eluted. Triangle and hexagon proteins (pI<7.0) are still bound to the column because they are negatively charged at pH 7.0. (B), When the second buffer (pH 4.0) was added, the pH in the column becomes lower, and this is halfway through the elution. (C), When the second buffer (pH 4.0) was continuously loaded, the pH in the column became pH 4.0 throughout the column, and then the hexagon proteins (pI=4.0) were eluted at the end of the elution.

3.1.2 Method

Rat plasma samples were collected at 24 h, 48 h, 72 h, 5 d, or 7 d after two-thirds hepatectomy or sham operation and stored at -80°C until use. Each plasma sample (200 μg/100 μl) was applied to chromatofocusing using a Mono P5/50 GL column (GE

Healthcare Inc.) and a fast protein liquid chromatograph (FPLC; AKTA Purifier, GE Healthcare, Inc). The starting buffer was 0.025 M bis-Tris, pH 7.1, and the elution buffer was 10% Polybuffer74, pH 4.0. Buffers were filtered through 0.22 μm filters under vacuum and degassed. Samples were adjusted to the pH of the starting buffer, or exchanged by dialysis into the starting buffer. The column was pre-equilibrated with the starting buffer until the eluent was applied to the column, which was the same pH as that of starting buffer. After samples were applied to FPLC, the elution buffer was added. Samples were separated by the pI of each component protein in the column, and sequentially eluted from the column. The effluent was continuously monitored by absorbance at 280 nm and divided by 1 mL/tube. After the pH gradient ended, the column was washed with two column volumes of 2M NaCl to elute the molecules still bound to the column. Finally, the column was re-equilibrated with five column volumes of starting buffer until UV absorbance and pH values reached a plateau. An aliquot (10 μL) of each fraction eluted from the column was subjected to SDS-PAGE on a 7% polyacrylamide gel, and the separated protein bands were electro transferred to a polyvinylidene difluoride membrane and detected with specific antibodies to vitronectin. Membranes were developed with ECL (GE Healthcare Inc.).

3.1.3 Results

An example of the elution pattern of rat plasma is shown in Fig. 9. As the buffer flowed through the chromatofocusing column, the pH decreased and a descending pH gradient was generated. As shown in Fig. 10, when vitronectin in rat plasma was separated by chromatofocusing, NO-VN was eluted at pH 4.7–4.5, while PH–VN (24 h) was eluted at pH 5.2-4.8.

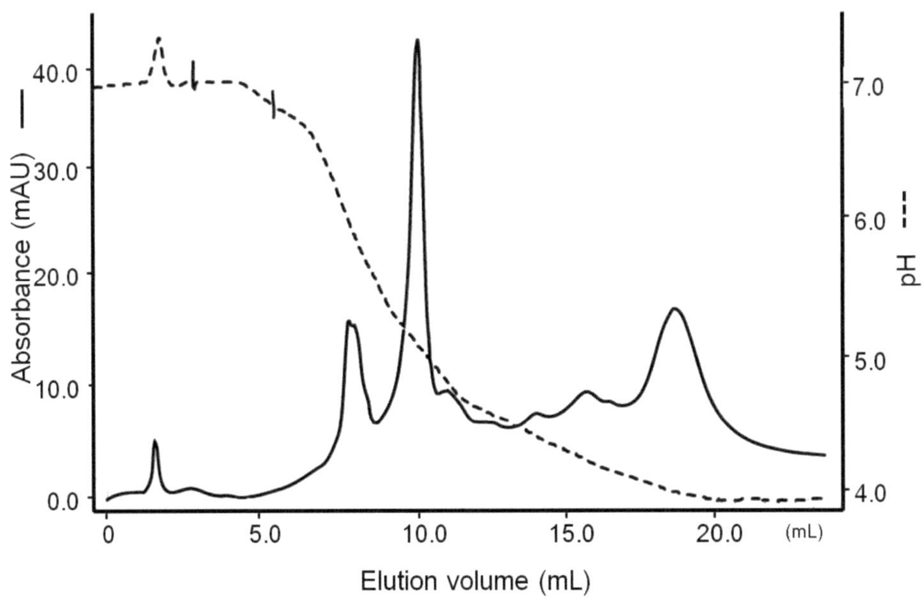

Fig. 9. Chromatogram of non-operated rat plasma. Dotted line, pH; solid line, absorbance at 280 nm. Flow rate: 0.5 mL/min, temperature: 4°C. Other conditions are described in the text.

Rat plasma vitronectin at 24 h after partial hepatectomy was shifted to a higher pI than that of NO-VN on two-dimensional PAGE, as shown in Fig. 7. The result of chromatofocusing combined with immunodetection suggests that sialylation of vitronectin remained low with pI over pH 5 for the period from 24 to 72h after partial hepatectomy. The pI of PH-VN after 7d had recovered to pH 4.7–4.5 (data not shown).

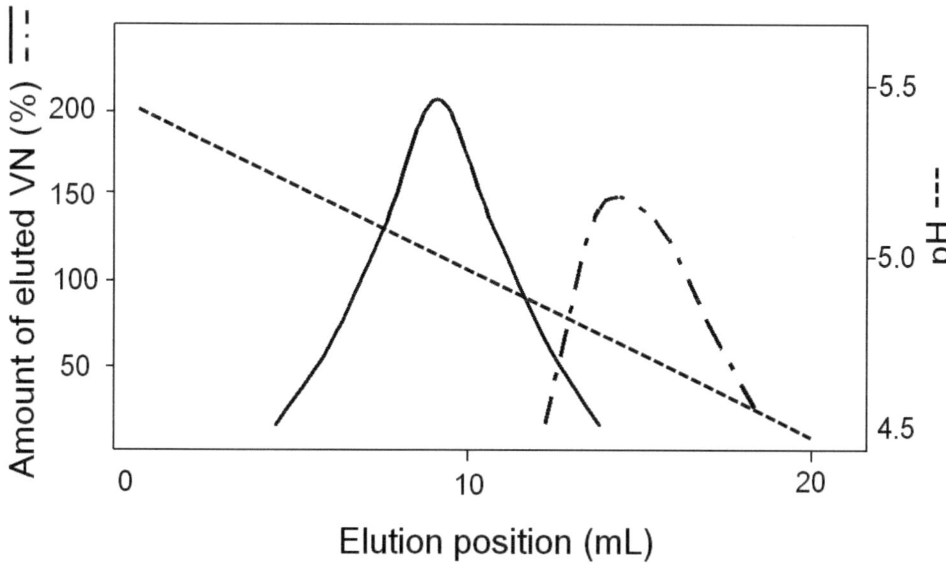

Fig. 10. The elution pattern of vitronectin with descending elution pH. The amount of vitronectin was measured by the band intensity of an immunoblot of the eluted fraction and is expressed as %, when taking that of 1 μg of purified NO-VN as 100%. Solid line: PH-VN (24 h), chain line: NO-VN, dotted line: pH of the eluted fraction.

3.1.4 Discussion

The elution pattern on chromatofocusing was very reproducible, and the pIs of NO-VN and PH-VN were found to be 4.7–4.5 and 5.2–4.8, respectively (Fig. 10). The immunostaining of PH-VN after two-dimensional PAGE (Fig. 7) indicated the presence of two components (pI 4.6 and 6.0), whereas PH-VN was eluted by chromatofocusing at intermediate pH, although the two methods agreed on the tendency for the pI of PH-VN to be considerably shifted toward alkalinity compared to that of NO-VN. This discrepancy may be due not only to the difference in the plasma sample lots but also because parts of the vitronectin eluted by chromatofocusing are multimerized in the buffer like the vitronectin in physiological plasma, which shows a broad pH range of elution peaks indicating variation in sialylation of the vitronectin molecules contained. Chromatofocusing has the advantage of analyzing the isoelectric property under physiological conditions, especially in detection of changes in sialylation of glycoproteins, and the eluted fractions are utilizable for activity measurements (Kneba, M., et al. 1983).

3.1.5 Summary

Each plasma sample was subjected to chromatofocusing using FPLC, which can separate molecules by their isoelectric point (pI). Fractions eluted from the column were subjected to

SDS-PAGE, electrotransferred to a PVDF membrane, and detected with specific antibodies to vitronectin or fibronectin.

Key results: The chromatofocusing and immunoblotting of plasma before and after sialidase treatment enabled us to demonstrate when and what alteration of sialylation occurs in each glycoprotein at each different stage during liver regeneration. The changes in pI essentially coincided with that of 2D-PAGE; however, we can determine the pI of a sample under physiological condition by using the chromatofocusing technique. Furthermore, this method is facile, quick, and applicable to recovered samples for activity analyses because they are non-denatured and separated by pI.

4. Conclusion

This study proposes that alterations of glycosylation, especially decreased sialylation of vitronectin, modulate tissue remodeling processes in multiple steps, especially HSC spreading and survival. Our findings suggest that the removal of sialic acid from vitronectin suppressed activation of stellate cells, indicating the possibility of a new treatment for or method to prevent liver cirrhosis. This will open new windows to the paradigm of glyco-regenerative medicine, which is based on the modulatory functions of glycans of ECM glycoproteins such as vitronectin.

5. Acknowledgements

This work was supported in part by the Japan Society for the Promotion of Science Grant 12995 (to K. S.), a fellowship of the Hayashi Memorial Foundation for Female Natural Scientists (to K. S.), a research grant from The Uehara Memorial Foundation (to H.O.), and Grants-in-aid for Scientific Research on Priority Areas 15040209, 17046004, and Research (C) on 17570109 (HO) and 22570111 (HO) from the Ministry of Education, Culture, Sports, Science, and Technology.

We thank K. Ono for editing the English.

6. References

Bauer, C.H., Hassels, B.F., & Reutter, W. G. (1976) Galactose metabolism in regenerating rat liver. *Biochem. J.*, 154, 141-147.

Benyon, R.C. and Arthur, M.J. (2001) Extracellular matrix degradation and the role of hepatic stellate cells. Semin. *Liver Dis.*, 21, 373-384.

Black, J.A. and Harkins, R.N. (1977) Amino acid compositions and evolutionary relationships with protein families. *J Theor Biol*, 66, 281-295.

Dalziel, M., Lemaire, S., Ewing, J., Kobayashi, L. & Lau, J. T. (1999) Hepatic acute phase induction of murine beta-galactoside alpha 2,6 sialyltransferase (ST6Gal I) is IL-6 dependent and mediated by elevation of exon H-containing class of transcripts. *Glycobiology*, 9, 1003-1008.

DeClerck, Y.A., Mercurio, A.M., Stack, M. S., Chapman, H. A., Zutter, M. M., Muschel, R. J., Raz, A., Matrisian, L. M., Sloane, B. F., Noel, A., Coussens, L. & Padarathsingh, M.

(2004) Proteases, extracellular matrix, and cancer: a workshop of the path B study section. *Am. J. Pathol.*, 164, 1131-1139.

Diehl, A.M. and Rai, R.M. (1996) Liver regeneration 3: Regulation of signal transduction during liver regeneration. *Faseb. J.*, 10, 215-227.

Finne, J., Krusius, T. & Rauvala, H. (1977) Occurrence of disialosyl groups in glycoproteins. *Biochem. Biophys. Res. Commun.*, 74, 405-410.

Finne, J., Krusius, T., Rauvala, H. & Hemminki, K. (1977) The disialosyl group of glycoproteins. Occurrence in different tissues and cellular membranes. *Eur. J. Biochem.*, 77, 319-323.

Gebb, C., Hayman, E.G., Engvall, E. & Ruoslahti, E. (1986) Interaction of vitronectin with collagen. *J. Biol. Chem.*, 261, 16698-16703.

Hayman, E.G., Pierschbacher, M.D., Ohgren, Y. & Ruoslahti, E. (1983) Serum spreading factor (vitronectin) is present at the cell surface and in tissues. *Proc. Natl. Acad. Sci. U S A*, 80, 4003-4007.

Holmes, R. (1967) Preparation from human serum of an alpha-one protein which induces the immediate growth of unadapted cells in vitro. *J Cell Biol*, 32, 297-308.

Hughes, R.C. (1997) *Adhesive glycoproteins and receptors*. Elsevier Sciences B.V., New York.

Ip, C. (1979) Effect of partial hepatectomy and hydrocortisone administration on liver and serum sialyltransferase activities. *Biochim. Biophys. Acta*, 583, 14-19.

Ishii, I., Takahashi, N., Kato, S., Akamatsu, N. & Kawazoe, Y. (1985) High-performance liquid chromatographic analysis of changes of asparagine-linked oligosaccharides in regenerating rat liver. *J. Chromatogr.*, 345, 134-139.

Izumi, M., Yamada, K.M., & Hayashi, M. (1989) Vitronectin exists in two structurally and functionally distinct forms in human plasma. *Biochim. Biophys. Acta*, 990, 101-108.

Kato, S. and Akamatsu, N. (1984) Alterations in N-linked oligosaccharides of glycoproteins during rat liver regeneration. *Biochim Biophys Acta*, 798, 68-77.

Kato, S. and Akamatsu, N. (1985) Alterations in fucosyl oligosaccharides of glycoproteins during rat liver regeneration. *Biochem. J.*, 229, 521-528.

Kato, S., Otsu, K., Ohtake, K., Kimura, Y., Yashiro, T., Suzuki, T.& Akamatsu, N (1992) Concurrent changes in sinusoidal expression of laminin and affinity of hepatocytes to laminin during rat liver regeneration. *Exp Cell Res*, 198, 59-68.

Kneba, M., Krieger, G., Kehl, A., Bause, I.& Nagel, G. A. (1983) Chromatofocusing combined with the ELISA technique. A sensitive method for the analysis of immune complexes. *J Immunol Methods*, 61, 233-243.

Kobayashi, J., Yamada, S. & Kawasaki, H. (1994) Distribution of vitronectin in plasma and liver tissue: relationship to chronic liver disease. *Hepatology*, 20, 1412-1417.

Miyoshi, E., Ihara, Y., Nishikawa, A., Saito, H., Uozumi, N., Hayashi, N., Fusamoto, H., Kamada, T., & Taniguchi, N. (1995) Gene expression of *N*-acetylglucosaminyltransferases III and V: a possible implication for liver regeneration. *Hepatology*, 22, 1847-1855.

Oda-Tamai, S., Kato, S., Hara, S. & Akamatsu, N. (1985) Decreased transfer of oligosaccharide from oligosaccharide-lipid to protein acceptors in regenerating rat liver. *J. Biol. Chem.*, 260, 57-63.

Ogawa, H., Yoneda, A., Seno, N., Hayashi, M., Ishizuka, I., Hase, S. & Matsumoto, I. (1995) Structures of the *N*-linked oligosaccharides on human plasma vitronectin. *Eur. J. Biochem.*, 230, 994-1000.

Okamoto, Y. and Akamatsu, N. (1977) Synthesis in vitro of glycoprotein in regenerating rat liver. *Biochim. Biophys. Acta*, 498, 272-281.

Okamoto, Y., Ito, E. & Akamatsu, N. (1978) UDP-*N*-acetylglucosamine-glycoprotein *N*-acetylglucosaminyltransferase in regenerating rat liver. *Biochim. Biophys. Acta*, 542, 21-27.

Preissner, K.T. (1991) Structure and biological role of vitronectin. *Annu. Rev. Cell Biol.*, 7, 275-310.

Preissner, K.T. and Muller-Berghaus, G. (1987) Neutralization and binding of heparin by S protein/vitronectin in the inhibition of factor Xa by antithrombin III. Involvement of an inducible heparin-binding domain of S protein/vitronectin. *J Biol Chem*, 262, 12247-12253.

Preissner, K.T. and Seiffert, D. (1998) Role of vitronectin and its receptors in haemostasis and vascular remodeling. *Thromb. Res.*, 89, 1-21.

Pretzlaff, R.K., Xue, V.W. & Rowin, M. E. (2000) Sialidase treatment exposes the beta1-integrin active ligand binding site on HL60 cells and increases binding to fibronectin. *Cell Adhes. Commun.*, 7, 491-500.

Rutishauser, U. (1998) Polysialic acid at the cell surface: biophysics in service of cell interactions and tissue plasticity. *J Cell Biochem*, 70, 304-312.

Sano, K., Asahi, M., Yanagibashi, M., Hashii, N., Itoh, S., Kawasaki, N.& Ogawa, H. (2008) Glycosylation and ligand-binding activities of rat plasma fibronectin during liver regeneration after partial hepatectomy. *Carbohydr. Res.*, 343, 2329-2335.

Sano, K., Asanuma-Date, K., Arisaka, F., Hattori, S. & Ogawa, H.. (2007) Changes in Glycosylation of Vitronectin Modulate Multimerization and Collagen Binding During Liver Regeneration. *Glycobiology*, 17, 784-794.

Sano, K., Miyamoto, Y., ,Kawasaki, N., Hashii, N., Itoh, S., Date, K., Yokoyama, M., Sato, C., Kitajima, K.& Ogawa, H. (2010) Survival signals of hepatic stellate cells in liver regeneration are regulated by glycosylation changes in rat vitronectin, especially decreased sialylation. *J. Biol. Chem.*, 285, 17301-17309.

Schvartz, I., Seger, D. & Shaltiel, S. (1999) Vitronectin. *Int. J. Biochem. Cell Biol.*, 31, 539-544.

Seiffert, D. (1997) Constitutive and regulated expression of vitronectin. *Histol. Histopathol.*, 12, 787-797.

Seiffert, D., Keeton, M., Eguchi, Y., Sawdey, M. & Loskutoff, D. J. (1991) Detection of vitronectin mRNA in tissues and cells of the mouse. *Proc. Natl. Acad. Sci. U S A*, 88, 9402-9406.

Serafini-Cessi, F. (1977) Sialyltransferase activity in regenerating rat liver. *Biochem. J.*, 166, 381-386.

Shan, L. and Anderson, D.J. (2001) Effect of buffer concentration on gradient chromatofocusing performance separating protiens on a high-performance DEAE column. *J Chromatogr A*, 909, 191-205.

Sluyterman, L.A. and Wijdenes, J. (1978) Chromatofocusing : isoelectric focusing on ion excharge columns. II. Experimental verification. . *J. Chromatogr.* , 150, 31-44.

Sluyterman, L.A.A.E., O (1978) Chromatofocusing : isoelectric focusing on ion excharge columns. I. General principles. *J. Chromatogr.*, 150, 17-30.

Suzuki, R., Yamada, S., Uchibori-Iwaki, Haruhi, Oda-Tamai, S., Kato, S., Akamatsu, N., Yoneda, A. & Ogawa, H. (2001) Changes in glycosylation and collagen binding of vitronectin in liver cirrhosis. *International Congress Series*, 1223, 103-107.

Uchibori-Iwaki, H., Yoneda, A., Oda-Tamai, S., Kato, S., Akamatsu, N., Otsuka, M., Murase, K., Kojima, K., Suzuki, R., Maeya, Y., Tanabe, M., Ogawa, H. (2000) The changes in glycosylation after partial hepatectomy enhance collagen binding of vitronectin in plasma. *Glycobiology*, 10, 865-874.

Uchibori, H., Ogawa, H., Matsumoto, I. & Seno N. (1992) Contribution of the sugar chains to the stability for proteolysis of cell adhesive glycoprotein, vitronectin. *Connective Tissue*, 23, 117-124.

Ueda, H., Kojima, K., Saitoh, T. & Ogawa, H. (1999) Interaction of a lectin from *Psathyrella velutina* mushroom with *N*-acetylneuraminic acid. *FEBS Lett.*, 448, 75-80.

Ueda, H., Matsumoto, H., Takahashi, N. & Ogawa, H. (2002) *Psathyrella velutina* mushroom lectin exhibits high affinity toward sialoglycoproteins possessing terminal *N*-acetylneuraminic acid 2,3-Linked to penultimate galactose residues of trisialyl *N*-glycans. Comparison with other sialic acid-specific lectins. *J. Biol. Chem.*, 277, 24916-24925.

Varki, A. (1993) Biological roles of oligosaccharides: all of the theories are correct. *Glycobiology*, 3, 97-130.

Varki, A., Commings, R.D., Esko, J. D., Freeze, H. H.,Stanley, P., H., Bertozzi, C.R., Hart, G.W., Etzler, M.E. eds. (2009) *Essentials of Glycobiology*. Cold Spring Harbor (NY): Cold Spring Harbor Laboratory Press.

Yamada, S., Kobayashi, J. & Kawasaki, H. (1997) Plasma collagen-binding vitronectin activated by heparin and dextran sulfate in chronic liver disease. *Res Commun Mol Pathol Pharmacol*, 97, 315-324.

Yamada, S., Kobayashi, J., Murawaki, Y., Suou, T. & Kawasaki, H. (1996) Collagen-binding activity of plasma vitronectin in chronic liver disease. *Clin Chim Acta*, 252, 95-103.

Yasukawa, Z., Sato, C. & Kitajima, K. (2005) Inflammation-dependent changes in alpha2,3-, alpha2,6-, and alpha2,8-sialic acid glycotopes on serum glycoproteins in mice. *Glycobiology*, 15, 827-837.

Yasukawa, Z., Sato, C. & Kitajima, K. (2007) Identification of an inflammation-inducible serum protein recognized by anti-disialic acid antibodies as carbonic anhydrase II. *J. Biochem. (Tokyo)*, 141, 429-441.

Yasukawa, Z., Sato, C., Sano, K., Ogawa, H. & Kitajima, K. (2006) Identification of disialic acid-containing glycoproteins in mouse serum: a novel modification of immunoglobulin light chains, vitronectin, and plasminogen. *Glycobiology*, 16, 651-665.

Yoneda, A., Ogawa, H., Kojima, K. & Matsumoto, I. (1998) Characterization of the ligand binding activities of vitronectin: interaction of vitronectin with lipids and identification of the binding domains for various ligands using recombinant domains. *Biochemistry*, 37, 6351-6360.

Yoneda, A., Ogawa, H., Matsumoto, I., Ishizuka, I., Hase, S. & Seno, N. (1993) Structures of the *N*-linked oligosaccharides on porcine plasma vitronectin. *Eur. J. Biochem.*, 218, 797-806.

6

The Protective Effect of Antioxidants in Alcohol Liver Damage

José A. Morales González[1], Liliana Barajas-Esparza[1], Carmen Valadez-Vega[1], Eduardo Madrigal-Santillán[1], Jaime Esquivel-Soto[2], Cesar Esquivel-Chirino[2], Ana María Téllez-López[1], Maricela López-Orozco[1] and Clara Zúñiga-Pérez[1]
[1]*Instituto de Ciencias de la Salud, UAEH,*
[2]*Facultad de Odontología, UNAM*
México

1. Introduction

The term antioxidant was originally utilized to refer specifically to a chemical product that prevented the consumption of oxygen (Burneo, 2009); thus, antioxidants are defined as molecules whose function is to delay or prevent the oxygenation of other molecules. The importance of antioxidants lies in their mission to end oxidation reactions that are found in the process and to impede their generating new oxidation reactions on acting in a type of sacrifice on oxidating themselves. There are endogenous and exogenous antioxidants in nature. Some of the best-known exogenous antioxidant substances are the following: β-carotene (pro-vitamin A); retinol (vitamin (A); ascorbic acid (vitamin C); α-tocopherol (vitamin E); oligoelements such as selenium; amino acids such as glycine, and flavonoids such as silymarin, among other organic compounds (Venereo, 2002).

Historically, it is known that the first investigations on the role that antioxidants play in Biology were centered on their intervention in preventing the oxidation of unsaturated fats, which is the main cause of rancidity in food (Wolf, 2005). However, it was the identification of vitamins A, C, and E as antioxidant substances that revolutionized the study area of antioxidants and that led to elucidating the importance of these substances in the defense system of live organisms (Jacob, 1996).

Due to their solubilizing nature, antioxidant compounds have been divided into hydrophilics (phenolic compounds and vitamin C) and lipophilics (carotenoids and vitamin E). The antioxidant capacity of phenolic compounds is due principally to their redox properties, which allow them to act as reducing agents, hydrogen and electron donors, and individual oxygen inhibitors, while vitamin C's antioxidant action is due to its possessing two free electrons that can be taken up by Free radicals (FR), as well as by other Reactive oxygen species (ROS), which lack an electron in their molecular structure. Carotenoids are deactivators of electronically excited sensitizing molecules, which are involved in the generation of radicals and individual oxygen, and the antioxidant activity of vitamin A is characterized by hydrogen donation, avoiding chain reactions (Burneo, 2009).

The antioxidant defense system is composed of a group of substances that, on being present at low concentrations with respect to the oxidizable substrate, delay or significantly prevent oxygenation of the latter. Given that FR such as ROS are inevitably produced constantly during metabolic processes, in general it may be considered as an oxidizable substrate to nearly all organic or inorganic molecules that are found in living cells, such as proteins, lipids, carbohydrates, and DNA molecules. Antioxidants impede other molecules from binding to oxygen on reacting or interacting more rapidly with FR and ROS than with the remainder of molecules that are present in the microenvironment in which they are found (plasma membrane, cytosol, the nucleus, or Extracellular fluid [ECF]). Antioxidant action is one of the sacrifices of its own molecular integrity in order to avoid alterations in the remainder of vitally functioning or more important molecules. In the case of the exogenic antioxidants, replacement through consumption in the diet is of highest importance, because these act as suicide molecules on encountering FR, as previously mentioned (Venereo, 2002).

This is the reason that, for several years, diverse researchers have been carrying out experimental studies that demonstrate the importance of the role of antioxidants in protection and/or hepatic regeneration in animals. Thus, in this chapter, the principal antioxidants will be described that play an important role in the regeneration of hepatic cells and in the prevention of damage deriving from alcohol (Burneo, 2009; Venereo 2002).

2. Retinol (Vitamin A)

Vitamin A, also called trans-retinol, is an isoprenoid alcohol that that performs several important functions in the organism, is essential in vision, in addition to being necessary for epithelial tissue regulation and differentiation, as well as for bone growth, reproduction, and embryonic development. Together with some carotenoids, vitamin A increases the immunitary function, reduces the consequences of infectious diseases (Goodman & Gilman, 1996), and, more recently, it has been observed that it provides certain protection against malignant diseases such as cancer (Morales-González 2009).

Vitamin A belongs to a family of similarly structured molecules that are generically denominated retinoids (low-molecular-weight molecules, derived from the hydrophobic molecules of vitamin A) (Mathews-Van Holde, 1998).

The activity of vitamin A in mammals is due not only to retinoids, but also to certain carotenes that are widely distributed in the majority of vegetables. Carotenes do not possess intrinsic vitamin A activity, but are converted into vitamin A by means of enzymatic actions that take place in the intestinal mucosa and in the liver (Morales-González 2009).

2.1 Structure

Vitamin A can present as a free alcohol, as a fatty acid ester, as an aldehyde, and as an acid (Figure 1). In this structure, on replacing the alcohol group, it obtained retinal, the principal functional form of rods and cones in the retina and, by an acid group, retinoic acid, the main functional form in cellular regulation and differentiation (Morales-González 2009; Mathews Van-Holde 1998).

2.2 Digestion, absorption, and metabolism

Because vitamin A is a liposoluble vitamin, retinol digestion and absorption is intimately linked to that of lipids. Retinol esters dissolved in fat from the diet arrive in the small intestine, forming micelles with the aid of bile salts. Later, hydrolysis is produced in which the pancreatic lipase enzyme participates, acting on formed micelles, causing the absorption of 90% of dietary fats. Vitamin A, together with the additional products of enzymatic hydrolysis, enter the enterocyte after passing through the cellular membrane, whether by facilitated diffusion or passively depending on the concentrations present (Morales-González 2009).

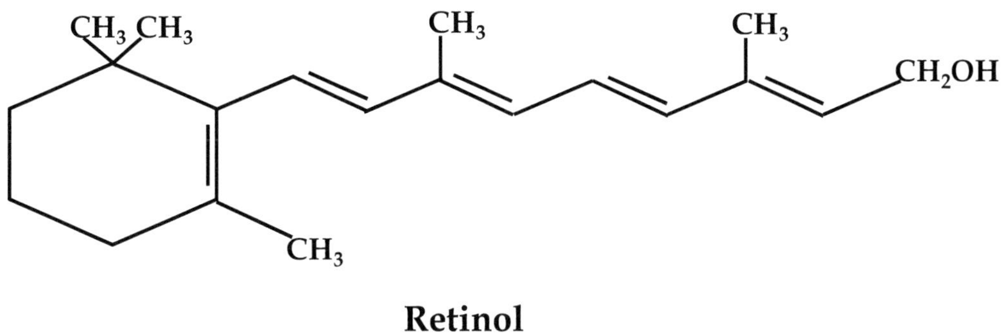

Fig. 1. The chemical structure of vitamin A is made up of a 6-carbon-atom cyclic nucleus with an 11-carbon side-chain.

Carotenes as such are absorbed passively, and once in the cytoplasm, are transformed into retinol. Within the intestinal cell, the greater part of the retinol is esterified with saturated fatty acids such as palmitic acid and is incorporated in lymphatic kilomicrons, which enter into the bloodstream and are transported to the liver, where it is stored in parenchymatous cells and in the adiposites in the form of retinyl ester. The greater part of this is taken up by the hepatocytes of kilomicron fragments and is transferred in the form of light retinol to the Retinol binding protein (RBP) and toward the Kupffer cells, whose main function appears to be storage of these. When the tissues require retinol, this is transported by means of RBP and Transthyretin (TTR, prealbumin) for transport in the circulation of the target cells. The tissues are capable of taking this up through surface receptors, where the retinol is transferred to a retinol membrane binding protein and becomes a retinyl ester. Later, a hydrolase related with the membrane unfolds the latter. RBP exists in nearly all tissues; the exceptions comprise cardiac and skeletal muscle. In addition to its uptake of retinol, the RBP functions as a reservoir for cellular retinol and releases the vitamin to the appropriate sites for its conversion into active compounds. In the retina, retinol becomes 11-cis-retinal, which is incorporated into the rhodospin. In other target tissues, retinol apparently is oxidized into retinoic acid, which is transported to the nucleus. It is noteworthy that the RBP plasma concentration is crucial for regulation of the retinol in plasma and its transport to the tissues (Morales-González, 2009). In general, within the organism retinol can follow three processes; esterification and storage in the liver; conversion into active metabolites (retinal), and/or catabolism and excretion as retinoic acid (Allende-Martínez 1997).

2.3 Antioxidant action

Vitamin A is a natural antioxidant that prevents cellular aging, eliminates FR, and protects the DNA in its mutagenic action. β-carotene is also a powerful antioxidant. We must clarify that this antioxidant function is only obtained in foods that were submitted to cooking for at least 5 minutes. Some studies have demonstrated that β-carotene supplemented in the diet has shown some evidence of antitumor action (Allende-Martínez 1997).

On the other hand, in the case of its participation in the regeneration of hepatic cells in alcohol-induced damage, a positive result is obtained, because vitamin A has shown a lesser protector effect against the formation of FR in the reversion of the hepatic regeneration inhibition caused by ethanol consumption, due to that retinol is the principal component of vitamin A, that it is an alcohol, and that retinol as well as ethanol utilize the same enzymatic pathways; thus, storage of vitamin A and of ethanol in hepatic cells is altered. Therefore, an excess of vitamin A and its interaction with the alcohol increase the capacity to produce fibrous tissue that, in the long term, can cause a cirrhosis (Ramírez-Farías et al., 2008). Thus, it is considered that vitamin A can generate, instead of a benefit, a hepatotoxicity with subsequent inflammation, necrosis, and the increase of some serum enzymes (Morales-González, 2009).

3. Ascorbic acid (Vitamin C)

L-Ascorbic acid (AA), commonly known as vitamin C, is considered one of the organism's most powerful antioxidant agents due to its capacity to donate two electrons from its double link, that of positions two and three; thus, it interacts with the FR, blocking their harmful effect. The human body is not capable of obtaining vitamin C exogenously through foods; it is found concentrated in certain organs such as eye, liver, spleen, suprarenal glands, and thyroids. It is an essential vitamin, in that it participates in reactions such as the synthesis of molecules such as carnitine or thrysine acid, which are fundamental for good bodily function; it participates in iron absorption and presents immunological and anti-inflammatory actions such as the synthesis of neurotransmitters and hormones, playing an important role in collagen synthesis (Morales-González, 2009; Mathews-Van Holde 1998).

AA ($C_6H_8O_6$) is an essential vitamin that is chemically synthesized from glucose by a series of enzyme-catalyzed actions, the last enzyme involved in its synthesis being L-gulono-gamma-lactone oxidase (GLO); it is hydrosoluble and possesses acidic and strongly reductive properties. These properties are due to its enediol structure and to the possibility of ionizing the hydroxyl situated on carbon 3, forming an anion that remains stabilized by resonance. Eventually, it can dissociate the carbon-2 hydroxyl, forming a dianion, although it does not acquire the same stability as that of carbon 3. In nature, two isomers are found with nutritive properties (Figure 1): the L- isomer (L-ascorbic acid), and the D-isomer (L-dehydroascorbic acid). The functions of vitamin C apparently reflect its redox capacity. Thus, it participates in some hydroxylation reactions in which it maintains optimal enzymatic activity by means of electron donation. Vitamin C also increases the absorption of no-heme iron and serves as an important mechanism for inactivating highly reactive radicals in tissue cells. Similarly, it delays the formation in the body of nitrosamines , which are possible carcinogens. The accumulated evidence links ascorbic acid with many elements of the immunitary system (Morales-González, 2009).

3.1 Absorption

Vitamin C absorption is carried out in the small intestine by sodium-dependent active transport (faster and more efficient), or by passive transfusion through a glucose transporter (insulin). L-dehydroascorbic acid is more easily absorbed than L-ascorbic acid. This characteristic is attributed to that the oxidized form of the vitamin remains with ionization to the physiological pH and to that the molecule is slightly hydrophobic, which permits it to penetrate into the membranes, Once inside, it is reduced to ascorbate; in this manner, it circulates mainly in plasma and in suprarenal glands and hypophysis. The effectiveness of the absorption depends on the dose administered, because it has been observed that on increasing the dose, the effectiveness of the absorption diminishes (Morales-González, 2009).

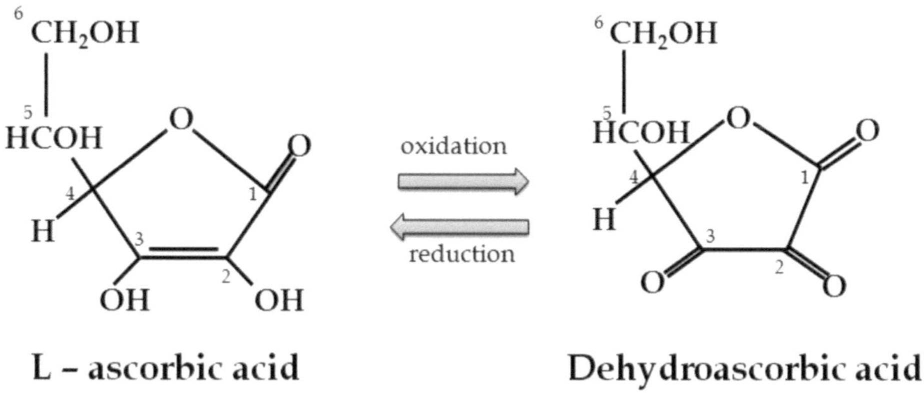

Fig. 2. The chemical structure of vitamin C corresponds to that of a lactone; L-ascorbic acid presents a double connection link between the 2 and 3 carbon; this characteristic allows the molecule to donate electrons from this double link, forming L-dehydroascorbic acid, or in a reversible reaction.

3.2 Antioxidant action

Similar to vitamin A, vitamin C is a natural antioxidant characterized by the capacity to donate two electrons from its double link at positions two and three, in such as way that it interacts with FR, blocking their harmful effect; consequently, it is oxidized. In addition to exerting an effect on FR, it also acts by regenerating oxidized antioxidants such as α-tocopherol and β-carotene. In a study published by Ramírez-Farías et al., in 2008, the author concluded that administration of vitamins C and E provided a protector effect against liver damage; they attenuate lipid peroxidation, and both vitamins present a significantly greater effect than vitamin A against ethanol-mediated toxic effects during hepatic regeneration.

4. Vitamin E

Vitamin E ($C_{29}H_{50}O_2$) is known as the generic of a derived set of tocols, with α-tocopherol the most active form for humans. Tocopherols possess a functional phenolic group in a chromanol ring and an isoprenoid side- chain of 16 carbons; it is saturated with three double links. There are two groups of compounds with vitamin E activity, and the tocopherols (with a 16-carbon isoprenoid side-chain) and the tocotrienols (with the same 16-carbon chain, but with three double links); both groups present vitameres that differ in the number and position of the methyl groups in the chromanol ring, designated as α, β, ∂, and δ (Figure 3) (Morales-González, 2009).

This vitamin forms part of the essential vitamins; thus, it should be acquired by means of its consumption in the daily diet. The distinct forms of vitamin E are not interchangeable among themselves in humans; in addition, they present distinct metabolic behaviors; therefore, other existing forms do not convert into α-tocopherol at any time and do not contribute to covering the vitamin E requirement. One of the main functions of α-tocopherol is that of its being a lipid antioxidant; on the other hand, it diminishes the production of thromboxanes and prostaglandins, is capable of inducing apoptosis directly or indirectly in tumor cells, modulates microsomal enzyme activity, inhibits protein kinase C activity, functions as a genetic regulator at the messenger RNA (mRNA) level, modulates the immunitary response during oxidative stress, and intervenes in the processes of fetal development and gestation, as well as in the processes of the formation of elastic and collagen fibers of the connective tissues, and in addition promotes the normal formation of erythrocytes; thus, the importance of this vitamin (Morales-González, 2009; Sayago, 2007).

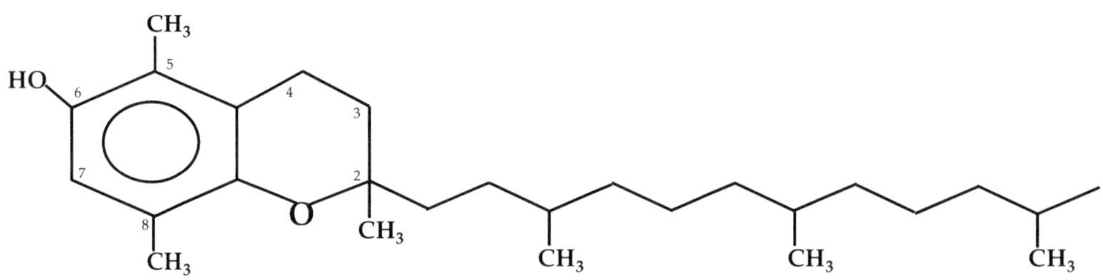

α- Tocopherol

Fig. 3. In tocopherol structure, the alpha (α) homologue possesses four methyl groups in positions 2, 5, and 7.

4.1 Absorption and metabolism

Vitamin E is principally absorbed in the small intestine, with biliary secretion and micelle solubilization as indispensible. The bile acids, proceeding from the liver and segregated in the small intestine, favor the formation of micelles and facilitate the action of pancreatic lipases on lipids. Absorption of vitamin E within the erythrocyte is a passive process; α-tocopherol and non-esterized δ-tocopherol are incorporated into the kilomicrons and for transport to the liver, in which the alpha-Tocopherol transfer protein (α-TTP) binds to the natural α-tocopherol stereoisomer or to the Golgi apparatus to incorporate it into Very-low-density lipoproteins (VLDL), from which is transferred to other circulating lipoproteins such as High-density (HDL) and Low-density lipoproteins (LDL) during their catabolism by the Lipoprotein lipase (LPL). LPL can also act on HDL and LDL in order for vitamin E to be able to accede to the peripheral tissues (Morales-González, 2009; Sayago, 2007).

4.2 Antioxidant activity

α-Tocopherol inhibits lipid oxidation by means of two mechanisms. On the one hand, it eliminates the FR produced during peroxidation, thus inhibiting the oxidation chain reaction and, on the other hand, it acts as a singlet oxygen chelator. Polyunsaturated fatty acids (PUFA) have methylene groups localized between two double links; this renders PUFA sensitive to auto-oxidation; the initial reaction produces the lipid radical, a conjugated diene radical (L•), which reacts rapidly with the oxygen molecule, giving rise to a peroxyl free radical L-OO. In turn, this FR can act upon another PUFA, which will re-initiate the entire process and which would give rise to the peroxidation chain. The FR chain reaction is broken when α-tocopherol, present in the membrane, transfers a hydrogen to a peroxyl FR, transforming it into a hydroperoxide and giving rise to a tocopherol radical. The latter can react with different electron donors, among which ascorbate (vitamin C) is highlighted, for regenerating the tocopherol. In general, each α-tocopherol molecule can react with two peroxyl radicals (Morales-González, 2009; Sayago, 2007).

5. Oligoelements

Oligoelements comprise nine micronutrients that are found in the organism in amounts of <0.01% of body weight, and are the following: iron; zinc; selenium; manganese; iodine; chromium; fluorine; copper, and molybdenum. Oligoelements are very important for the organism because they perform functions such as serving as co-factors in enzymatic systems or as vital molecular components (Morales-González, 2009).

5.1 Selenium (Se)

Selenium is localized within the group of micronutrients constituting a trace element or an essential micronutrient for all mammals. Selenium is defined as a non-volatile micromineral that fulfills numerous biological functions; thus, its best known function is its role as part of the glutathione peroxidase enzyme that protects cells from oxidative damage. Its presence in tissues such as liver, heart, lung, and pancreas is essential because it promotes the breakdown of toxic peroxides formed during metabolism, impeding cell membrane damage. Selenium protects against toxicity by means of mercury, cadmium, and silver (Morales-González, 2009; Manzanares-Castro, 2007).

5.1.1 Absorption and metabolism

Absorption is mainly carried out in the duodenum. This element enters into the body in two principal ways depending on its source: selenocysteine (in animals), and selenomethionine (in plants); once in the organism, sulfur replacement takes place in cysteine and methionine for the formation of amino acids and selenoproteins. It is excreted through the urine and when consumed in high quantities, it also can be eliminated through the breath (Manzanares-Castro, 2007).

5.1.2 Antioxidant action

Selenium (Se) possesses a potent antioxidant power, which is associated with the so-called selenoenzymes. To date, approximately 35 selenoenzymes have been described; these are proteins contain a selenocysteine residue in their active site and in which Se constitutes their enzymatic co-factor. Among the selenoenzymes, the best characterized and studied are Glutathione peroxidase (GPx) and Selenoprotein P (SePP). Another two very important enzymes are thioredoxin reductase, whose function is to reduce nucleotides during DNA synthesis, and iodothyronine deiodinase, which is responsible for the peripheral conversion of T_4 in active T_3. Selenium, together with vitamin E, protects cells from peroxidation because the former destroys peroxides through the cytoplasm, while the latter prevents peroxide formation (Manzanares-Castro, 2007).

The biological role that selenium presents is based on two fundamental properties: the antioxidant protector function from oxidative damage, and immunomodulation; in this case, it had the antecedent of a study carried out in rats in whom oxidative damage was induced and by means of Selenium (Se); a significant diminution of the enzymes aspartate aminotransferse (AST), Alkaline phosphatase (ALT), and Alanine aminotransferase (ALP) was observed, as well as an improvement in the antioxidant state. The results suggested that administration of Se to hepatic tissue protects against intoxication due to its antioxidant properties (Mahfoud et al., 2010).

6. Amino acids

Amino acids are nutrients that function as raw material for protein formation. According to their classification due to their requirement in the diet, they are classified as essential, non-essential, and semi-essential. Their most important function is the formation of peptides, structural proteins, enzymes, transporter proteins, immunoproteins, and hormones. However, each has special chemical functions in which they are exchanged or cede methyl or sulfhydroxyl groups in choline synthesis or substance detoxification. Some of these, such as glycine, cysteine, glutamic acid, or taurine, assume the role of antioxidants, and under extreme conditions when other energy sources are insufficient, these are utilized to produce energy through glyconeogenesis; each amino acid possesses different specific and concrete functions (Morales-González, 2009).

6.1 Glycine (Gli)

This is the simplest amino acid of all, and it is one of the so-called non-essential amino acids; thus, no minimal nutritional contribution is required, given that there are substances available in the organism for its synthesis. Glycine (C2H5NO2) is produced in hepatocyte mitochondria from 3-D-phosphoglycerate, giving rise to serine, and in the presence of pyridoxal phosphate, serine hydroxymethyltransferase removes one carbon atom, thus producing glycine. It can also be constituted from carbon dioxide, ammonium, and from N5N10-methylenoTetrahydrofolate (TFH) in the same manner as in the mitochondria (Morales-González, 2009; Mathews-Van Holde, 1998).

Glycine is found at high concentrations in the organism, functioning as an important carbon donor for the formation of numerous essential compounds. It also functions in the biosynthesis of multiple compounds, such as the heme group, purines, proteins, nucleotides, nucleic acids, creatinine, conjugated bile salts, and porphyrins, or it can be degraded and converted into serine. In the brain, it functions as a neurotransmitter inhibitor; in addition, it serves as an extracellular communications molecule; therefore, it possesses different antioxidant protector effects (Morales-González, 2009).

6.1.1 Antioxidant activity

Glycine possess a protector effect due to that it prevents due to that it prevents the decrease of antioxidant hepatic enzyme activity after hemorrhagic processes; this effect can be due to that glycine blocks the activation of Kupffer cells, which produce FR. Likewise, glycine exercises an ascorbate oxidation protector effect by means of cupric ions; consequently, it diminishes hydroxyl radical generation. In addition to this, glycine forms part of glutathione, tripeptidic and intracellular, that combats FR and maintains some essential biological molecules in a reduced chemical state (Morales-González, 2009).

A group of researchers demonstrated the hepatoprotector effect of glycine and vitamin E in a study conducted in rats in which Partial hepatoctomy (PH) was practiced with subsequent administration of these antioxidants; finally, it was observed that treatment with either of the two antioxidants causes an increase in the peroxidase dismutase enzyme; it diminishes Thiobarbituric acid (TBARS) levels, exhibiting the protector effect in hepatic regeneration (Parra-Vizuet et al., 2009).

7. Flavonoids

Flavonoids are compounds that make up part of the polyphenols and are also considered essentials nutrients. Their basic chemical structure consists of two benzene rings bound by means of a three-atom heterocyclic carbon chain. Oxidation of the structure gives rise to several families of flavonoids (flavons, flavonols, flavanons, anthocyanins, flavanols, and isoflavons), and the chemical modifications that each family can undergo give rise to >5,000 compounds identified by their particular properties (Morales-González, 2009).

Flavonoid digestion, absorption, and metabolism have common pathways with small differences, such as, for example, unconjugated/non-conjugated flavonoids can be absorbed at the stomach level, while conjugated flavonoids are digested and absorbed at the intestinal level by extracellular enzymes on the enterocyte brush border. After absorption, flavonoids are conjugated by methylation, sukfonation, ands glucoronidation reactions due to their biological activity, such as facilitating their excretion by biliary or urinary route. The conjugation type the site where this occurs determine that metabolite's biological action, together with the protein binding for its circulation and interaction with cellular membranes and lipoproteins. Flavonoid metabolites (conjugated or not) penetrate the tissues in which they possess some function (mainly antioxidant), or are metabolized (Morales-González, 2009).

7.1 Silymarin

Silymarin is a compound of natural origin extracted from the *Silybum marianum* plant, popularly known as St. Mary's thistle, whose active ingredients are flavonoids such as silybin, silydianin, and silycristin. This compound has attracted attention because of its possessing antifibrogenic properties, which have permitted it to be studied for its very promising actions in experimental hepatic damage. In general, it possesses functions such as its antioxidant one, and it can diminish hepatic damage because of its cytoprotection as well as due to its inhibition of Kupffer cell function (Sandoval, 2008).

7.1.1 Antioxidant and hepatoprotector action

Silymarin is an active principle that possesses hepatoprotector and regenerative action; its mechanism of action derives from its capacity to counterarrest the action of FR, which are formed due to the action of toxins that damage the cell membranes (lipid peroxidation), competitive inhibition through external cell membrane modification of hepatocytes; it forms a complex that impedes the entrance of toxins into the interior of liver cells and, on the other hand, metabolically stimulates hepatic cells, in addition to activating RNA biosythesis of the ribosomes, stimulating protein formation. In a study published by Sandoval et al., in 2008, the authors observed that silymarin's protector effect on hepatic cells in rats when they employed this as a comparison factor on measuring liver weight/animal weight % (hepatomegaly), their values always being less that those of other groups administered with other possibly antioxidant substances; no significant difference was observed between the silymarin group and the silymarin-alcohol group, thus demonstrating the protection of silymarin. On the other hand, silymarin diminishes Kupffer cell activity and the production of glutathione, also inhibiting its oxidation. Participation has also been shown in the increase of protein synthesis in the hepatocyte on stimulating polymerase I RNA activity. Silymarin reduces collagen accumulation by 30% in biliary fibrosis induced in rat (Boigk, 1997). An assay in humans reported a slight increase in the survival of persons with cirrhotic alcoholism compared with untreated controls (Ferenci, 1989).

8. Ethanol metabolism

Ethanol is absorbed rapidly in the gastrointestinal tract; the surface of greatest adsorption is the first portion of the small intestine with 70%; 20% is absorbed in the stomach, and the remainder, in the colon. Diverse factors can cause the increase in absorption speed, such as gastric emptying, ingestion without food, ethanol dilution (maximum absorption occurs at a 20% concentration), and carbonation. Under optimal conditions, 80–90% of the ingested dose is completely absorbed within 60 minutes. Similarly, there are factors that can delay ethanol absorption (from 2-6 hours), including high concentrations of the latter, the presence of food, the co-existence of gastrointestinal diseases, the administration of drugs, and individual variations (Goldfrank et al., 2002).

Once ethanol is absorbed, it is distributed to all of the tissues, being concentrated in greatest proportion in brain, blood, eye, and cerebrospinal fluid, crossing the feto-placentary and hematocephalic barrier (Téllez & Cote, 2006). Gender difference is a factor that modifies the distributed ethanol volume; this is due to its hydrosolubility and to that it is not distributed in body fats, which explains why in females this parameter is found diminished compared with males.

Ethanol is eliminated mainly (> 90%) by the liver through the enzymatic oxidation pathway; 5-10% is excreted without changes by the kidneys, lungs, and in sweat (Goldfrank et al., 2002). The liver is the primary site of ethanol metabolism through the following three different enzymatic systems:

8.1 Alcohol dehydrogenases (ADH)

Alcohol dehydrogenases (ADH) are cytoplasmic enzymes with numerous isoforms in the liver of humans, with high specificity for ethanol as substrate; these are codified by three separate genes designated as *ADH1*, *ADH2*, and *ADH3*; these genes translate into peptide subunits denominated alpha, beta, and gamma. Variations in ADH isoforms can explain the significance in alcohol elimination levels among ethnic groups (Feldman et al., 2000).

This enzyme utilizes Nicotinamide adenin dinucleotide (NAD^+) as a Hydrogen (H^+) receiver for oxidizing ethanol into acetaldehyde. In this process, H^+ is transferred from the substrate (ethanol) to the co-factor (NAD^+), converting it into its reduced form, NADH; likewise, H^+ acetaldehyde is transferred from acetaldehyde to NAD^+. Later, the acetaldehyde oxidizes into acetate by means of the reduced Aldehyde-dehydrogenase enzyme (ALDH). Under normal conditions, acetate is converted into acetyl coenzyme A (acetyl-CoA), which enters the Krebs cycle and is metabolized into carbon dioxide and water (Goldfrank et al., 2002).

8.2 Microsomal ethanol oxidation system (MEOS)

This system is localized at hepatocyte smooth reticulum cisterns; this cytochrome P-450 2E1-dependent enzymatic system contributes to 5–10% of ethanol oxidation in moderate drinkers, but its activity increases significantly in chronic drinkers by up to 25% (Roldán et al., 2003).

A critical component of MEOS is Cytochrome P-450 2E1 (CYP2E1); this enzyme catalyzes not only ethanol oxidation, but also the metabolism of other substances such as paracetamol, the barbiturates, the haloalkalines, and the nitrosamines, among others (Feldman et al., 2000).

CYP2E1 utilizes Nicotinamide adenin dinucleotide phosphate ($NADP^+$) as a (H^+) receiver for oxidizing ethanol into acetaldehyde. In this process, H^+ is transferred from the substrate (ethanol) to the co-factor ($NADP^+$), converting this into its reduced form, NADH; Similarly, the acetaldehyde is as the H^+ transfers from acetaldehyde to $NADP^+$ (Goldfrank et al., 2002).

While MEOS participation is more active in the ethanol metabolism of chronic than in occasional drinkers and its relative contribution in comparison with ADH is difficult to determine, notwithstanding this, MEOS is important in the pathogeny of ethanol consumption-associated hepatic lesions because oxidation of this CYP2E1-mediated substance produces reactive oxygen intermediaries as subproducts (Feldman et al., 2000).

8.3 Catalase system

This presents in the peroxisomes and utilizes Hydrogen peroxide (H_2O_2) for ethanol oxidation; its contribution is minimal (Roldán et al., 2003). This system exists in a tight relationship with the reduced-oxidase-glutathione system and, like MEOS, is induced by chronic ethanol consumption. Ethanol oxidizes into acetaldehyde, utilizing H_2O_2 as co-enzyme; this metabolite continues the same course as for converting into acetate through the ALDH enzyme (Morales-González et al., 2001; Morales-González et al., 1998).

Any of the three ethanol pathways transforms it into acetaldehyde, which afterward is oxidized into acetate by the Aldehyde-dehydrogenase (ALDH) enzyme. Aldehyde is a highly reactive compound and is potentially toxic for the hepatocyte.

9. Hepatic regeneration

Hepatic regeneration (HR) is a process arising throughout evolution to protect animals from the catastrophic results of hepatic necrosis caused by the effect of the toxins of plants that serve them as food; this extraordinary process has been the object of the curiosity of scientists of all times. In ancient Greece, the myth of the chained Prometheus in Caucasus mountains of the Caucasus while an eagle daily devoured his entrails, which regenerated during the night, is based on the recognition from those times of hepatic regeneration (Michalopoulos & DeFrances, 1997).

Several terms such as replacement of lost parts, restitution, or repair have been employed to designate "regeneration", which is indicative of the active cell proliferation that invariably precedes differentiation. After partial liver extraction, cell proliferation does not occur at the level of the incision, new lobules do not develop instead of the removed part; ascribing to cell migration and differentiation results in the formation of new lobules, and the liver's preoperative weight is rapidly restored (Higgins & Anderson, 1931).

The liver in humans as well as in rodents possess a noteworthy capacity of regenerative response to several stimuli, including massive destruction of hepatic tissue by toxins, viral agents, or by surgical desertion; liver regeneration depends on the ability of the hepatocytes to be submitted to cell division, which is strictly controlled by intra- and extrahepatic factors (Gutiérrez-Salinas et al., 1999).

HG is a physiological process that includes hypertrophy (increase in cell size or in protein content in the replicative phase) and hyperplasia (increase in cell number), the latter governed to a greater degree by functional than by anatomical needs (Palmes & Spiegel, 2004). Studies with hepatic resections in large animals (dogs and primates) and in humans have established that the regenerative response is proportional to the amount of the liver removed (Michalopoulos & DeFrances, 1997).

Partial hepatectomy (PH) is the best studied animal hepatic regeneration model and was proposed in 1931 by Higgins and Anderson and consists of the surgical extraction of 70% of hepatic mass; hepatic regeneration is induced at an important stage when all of the hepatocytes are virtually found in phase G_0 of the cell cycle. Stimulated by PH, all of the hepatocytes in synchronized fashion enter into the cell cycle; initiation of maximum Deoxyribonucleic acid (DNA) synthesis takes place 24 hours after PH and 7–14 days after the removed hepatic mass is restored (Palmes & Spiegel, 2004).

In contrast with other regenerating tissues (bone marrow and skin) hepatic regeneration does not depend on a small population of progenitor cells; liver regeneration after PH is carried out by the proliferation of all of the existing mature call populations that comprise the intact organ; this includes the hepatocytes, biliary endothelial cells, fenestrated endothelial cells, Kupffer cells and Ito cells; all of these cells divide during hepatic proliferation, hepatocytes the first to do this. Cell proliferation kinetics differs slightly from one species to another; the first DNA synthesis peak in hepatocytes occurs at 24 hours, with a second peak between 36 and 48 hours; when three quarters of the hepatic tissue is removed, restoration of the original number of hepatocytes theoretically requires 1.66 cell cycles per residual hepatocyte (Michalopoulos & DeFrances, 1997).

10. Ethanol-derived hepatic damage

The liver is the main target organ of ethanol toxicity; it has been demonstrated experimentally that chronic ethanol ingestion leads to an increase of lipid peroxidation products and a decrease of antioxidant factors such as glutathione (GSH) and derived enzymes. Likewise, oxidative stress has been related as the main factor implicated in alterations derived from its chronic consumption, from the Central nervous system (CNS) as well as from the peripheral nervous system (Díaz et al., 2002).

There are several mechanisms involved in ethanol consumption-related liver damage; the former results from the effects of alcohol dehydrogenase (ADH) mediated by the excessive generation of NADH and acetaldehyde, which generates the formation of FR (Lieber, 2004). Acetaldehyde increases the production of alkanes and causes an imbalance in potential cytosolic redox on altering the NAD^+/NADH relationship, as occurs in the mitochondria; this redox alteration favors the production of lipoperoxides, which increase damage to the cell (Gutiérrez-Salinas & Morales-González, 2004).

Acetaldehyde inactivates the enzymes, diminishing DNA repair, antibody production, and glutathione depletion, and increasing mitochondrial toxicity, endangering oxygen utilization, and increasing collagen synthesis (Lieber, 2000).

Another probable mechanism of hepatic damage associated with chronic ethanol consumption is the increase in the synthesis of fatty acids and triglycerides and a decrease of the oxidation of the former, generating hyperlipidemias that leads to the development of fatty liver, in addition to inhibiting fatty acids utilization and the availability of precursors, which stimulate the hepatic synthesis of triglycerides (Téllez & Cote, 2006).

Previous studies have suggested that Kupffer cells are involved in hepatic damage caused by ethanol consumption; this is due to that there are reports that ethanol alters the functions of these cells, such as phagocytosis, bactericide activity, and the production of inflammatory cytokines such as Tumor growth factor alpha (TNF)-α, Interleukin 1 (IL-1), and IL-6, among others, which result in hepatic cell toxicity (Thurman et al., 1998). It has been reported that TNF-α and IL-1 inhibit protein synthesis in hepatocytes in rat, in addition to stimulating neutrophil migration and activation, as well as protease induction and FR release (Thurman, 1998).

Cytokines and chemokines originated by the Kupffer cells employ autoparacrine as well as paracrine effects that initiate the defensive response in the liver, but that also promote the infiltration of inflammatory leukocytes and activate the oxidative attack response, accompanied by strong damage originating from degrading cytokines and proteins. Cellular infiltration of activated neutrophils produces oxygen FR and secretes other toxic mediators; additionally, these can increase the inflammatory response, causing damage and cell death (Thurman, 1998).

Ethanol consumption induces changes in the mitochondrial membrane, such as Mitochondrial permeability transition (MPT), which is associated with loss of mitochondrial energy, mitochondrial matrix inflammation, and external membrane rupture; this is accompanied by the release of numerous proapoptotic factors; in addition, TNF-α activates different hepatic cell cascades, resulting in the stimulation of mitotic genes such as *p38* of Mitogen-activated protein kinases (MAPK) and the Jun N-terminal kinase (JNK), which affect mitochondrial sensitivity to the proapoptotic stimulus as follows: JNK by the phosphorylation of proapoptotic proteins such as Bcl-2 and Bcl-X_L, and *p38* MAPK by re-enforcing the effects of the Bcl-2 proapoptotic Bax Protein family.

Acute ethanol consumption can produce a hypermetabolic state in the liver that is characterized by increase in mitochondrial respiration, which is driven by the great demand for NADH reoxidation produced during ethanol metabolism by cytosolic ADH (Adachi & Ishii, 2002); in addition, it alters hepatic microcirculation by stimulating endothelial-1 production (Thurman, 1998); similarly, the acetaldehyde generated by ethanol metabolism causes hypoxia on chemically reacting with free sulfate groups such as glutathione, in such as way as to alters the reaction of this metabolite, which activates the xantine oxidase and xantine dehydrogenase enzymes, in order to finally diminish the NAD^+/NADH equilibrium (Gutiérrez-Salinas & Morales-González, 2004).

Depletion of antioxidant levels, above all that of hepatic glutathione, caused by acute as well as by chronic ethanol consumption, increases oxidative stress, which induces changes in the mitochondrial membrane, such as diminution of the mitochondrial membrane potential in hepatocytes and MPT, both inhibited by the antioxidants or by an ADH inhibitor (Adachi & Ishii 2002).

11. Inhibition of hepatic regeneration by ethanol

When there is an important, functional hepatic mass loss, such as occurs in PH, the remnant tissue undergoes a regeneration process in which the removed tissue is replaced in its totality; during this process, DNA synthesis increases notably, reaching a maximal peak of 23 at 25 hours postsurgery. After PH, hepatic tissue become more vulnerable to the damage caused by consumption of xenobiotics, particularly ethanol administration, which causes damage to HR, above all in the early regenerative process phase (Morales-González et al., 2001; Morales-González et al., 1999).

Studies performed in animals in which PH was carried out suggest that the acute ethanol administration rapidly inhibits the result of the HR after surgery; this has been assessed by a frank diminution of the cell proliferation parameters in the remnant liver. Although the exact mechanism by which ethanol inhibits HR, it is reasonable to assume that this hepatotoxicity could alter the total metabolism of the regenerating liver, which includes ethanol oxidation into acetaldehyde, catalyzed by ADH, and the later conversion of this into acetate by means of the mitochondrial ALDH (Gutiérrez-Salinas, 1999).

Acute ethanol administration produces structural and biochemical changes such as partial inhibition of protein and DNA synthesis, which indicates the diminution of the mitotic index, transitory accumulation of fat, the presence of inflammation, modifications in hepatocellular organization, diminution of weight gain in the regenerating liver, and inhibition of hepatic regeneration (Morales-González et al., 2001).

Some physiological processes that are altered by ethanol are metabolite levels in serum (glucose, triglycerides, albumin, and bilirubin), in addition to causing modification of the serum activity of enzymes that reflect liver integrity (alanine and aspartate aminotransferase, lactate dehydrogenase, ornithine carbamoyltransferase, and glutamate dehydrogenase); also, on inhibiting DNA synthesis and the activity of enzymes intimately related with this process, such as Thymidine synthetase (TS) and Thymidine kinase (TK), in addition to diminution of the mitotic index (Morales-González et al., 2001).

A sole dose of ethanol is capable of significantly inhibiting the synthesis of the protein ornithine decarboxylase, in addition to causing thyrosine aminotransferase degradation, which suggests that acute ethanol consumption inhibits protein synthesis and regenerating liver activity on transcriptional levels, interfering with RNA synthesis in the nucleus (Morales-González et al., 2001; Morales-González et al., 1999).

Investigations that have been conducted have demonstrated that the acute as well as the chronic ethanol administration jeopardize the incorporation of thymidine into the DNA of hepatocytes of rats on which PH had been performed with or without diminution of the DNA contents, in addition to reporting that chronic consumption of this substance inhibits regeneration 24 hours after the PH due to delay in the induction of ornithine carbamoyltransferase activity (Yoshida et al., 1997).

It has been suggested that damage cause by FR as the product of ethanol consumption occurs at the early phase of HR; on the other hand, a transcending increase has been reported in mitochondrial lipoperoxidation of the liver in rats after PH. In the same study, a diminution was also observed in the early HR phase of mitochondrial glutathione levels (Guerrieri et al., 1998).

12. Ethanol and free radicals

One factor that suggested that ethanol causes cell damage due to its hepatic metabolism is the excessive generation of FR, which can be the result of a state denominated oxidative stress; this is because any of ethanol's metabolic pathways, principally MEOS, is made up of chemicals oxido-reduction reactions, which produce highly unstable molecules called Reactive oxygen species (ROS), such as the superoxide anion (O_2^-), Hydrogen peroxide (H_2O_2), and the hydroxyl radical ($OH^\bullet$) (Nanji & French, 2003).

FR can perform four main reactions (Wu & Cederbaum, 2003):

1. Hydrogen abstraction, in which FR interact with another molecule that acts as the donor of an atom of H^+. As a result, FR bind to H^+ and become more stable, while the donor is converted into a FR.
2. Addiction. Because the FR binds to a more stable molecule, which converts a receiver molecule into a FR.
3. Termination. In this, two FR react between themselves to form a more stable compound.
4. Disproportion. This consists of two LR that are identical to each other react between themselves. In this reaction, one FR acts as an electron donor and the other, as an electron receiver. In this manner, they become two more stable molecules.

Free radicals are chemical species that possess an unpaired electron in their last layer, which allows these to react with a high number of molecules of all types, first oxidizing these and afterward attacking their structures. If lipids (polyunsaturated fatty acids) are involved, they damage the structures rich in the latter, such as cell membranes and lipoproteins (Rodríguez et al., 2001). Within this generic concept, the partially reduced forms of oxygen are denominated Reactive oxygen species (ROS). This is a collective term that includes not only oxygen free radicals, but also some reactive non-radical oxygen derivatives.

The oxidating mechanism of FR is intimately linked to their origin, which follows a sequence of chain reactions; in these reactions, a very reactive molecule is capable of reacting with another, non-reactive molecule, inducing in the latter the formation of a FR ready to initiate a new neutrophilic attack, and so on successively.

The greatest source of ROS production in the cell is the mitochondrial respiratory chain, which utilizes approximately 8090% of the O_2 that a person consumes; another important source of ROS, especially in the liver, is a group of Mixed function oxidase (MFO) Cytochrome P-450 enzymes. In addition to the ROS generation that takes place naturally in the organism, humans are constantly exposed to environmental FR including ROS in the form of radiation, Ultraviolet (UV) light, smoke, tobacco smoke, pesticides, and drugs utilized in the treatment of cancer (Wu & Cederbaum, 2003).

The increase in O_2^- and H_2O_2 formation is justified with the finding that in aging, electron flow conditions are modified in the transport chain of these, which is the last stage of high energy proton production, and whose passage through the internal mitochondrial membrane generates an electrical gradient that provides the energy necessary for forming ATP (5-Adenosin triphosphate). Researchers postulate that the ROS generated can produce damage to the internal mitochondrial membrane as well as to electron transport chain components or to mitochondrial DNA, which further increases ROS production and consequently, more damage to the mitochondria and an increase of oxidative stress due to increased oxidant production. FR produced during aerobic stress cause oxidative damage that accumulates and results in a gradual loss of the homeostatic mechanisms, in an interference of genetic expression patterns, and loss of cell functional capacity, which leads to aging and death.

ROS generation promotes the decrease of intracellular glutathione, elevation of cytoplasmic calcium, lipid peroxidation of the membranes, and a series of chain reactions that are accompanied by the disappearance of glycogen, decrease of ATP, and the descent of the energy state of hepatic cells; these events are the origin of membrane destruction and cell death (Thurman et al., 1999). Sustained elevation of cytoplasmic calcium is associated with activation of calcium-dependent enzymes such as phospholipase A_2, glycogen phosphorylase, and the endonucleases, which cause plasmatic membrane ruptures and DNA molecule fragmentation. On the other hand, the transitory elevation of intracellular calcium intervenes in the progression of cell division in G_1-to-S transitions and in those of the G_2-to-M phase. Immediately after cell death, hepatocellular proliferation begins in order to re-establish cell populations that have been destroyed, thus restoring hepatic function (Andrés & Cascales, 2002).

The OH^- radical is highly toxic for the hepatocytes, which do not possess a direct system for its elimination; this FR is produced intracellularly by two reactions that occur spontaneously and that are catalyzed by a transition metal, generally iron (Fe), which are termed the Fenton reaction and the Haber-Weiss reaction (Boveris et al., 2000).

In this reaction, Hydrogen peroxide (H_2O_2) in the presence of Fe as catalyzer produce the hydroxyl radical ($OH^\bullet$).

The superoxide radical (O_2^-) leads to the formation of Hydrogen peroxide (H_2O_2) and both products of the partial reduction of oxygen (O_2) produce the hydroxyl radical ($OH^\bullet$).

13. Effect of glycine in hepatic regeneration in the presence of ethanol damage

Recent studies have reported the beneficial effects of glycine, including protection against toxocity induced by anoxia and oxidative stress caused by several toxic agents for the cell, including ethanol. These investigations have demonstrated that glycine blocks the increase of Calcium (Ca^{2+}) in hepatocytes, caused by agonists released during stress, such as Prostaglandin E_2 (PGE_2) and the adrenergic hormones (Qu et al., 2002).

It is known that the early, ethanol consumption-related hepatic-damage phase is characterized by steatosis, inflammation, and necrosis, which are principally mediated by resident Kupffer cells in liver macrophages (Yin et al., 1998). Senthilkumar and colleagues (2004) in a model of ethanol-intoxicated rats administered with glycine, observed a significant diminution of total free fatty acid levels in the liver, in addition to a decrease in steatosis and necrosis after glycine administration (Senthilkumar & Nalini, 2004). In another study conducted by the same researchers, the latter demonstrate that treatment with glycine offers protection against FR-mediated oxidative stress in the membrane of erythrocytes, plasma, and hepatocytes of rats with ethanol-induced hepatic damage (Senthilkumar et al., 2004).

It has been reported that glycine inhibits macrophage activation and TNF-α release; this amino acid prevents liver damage after chronic exposure to ethanol and attenuates the lipoperoxidation and glutathione depletion induced by diverse xenobiotics (Mauriz et al., 2001). Kupffer cells in liver constitute 80% of the resident macrophages in the organism. Glycine blocks the systemic inflammatory process arising in a broad variety of pathological states, due to the activation of macrophages that release potent inflammatory mediators such as toxic cytokines and eicosanoids, which perform an important role in the progressive inflammatory response (Matilla et al., 2002).

Prior studies have demonstrated that glycine administration prevents several forms of hepatic lesions; because it prevents the necrosis and the inflammation developed in the early phase during chronic ethanol administration, glycine's mechanism of protection against damage involves the Kupffer cells (Ishizaki et al., 2004). As mentioned previously, Kupffer cell activation releases active substances such as NO, TNF-α, ROS, Interleukin β (IL-β), IL-6, and TGF, which cause potential damage to the liver; in particular, TNF-α, as a product of Lipopolysaccharide (LPS), plays an important role in the induction of hepatic damage on activating gene transcription factors of nuclear regulator cytokines such as NF-$_{\kappa}$B, which is related with the inflammatory response and which plays an important role in the activation of hepatic stellate cells (Mauriz et al., 2001).

One probable hepatoprotector mechanism of glycine is due to that this amino acid activates the chloride channels, causing a diminution of intracellular Calcium,(Ca^{2+}) concentration in the Kupffer cells, which hyperpolarize the cell membrane, rendering the opening of voltage-dependent Ca^{2+} channels difficult; in addition to inhibiting macrophage activation and TNG-α release, glycine prevents liver damage after chronic exposure to ethanol and attenuates lipoperoxidation and hepatotoxin-induced glutathione depletion (Senthilkumar & Nalini, 2004). It is well known that intracellular Ca^{2+} is important for activation and release of inflammatory cytokines. TNF-α production by Kupffer cells has been associated with an increase of Ca^{2+} (Ishizaki et al., 2004).

14. Effect of vitamin E on liver regeneration in the presence of ethanol-derived damage

Vitamin E is well known for its antioxidant properties; these function as rupturing the antioxidant chain that prevents the propagation of FR reactions; this protects cells from oxidative stress. Vitamin E is particularly important, protecting cells against lipoperoxidation, during which FR attack the fatty acids, causing structural damage to the membrane, resulting in the formation of secondary cytotoxic products such as MDA, among others (Jervis & Robaire, 2004).

Since its discovery, α-tocopherol has shown to possess two important functions in the membrane: the first as a liposoluble antioxidant that acts to prevent FR damage to polyunsaturated acids, and the second, as a membrane stabilizing agent, which aids in preventing phospholipid-caused lesions. It has also shown to inhibit protein kinase C (Bradford et al., 2003).

Studies *in vitro* in animals and in artificial membranes have demonstrated that tocopherols interact with polyunsaturated-lipid acyl groups, stabilizing the membranes and eliminating ROS and the subproducts of oxidative stress (Sattler et al., 2004).

In vivo, vitamin E acts by rupturing the antioxidant chain and in this manner preventing propagation of damage to cell membranes that cause FR. This vitamin protects polyunsaturated fatty acids from biological membrane phospholipids and lipoproteins in plasma. When lipid peroxides oxidize and are converted into peroxyl radicals ($ROO^{\bullet}$), the phenolic OH^- tocopherol group reacts with an organic $ROO^{\bullet}$ radical to form the corresponding organic hydroxyperoxide and the tocopheroxyl radical (Vit E-$O^{\bullet}$) (Shils et al., 1999).

α-Tocopherol inhibits the activity of monocyte protein kinase C, followed by inhibition of phosphorylation and translocation of the p47 cytosolic factor, damaging the NADPH-oxidase assembly and O_2^- production; α-tocopherol exerts an important biological effect on inhibiting the release of proinflammatory cytokines, interleukin 1β, and inhibition of the 5-lipooxygenase pathway.

15. References

Adachi, M., & Ishii, H. (2002). Role of mitochondria in alcoholic liver injury. *Free Radical Biology & Medicine*. Vol. 3, pp. 487-491, ISSN: 0891-5849

Allende Martínez, LM (1997). Effects of retinol (vitamin A) in human T lymphocyte activation and therapeutic implications (Thesis) Madrid, Spain: Universidad Complutense de Madrid. available at: http//eprints.ucm.es/tesis/bio/ucem-t21864.pdf

Andrés, D., Cascales, M. (2002). Novel mechanism of Vitamin E protection against cyclosporine A cytotoxicity in cultured rat hepatocytes. *Biochemical Pharmacology*. Vol. 64, No. 2, (Jul), pp. 267-276, ISSN: 0006-2952

Boigk, G., Stroedter, L., Herbst, H., Waldschmith, J., & Riecken, EO. (1997) Sylimarin retards collagen accumulation in early and advanced biliary fibrosis secondary to complete bile duct obliteration in rats, In: Hepatology, December 2003, Disponible en: http://onlinelibrary.wiley.com/doi/10.1002/hep.510260316/pdf

Boveris, A. (2005). The evolving concept of free radicals in biology and medicine. *Ars Pharmaceutica*. Vol. 46, pp. 85-95, ISSN: 0004-2927

Bradford, A., Atkinson, J., Fuller, N., & Rand, R. (2033). The effect of Vitamin E on the structure of membrane lipid assemblies. *Journal of Lipid Research*. Vol. 44, pp. 1940-1945, ISSN: 0022-2275

Burneo Palacios, ZL. (2009). Determination of total phenolic compounds and antioxidant activity of total extracts of twelve native plant species in southern Ecuador (Thesis) Loja, Ecuador: Universidad Técnica Particular de Loja. available at: http://es.scribd.com/doc/43393190/TESIS-ANTIOXIDANTES

Ferenci P, Dragosics B, Dittrich H. Randomizedd controlled trials of sylimarin treatment in patients with cirrhosis of the liver. (1989) En: *Journal of Hepatology*. Disponible en: http://www.ncbi.nlm.nih.gov/pubmed/2671116

Díaz, J., España, M., Soriano-Romero, S., Marin, N., Muriach., et al. (2002). Oxidative stress in the rat optic nerve induced by chronic administration of ethanol. *Archivos de la Sociedad Española de Oftamología*. Vol. 77, pp. 263-268, ISSN: 0365-6691

Feldman, M., Schardnmidt, M., & Sleisenger, M. (2000). Gastrointestinal and liver disease. Physiology, diagnosis and treatment. 6 ed. Ed. Medical Panamericana, Mexico, DF., pp. 1282-1298.

Goldfrank, L., Flomenbaum, N., & Lewin, N. (2002). Goldrank´s Toxicology Emergencies. 7th. Ed. McGraw-Hill, USA, pp. 952-962.

Goodman Gilman, A. (1996). Vitamins, In: The Pharmacological Basis of Therapeutics. Marcus R & Coulston A.M., p.p. (1647 - 1675), The McGraw-Hill Interamericana ISBN 970-10-1133-3, México, D.F.

Guerrieri, F., Vendímiale, G., Grattagliano, I., Cocco, T., Pellecchia, G., & Altomare, E. (1999).Mitochondrial oxidative alterations followin partial hepatectomy. *Free Radical Biology & Medicine*. Vol. 32, pp. 487-491, ISSN: 0891-5849

Gutiérrez-Salinas, J., & Morales-González, JA. (2004). Production of oxygen-derived free radicals and damage to the hepatocyte. Medicina Interna de México. Vol. 20, pp. 287-95, ISSN: 0186-4866

Gutiérrez-Salinas, J., Miranda-Garduño, L., Trejo-Izquierdo, E., Diaz-Muñoz, M., Vidrio, S., Morales-González, JA., & Hernández-Muñoz R. (1999). Redox state and energy metabolism during liver regeneration. Alterations produced by acute ethanol administration. *Biochemical Pharmacology*. Vol. 58, pp. 1831–1839, ISSN: 0006-2952

Higgins, GM. & Anderson, RM. (1931). Experimental pathology of the liver. I. Restoration of the liver of the white rat following partial surgical removal. *Archive Pathology*. Vol. 12, pp. 186-202

Ishizaki, S., Sonakaa, Y, Takeib, Y., Ikejimab, K., & Satob, N. (2004). The glycine analogue, aminomethanesulfonic acid, inhibits LPS-induced production of TNF-a in isolated rat Kupffer cells and exerts hepatoprotective effects in mice. *Biochimical and Biophysical Research Communications*. Vol. 322, pp. 514–519, ISSN: 0006-291X

Jacob RA. Three eras of vitamin C discovery (1996). En: *Subcellular Biochemistry*, (citado el 3 de septiembre de 2011) Disponible en: http://www.ncbi.ncbi.nlm.nih.gov/pubmed/8821966

Jervis, K., & Robaire, B. (2004). The effects of long-term vitamin E treatment on gene expression and oxidative stress damage in the againg brown Norway rat epididymis. *Biology of Reproduction*. Vol. 71, pp. 1018-1095, ISSN: 0006-3363

Lieber, C. (2000). Alcohol liver disease: new insights in pathogenesis lead to the new treatments. *Journal Hepatology*. Vol. 32, pp. 113-128, ISSN: 0168-8278

Lieber, C. (2004). The discovery of the microsomal ethanol oxidizing system and its physiologic and pathologic role. *Drug Metabolism Reviews*. Vol. 3, pp. 511-529, ISSN: 0360-2532

Mathews CK & Van Holde KE (1998). *Bioquímica* (2° edición), McGraw-Hill Interamericana, ISBN: 0-8053-3931-0, Madrid, España.

Manzanares Castro W. Selenium in critically ill patients with systemic inflammatory response (2007). In: Nutrition Hospital (cited September 3, 2011). available at: http//sicelo.iscii.es/pdf/nh/v22n3/revisión2.pdf

Mauriz, J., Culebras, J., González, P., & González, J. (2001). Dietary glycine inhibits activation of nuclear factor Kappa B and prevents liver injury in hemorrhagic shoch in the rat. *Free Radical Biology & Medicine*. Vol. 31, pp. 1236-1244, ISSN: 0891-5849

Messarah, M., Klibet, F., Boumendjel, A., Abdennour, C., Bouzerna, N., Boulakoud, M. S., & El Feki A. (2010) Hepatoprotective role and antioxidant capacity of selenium on arsenic-induced liver injury in rats. In: Experimental and Toxicologic Pathology. September 2010. Disponible en: http://www.sciencedirect.com/science/article/pii/S094029931000134X

Michalopous, G., & DeFrances, M. (1997). Liver regeneration. *Science*. Vol. 276, pp. 60-66, ISSN: 0036-8075

Morales-González, JA. (2009). Antioxidant defenses, In: Antioxidants and chronic degenerative diseases, Miranda Martínez I, Gasca León MI, Aedo Santos MA, Cantoral Preciado AJ, Gallardo Wong I, p.p. (131-225), ISBN 978-607-482-052-2, Hidalgo, México.

Morales-González, JA., Gutiérrez-Salinas, J., & Hernández-Muñoz, R. (1998). Pharmacokinetics of the ethanol bioavalability in the regenerating rat liver induced by partial hepatectomy. *Alcoholism Clinical and Expimental Research*. Vol. 22, pp. 1557-1563, ISSN: 0145-6008

Morales-González, JA., Gutierrez-Salinas, J., & Piña, E. (2004). Release of Mitochondrial Rather than Cytosolic Enzymes during Liver Regeneration in Ethanol-Intoxicated Rats. *Archives of Medical Research*. Vol. 35, pp. 263-270, ISSN: 0188-4409

Morales-González, JA., Gutiérrez-Salinas, J., Yánez, L., Villagómez, C., Badillo, J. & Hernández, R. (1999). Morphological and biochemical effects of a low ethanol dose on rat liver regeneration. Role of route and timing of administration. *Digestive Diseases and Sciences*. Vol. 44, No. 10 (October), pp. 1963-1974, ISSN: 0163-2116

Morales-González, JA., Jiménez, L., Gutiérrez-Salinas, J., Sepúlveda, J., Leija, A. & Hernández, R. (2001). Effects of Etanol Administration on Hepatocellular Ultraestructure of Regenerating Liver Induced by Partial Hepatectomy.

Digestive Diseases and Sciences. Vol. 46, No. 2 (February), pp. 360–369, ISSN: 0163-2116

Nanji, A., & French, S. (2003). Animal models of alcoholic liver disease-focus on the intragastric feeding model. *Alcohol Liver Disease*. Vol. 27, pp-325-330.

Palmes, D., & Spiegel, H. (2004). Animals models of liver regeneration. *Biomaterials*. Vol. 25, pp.1601-1611, ISSN: 0142-9612

Parra-Vizuet, J., Camacho-Luis, A., Madrigal-Santillán, E., Bautista, M., Esquivel-Soto, J., Esquivel-Chirino, C., García-Luna, M., Mendoza-Pérez, JA., Chanona-Pérez, J., & Morales-González, JA. (2009). Hepatoprotective effects of glycine and vitamin E during the early phase of liver regeneration in the rat. *African Journal of Pharmacy and Pharmacology*. Vol.3, No. 8, pp. 384-390 (August, 2009), ISSN 1996-0816

Qu, W., Ikejima, K., Zhong, Z., Waalkes, P., & Thurman, R. (2002). Glycine blocks the increase in intracellular free Ca^{2+} due to vasoactive mediators in hepatic parenchymal cell. *American Journal of Physiology Gastrointestinal and Liver Physiology*. Vol. 283, pp. G1249-G1256, ISSN: 0193-1857

Ramírez-Farías, C., Madrigal-Santillán, E., Gutiérrez-Salinas, J., Rodríguez-Sánchez, N., Martínez-Cruz, M., Valle-Jones, I., Gramlich-Martínez, I., Hernández-Ceruelos, A., & Morales-González JA. (2009) Protective effect of some vitamins against the toxic action of ethanol on liver regeration induced by partial hepatectomy in rats. *World Journal of Gastroenterology* Vol. 14, No. 6, (2008 february), pp (899-907), ISSN: 1007-9327

Rodriguez, J., Menéndez, J., & Trujillo, Y. (2001). Free radicals in the biomedical and oxidative stress. *Revista Cubana de Medicina Militar*. Vol. 1, pp. 15-20, ISSN: 0138-6557

Roldán, J., Frauca, C., & Dueñas, A. (2003). Alcohols poisoning. *Anales del Sistema Sanitario de Navarra*. Vol. 26, pp. 129-139, ISSN: 1137-6627

Sandoval M, Lazarte K, Arnao I. Antioxidant hepatoprotection peel and seeds of Vitis vinifera L. (grape) (2008). In: Proceedings of the Faculty of Medicine (cited September 3, 2011). available at: http://www.scielo.org.pe/scielo.php?pid=S1025-55832008000400006&script=sci_arttext

Sattler, S., Gilliland, L., Lundback, M., Pollard, M., & Della Penna, D. (2004). Vitamin E is essential for seed longevity and for preventing lipid peroxidation during germination. *The Plant Cell*. Vol. 16, pp.1419-1432, ISSN: 1040-4651

Sayago A, Marin MI, aparicio R, Morales MT. Vitamin E and vegetable oils (2007). In: Fats and oils (cited September 3, 2011). available at: http://digital.csic.es/bitstream/10261/2470/1/Sayago.pdf

Senthilkumar, R. and Nalini N. (2004). Effect of glycine on tissue fatty acid composition in an experimental model of alcohol-induced hepatotoxicity. *Clinical and Experimental Pharmacology and Physiology*. Vol. 7, pp. 456-461, ISSN: 0305-1870

Senthilkumar, R., Sengottuvelan, M., & Nalini, N. (2004). Protective effect of glycine supplementation on the levels of lipid peroxidation and antioxidant enzymes in the erythrocyte of rats with alcohol-induced liver injury. *Cell Biochemistry and Function.* Vol. 22, pp. 123-128, ISSN: 1099-0844

Shils, M., Olson, J., Shike, M., & Ross, R. (1999). Modern nutrition in health and diseade. 9a ed. Ed. McGraw-Hill, USA, pp. 751-759.

Téllez, J., & Cote, M. (2006). Ethyl alcohol: a toxic risk to human health is socially acceptable. *Revista Facultad de Medicina de la Universidad Nacional de Colombia.* Vol 54, pp. 32-47, ISSN 0120-0011

Thurman, R. (1998). Mechanism of hepatic toxicity. II. Alcoholic liver injury involves activation of Kupffer cells by endotoxin. *American Journal of Physiology Gastrointestinal and Liver Physiology. Vol. 275, pp. G605-G611,* ISSN: 0193-1857

Venereo Gutiérrez, JR. (2002). Oxidative damage, free radicals and antioxidants. In: Journal of Military Medicine, February 2002, available at: http://bvs.sld.cu/revistas/mil/vol31_2_02/MIL09202.pdf

Wolf G. (2005) The discovery of the antioxidant function of vitamin E: the contribution of Henry A. Matiill. En: *The Journal of Nutrition* (citado el 3 de septiembre de 2011). Disponible en: http://jn.nutrition.org/content/135/3/363.long

Wu, D., & Cederbaum, A. (2003). Alcohol, oxidative stress, and free radical damage mechanism of injury. *Alcohol Research & Health.* Vol. 27, pp. 277-284, ISSN: 1535-7414

Yin, M., Ikejima, K., Arteel, G., Seabra, V., Bradford, B., Kono, H., Rusyn, I., & Thurman, R. (1998). Glycine accelerates recovery from alcohol-induced liver injury. *American Journal of Physiology Gastrointestinal and Liver Physiology.* Vol. 286, pp. G1014-G1019, ISSN: 0193-1857

Yoshida, Y., Komatsu, M., Ozeki, A., Nango, R., Tsukamoto, I. (1997). Ethanol represses thymidylate synthase and thymidine kinase at mRNA level in regenerating rat liver after partial hepatectomy. *Biochim Biophys Acta.* Vol. 1336, pp. 180-186, ISSN: 0006-3002

Section 2

Animal Models of Liver Regeneration

7

Analbuminemic Rat Model for Hepatocyte Transplantation

Katsuhiro Ogawa and Mitsuhiro Inagaki
Departments of Pathology & Surgery, Asahikawa Medical University, Japan

1. Introduction

Orthotropic whole and split liver transplantation are successful and well-established treatments for liver disease. However, the short supply of donor organs is a major obstacle to the widespread use of this therapy (Merion, 2010). Recent research has focused on cell transplantation as a therapy for liver disease. Although only a small number of hepatocytes can be transplanted into the liver by transfusion into the portal circulation, the transplanted cells can regenerate the normal hepatic tissue when liver growth is impaired as a result of continuous hepatic damage (Roy-Chowdhury & Roy-Chowdhury, 2011; Gilgenkranz, 2010). A number of animal models have demonstrated that much of the liver can be replaced via repopulation by a small number of transplanted hepatocytes, which restore normal liver functions and improve recipient survival. Hepatocyte transplantation is therefore expected to be used as a therapy for human liver disease. To optimize the therapeutic application of hepatocyte transplantation, however, number of problems such as the methods to monitor and evaluate the functionality of the transplanted cells and the suppression of host immunological responses against the transplanted cells require mitigation.

The cell transplantation model, which uses Fischer344 (F344) rats as donors and F344 congenic analbuminemic (F344-alb) rats as recipients, provides an excellent system for the investigation of cell transplantation, because no immunosuppressants are required for cell transplantation. In addition, the transplantability can be accurately evaluated with immunohistochemistry to stain for albumin, using PCR to detect albumin mRNA and genomic DNA, and by measuring serum albumin protein. In this review, we describe the hepatocyte transplantation system that uses F344 and F344-alb rats to study the intrahepatic hepatocyte transplantation.

2. Nagase analbuminemic rats (NARs)

Analbuminemic rats (Nagase analbuminemic rats, NARs) were first established from Sprague Dawley (SD) rats by Nagase et al. (1979). NARs show extraordinarily low serum albumin, hyperlipidemia and hormonal changes. However, their growth and reproduction rates do not differ from those of normal rats. The amount of total serum protein in NARs is similar to that seen in normal rats due to the increase in proteins other than albumin.

Analbuminemia in NARs is an autosomal recessive trait. Therefore, the serum albumin levels are nearly normal in the F1 hybrids of NARs and normal rats.

The molecular basis for analbuminemia in NARs is a deletion of base 5 to base 11 at the 5′ end of the 9th intron of the albumin gene (Figure 1A a) (Esumi et al., 1983). Most of the albumin mRNA in NARs shows a precise deletion of exon H, which is skipped during albumin pre-mRNA processing following the deletion in the 9th intron (Figure 1A b) (Shalaby & Shafritz, 1990). The deletion of exon H causes a frameshift in the mRNA and the occurrence of a translation termination signal at the 7th codon of exon I, preventing production of the albumin protein (Shalaby & Shafritz, 1990).

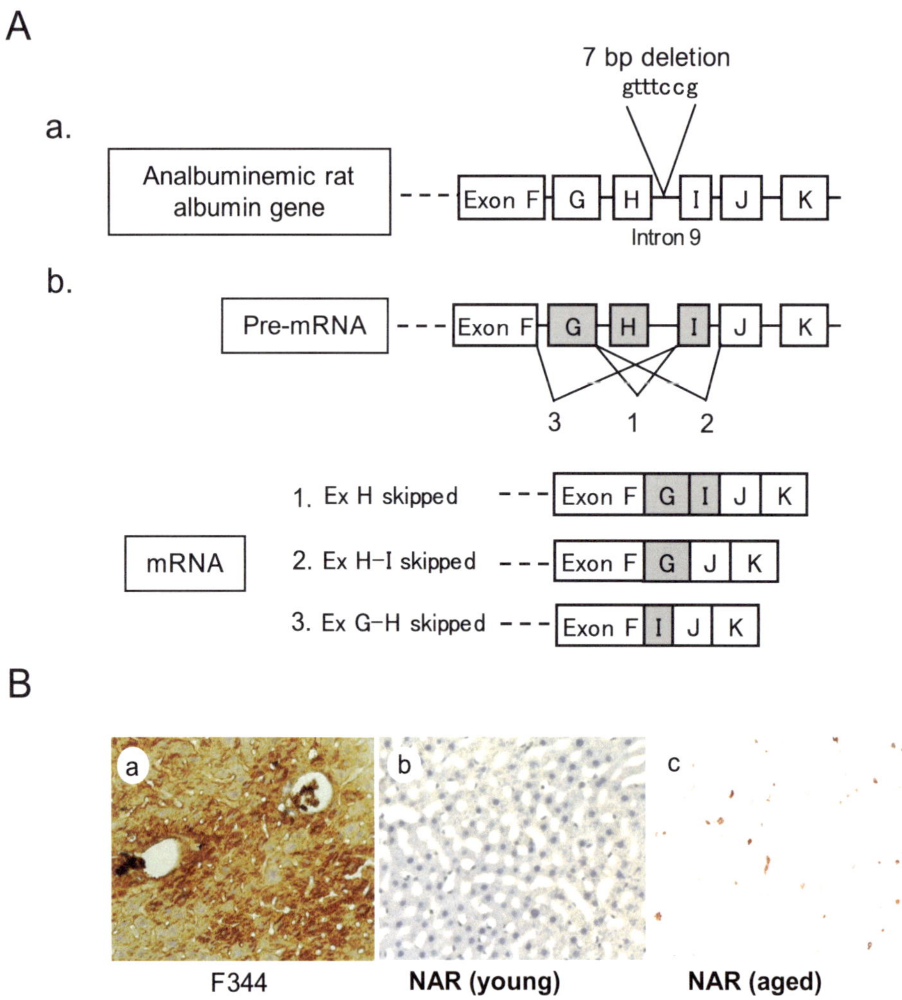

Fig. 1. A. The structures of the albumin gene and mRNA in the Nagase analbuminemic rat (NAR). a. Seven base pairs (the 5th to 11th base pairs from the start of the intron) are deleted between exons H and I (9th intron). b. Exon H is skipped in most albumin mRNAs during albumin pre-mRNA processing (1). With aging and treatment with hepatocarcinogens, the levels of albumin mRNA transcripts that skip exons H and I (2) and exons G and H (3)

increase. B. Immunostaining for albumin in an F344 rat liver (a) and in young (b) and aged NAR livers (c). Albumin-positive hepatocytes are visible in the aged NAR liver (c) and are usually present as single cells (inset) or two-cell clusters.

In contrast to the hepatocytes of normal rats (Figure 1B a), hepatocytes of NARs are generally negative for immunohisotochemical staining of albumin (Figure 1B b). However, although NARs display extremely low serum albumin, albumin-positive hepatocytes sometimes can be seen in the liver tissue of NARs at low frequency. The number of such albumin-positive hepatocytes increases with aging (Figure 1B c) and after treatment with hepatic carcinogens (e.g. 3′-methyl-4-dimethylaminoazobenzene) (Makino et al., 1986). These albumin-positive hepatocytes are present as single or double cells in cross sections (Figure 1B c inset) and rarely form clusters consisting of more than three cells. These cells remain as single or double cells after liver regeneration following two thirds hepatectomy (PH), suggesting that albumin-positive hepatocytes may have a low proliferative capacity. Under these conditions, the prevalence of albumin mRNAs missing exons G and H and exons H and I increase along with exon H-skipped albumin mRNA (Figure 1A b). In addition, aberrant 60-kD albumin is generated in the liver (Kaneko et al., 1991). This abnormal albumin may be a translation product of mRNA that skips exons H and I, and may accumulate in the cytoplasm because of defects in extracellular albumin secretion, resulting in positive albumin immunostaining (Kaneko et al., 1991).

3. F344-congenic analbuminemic rats (F344-alb rats)

Although NARs are derived from SD rats, the SD rats consist of out-bred strains. Therefore, SD rats may be genetically heterogeneous. We first tried to transplant SD rat hepatocytes into an NAR liver without immunosuppressants. However, this experiment was unsuccessful, most likely because the SD rats were immunogenetically heterogeneous. Co-cultured spleen cells from NARs and SD rats displayed a higher rate of cell proliferation compared to cells in NAR/NAR or SD/SD cultures (Yokota & Ogawa, 1978). These results indicate that the immunological rejection may occur after the transplantation of the SD hepatocytes into the NAR liver.

In addition to the variety of abnormalities in serum proteins, lipids and hormones, NARs differ from SD rats in their susceptibility to tumorigenesis induced by chemical carcinogens in various organs. Notably, NARs are highly susceptible to urinary bladder tumors that are induced by *N*-butyl-*N*-(4-hydroxybutyl)nitrosamine (BBN) (Kakizoe et al., 1982) and less susceptible to hepatocarcinogenesis induced by diethylnitrosamine (DEN) and 2-acetylaminofluorene (2-AAF) (Asamoto et al., 1989). It is not clear whether the difference in the susceptibility to chemical carcinogens is the result of the analbuminemia in NARs or of differences in the genetic backgrounds of NARs and SD rats. To reduce the genetic variability, Takahashi et al. (Takahashi et al., 1988) established F344-alb rats with the genetic background of F344 rats. F344-alb rats are highly susceptible to BBN-induced urinary tumors compared to F344 rats (Takahashi et al., 1988) and are equally sensitive to DEN-2-AAF-induced hepatocarcinogenesis (Ohta et al., 1994). Because the only genetic difference between F344-alb rats and F344 rats is the aforementioned 7-base-pair deletion in the albumin gene in the F344-alb rats, the pairing of F344 and F344-alb rats can be used for cell transplantation without using immunosuppressants.

4. Hepatocyte transplantation using F344 and F344-alb rats

To monitor the transplantability of hepatocytes in the liver, transplanted cells are detected using markers that are specifically expressed in the donor cells. For this purpose, various immunohistochemical and enzymatic histochemical procedures are used to visualize the proteins that are specifically expressed in the donor hepatocytes, such as *E. coli* β-galactosidase (β-gal) (ROSA26-transgenic mice) (Mao et al., 1999) or green fluorescence protein (GFP-transgenic mice) in the normal mouse liver (Chiocchetti et al., 1997), dipeptidyl peptidase IV (DPPIV) in the DPPIV$^{-/-}$ rat liver (Gupta et al., 1995) and fumarylacetoacetate hydrolase (FAH) protein in the FAH$^{-/-}$ mouse liver (Hamman et al., 2005). When combining cells from a male donor and a female recipient, Y chromosome-specific *in situ* hybridization techniques are also useful (Eckert et al., 1995).

In the F344/F344-alb model, albumin-positive donor F344 hepatocytes in the F344-alb liver are detected using immunohistochemical staining of albumin and PCR-based techniques detecting normal albumin mRNA and genomic DNA. In addition, the functionality of F344 hepatocytes can be repeatedly evaluated by taking blood samples and examining the elevation of serum albumin levels.

In immunohistochemical staining for albumin, the technique used for tissue fixation is important because albumin tends to diffuse out of hepatocytes during tissue fixation. We usually fix the liver tissues by perfusing periodate-lysine-paraformaldehyde (PLP) fixative through the portal vein. We then incubate the sliced hepatic tissues in the same fixative overnight at 4°C and then embed the tissues in paraffin (Ogawa et al., 1993). These procedures preserve the antigenicity of albumin and the morphological integrity of hepatic tissues. Antibodies that specifically recognize rat albumin are commercially available.

Donor F344 hepatocytes can also be detected by RT-PCR targeting normal albumin mRNA (Ohta et al., 1993a). Because albumin mRNA in F344-alb rats lacks exon H, RT-PCR with primers targeting exons G and I amplify both the normal and the short albumin mRNA from RNA that is isolated from recipient livers. In addition, primers targeting exons H and I specifically amplify normal albumin mRNA. Because the expression levels of exon H-skipped albumin mRNA in the F344-alb liver are low but relatively constant, the small amounts of normal albumin mRNA in the recipient F344-alb rat livers can be quantified using abnormal albumin mRNA as an internal standard (Ohta et al., 1993a).

The donor-derived F344 hepatocytes can also be detected by the amplification of the albumin genomic DNA sequences using DNA isolated from the recipient livers (Ogawa et al., 1993). Because of the 7-base-pair deletion in the 9th intron of the analbuminemic albumin gene, PCR amplification of amplicons spanning these sequences can differentially detect normal and abnormal genes, which can be used to quantify the expression of the normal albumin gene, using the abnormal albumin gene as an internal control (Ogawa et al., 1993). This method is applicable to hepatocytes and other cell types, such as bone marrow cells (Arikura et al., 2004; Inagaki et al., 2011).

The functionality of the transplanted F344 hepatocytes is evaluated based on the increased serum albumin levels in the recipients. Because the serum albumin level in F344-alb rats is extremely low, a small increase in albumin is detectable by sensitive methods such as the enzyme-linked immunosorbent assay (ELISA). However, conventional gel-electrophoretic assays are unsuitable because the serum of F344-alb rats contains proteins with molecular weights similar to that of albumin, and the levels of other proteins may increase to compensate for the lack of albumin (Ohta et al., 1993b). Therefore, conventional gel electrophoretic assays may falsely detect protein levels that are much higher than those of the actual albumin levels.

5. Changeability of phenotype of hepatocytes *in vivo* and *in vitro*

Cultured hepatocytes differ from hepatocytes *in vivo* with respect to many properties. They often lose specific functions such as the production of albumin and the activity of tyrosine aminotransferase and cytochrome P-450, and gain bile duct epithelium-specific functions such as cytokeratin 19 expression (Block et al., 1996). This difference may arise because gene expression in hepatocytes is strongly influenced by the culture environment, which may activate specific transcription factors (e.g. AP1 and NFκB). In addition, the signals mediated by the extracellular matrix influence gene expression in hepatocytes *in vitro* (Serandour et al., 2005; Fasset et al., 2006; Kim et al., 2003). We studied whether the altered phenotype of cultured hepatocytes reverts to that *in vivo* when the cells are reimplanted into the body (Nishikawa et al., 1994). To investigate this problem, two markers were used, one of which is newly expressed and the other is suppressed in cultured hepatocytes, by the F344/ F344-alb transplantation model.

F344 hepatocytes were cultured on hydrophobic plastic dishes to form spheroidal aggregates. Within 3 days of culture, the hepatocytes formed spheroidal aggregates of approximately 50 to 100 μm in diameter, most of which were detached from the bottom and floated freely in the medium. After 5 days of culture, the spheroidal hepatocyte aggregates were harvested and implanted into livers and spleens of F344-alb. The hepatocytes in the liver tissue of F344 rats are positive for p450 (CYP2C6), which is extensively expressed in normal rat hepatocytes (Figure 2C), as well as albumin (Figure 2A), but these cells were completely negative for the placental form of glutathione S-transferase (GST-P), which is one of the glutathione S-transferases that plays an active role in the detoxification of xenobiotics and noxious products generated after tissue damage (Sato, 1999) (Figure 4B). Five days after the start of culture, although albumin expression was maintained in the hepatocytes in spheroidal aggregates (Figure 2D), GST-P was strongly positive in the nucleus (Figure 2E), but P-450 was completely negative (Figure 2F).

After intrahepatic transplantation, the transplanted hepatocytes could be identified by albumin staining within the recipient livers (Figure 2G). On day 5-10 after transplantation, most of the transplanted F344 hepatocytes were located at the portal veins. These cells were firmly attached to their walls and covered by endothelial cells, while some were occasionally observed integrated into the interlobular connective tissue. Most transplanted hepatocyes became completely negative for GST-P staining (Figure 2H), while P-450 was detected in all of the transplanted hepatocytes at an expression level equivalent to that of the surrounding host hepatocytes (Figure 2I). After intrasplenic transplantation, most hepatocytes migrated into the red pulps on days 5-10. These hepatocytes were stained positive for albumin (Figure 2J), still weakly positive for GST-P (Figure 2K) and strongly positive for P-450 which generally gave a more intense signal than in the hepatocytes of F344 livers (Figure 2L).

These results indicated that the phenotype of cultured hepatocytes returned to that of hepatocytes *in vivo* after implantation into intrahepatic and intrasplenic environments.

However, GST-P expression in the cultured hepatocytes was more rapidly extinguished when these cells were transplanted into the liver than into the spleen, and P-450 was more intensely expressed in the spleen than the liver, possibly as a result of differences in the environmental factors between these two organs. Although hepatocyte transplantation would appear possible to various organs other than the liver, the liver is considered to the most likely site for transplanted hepatocytes to return to physiologic functioning.

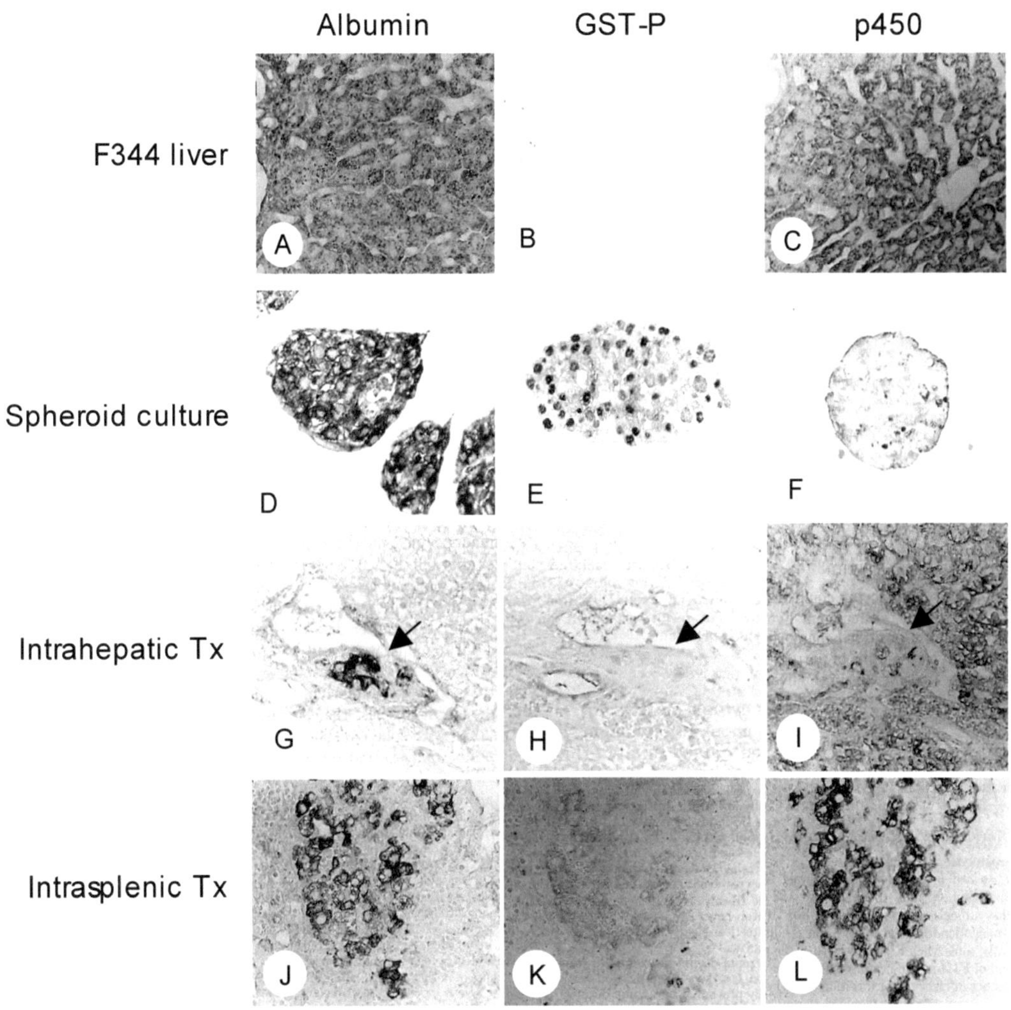

Fig. 2. Immunohistochemical staining for albumin, GST-P and P-450 (CYP2C6) in the hepatic tissue of F344 rat (A, B and C), spheroidal aggregates of F344 hepatocytes 5 days after the start of culture (D, E and F), and transplants of the cultured F344 hepatocytes in the liver (arrows in G, H and I) and spleen of F344-alb rats (J, K and L). A, D, G and J: albumin staining; B, E, H and K: GST-P staining; C, F, I and L; P-450 (CYP2C6) staining.

6. The effect of a continuous increase of serum albumin in analbuminemic rats

Albumin plays an important role in the maintenance of plasma osmotic pressure, and assists in the transport of various substances such as hormones, fatty acids, bilirubin, iron and other metals, and exogenous drugs. In spite of the absence of albumin, however, analbuminemic rats are generally healthy, showing none of the anticipated signs of disturbed physiology, such as edema and jaundice. This may be so fit because other serum proteins take the place of albumin. In analbuminemic rats, the total serum protein is usually normal because of increases in other serum proteins (Nagase et al., 1979). As the degree of increase is not uniform for all proteins, the electophoretic pattern of the non-albumin fractions is quite different from that of normal rats. The major elevated components involved are α1-trypsin, α2-macroglobulin, α-X protein, transferrin, ceruloplasmin, fibrinogen and immunoglobulins.

We investigated whether a continuous increase of serum albumin can normalize the unique patterns of serum proteins in analbuminemic rats (Ohta et al., 1993b). When carcinogen-induced preneoplastic hepatocytes isolated from hyperplastic hepatic nodules (HPN) of F344 rats were transplanted into F344-alb livers (Figure 3A), they could be induced to grow rapidly by application of the Solt and Farber regimen; sometimes these nodules progressed into hepatocellular carcinoma (HCC) (Ohta et al., 1994). As HPN and HCC cells can produce albumin and other serum proteins, this procedure allows the investigation of changes in serum protein content in analbuminemic rats under the condition in continuous albumin elevation.

The livers of all the F344-alb rats that had received F344 HPN cells contained a large number of HPN cells at 6 weeks (Figure 3B). At 12 months after transplantation, some recipients had HCC in addition to HPN. The transplanted cells were estimated to occupy aprpximately 6.0+1.4% and 30.0+10.0% of the total liver mass of the recipients at 6 weeks and 12 months, respectively. Serum albumin in F344-alb rats with HPN cell transplantation reached 7.0±2.4 mg/ml at 6 weeks and 38.8±7.9 mg/ml at 12 months; the latter number was comparable to or even higher than the values seen in normal F344 rats (34.0 ±0.2 mg/dl).

Although total serum protein in untreated F344-alb rats (6.8±0.4 g/dl) was similar to that in the untreated F344 rats (6.7±0.3 g/dl), electrophoresis of the serum proteins of the untreated F344-alb rats revealed that the α, β and γ fractions were elevated (Figure 3C a, b). On the other hand, although total serum protein in recipient F344-alb rats was almost at the normal level (7.0±0.4 g/dl) 6 weeks after transplantation, it was significantly higher at 12 months (10.4+2.1 g/dl) in these rats than in untreated F344 or F344-alb rats. Electrophoresis of the serum of F344-alb rats with HPN cell transplantation showed that albumin was clearly increased in the transplanted F344-alb cases, but the patterns of non-albumin fractions were unchanged compared with untreated F344-alb rats (Figure 3C b, c).

These results demonstrated that after transplantation of F344 HPN cells into F344-alb rats, although albumin was persistently elevated, the F344-alb-specific pattern of non-albumin fractions remained and was sometimes accompanied by an increase in total serum protein. The serum albumin level is determined by its synthesis, secretion, distribution and degradation in normal animals. When an excess of serum albumin is achieved by intravenous infusion of albumin in normal rats, both degradation and urinary excretion were increased (Rothschild et al., 1988). In contrast, the half-life of infused albumin is greatly extended in analbuminemic humans and rats, although the half-life of other plasma proteins is normal (Inoue, 1985). It is thus possible that the lack of albumin causes an involution of the ability to maintain a normal albumin levels in analbuminemic rats. Additionally, albumin produced by the transplanted HPN cells may accumulate because of slow degradation or impaired urinary excretion, causing hyperalbuminemia in some recipients.

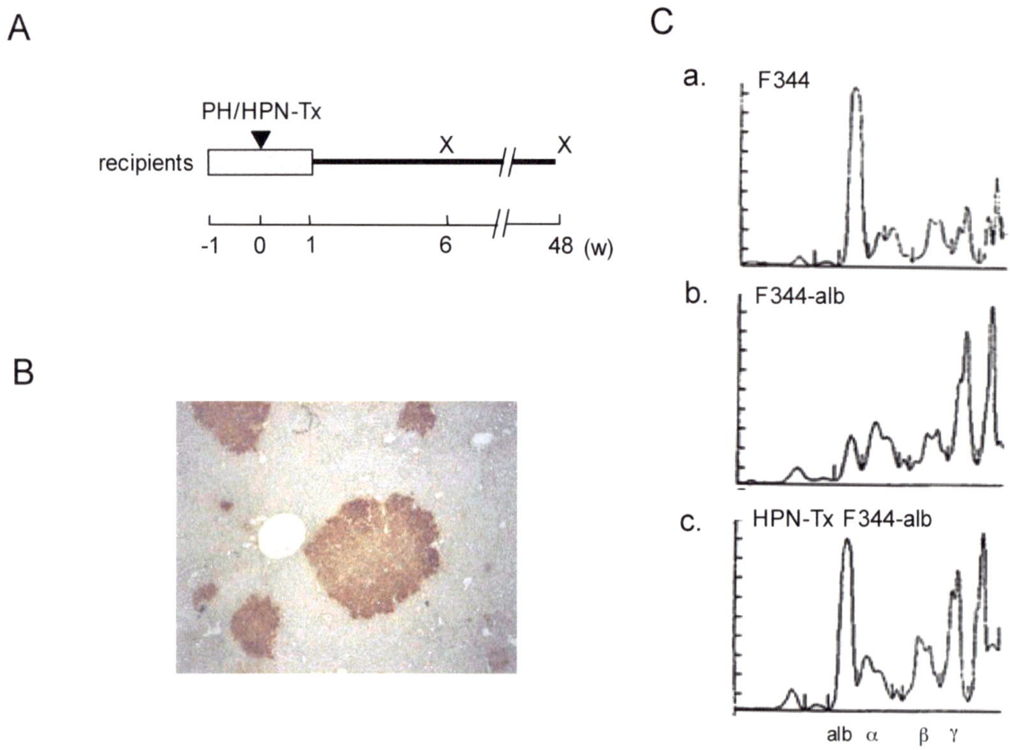

Fig. 3. The effect of continuous serum albumin elevation in F344-alb rats. A. Experimental procedure for transplantation of F344 HPN cells into the liver of F344-alb rats. Hatched box: dietary 2-acetylaminofluorene (2-AAF) treatment. PH/HPN-Tx: transplantation with HPN cells via the portal vein immediately after PH. X: sacrifice. B. Albumin immunostaining of a recipient F344-alb liver 6 weeks after F344 HPN cell transplantation. C. Densitometric analyses of electrophoretic patterns of serum proteins of untreated F344 and F344-alb rats, and as well as the F344-alb rats 12 months after HPN cell transplantation.

7. Hepatocytes derived from bone marrow cells

Hematopoietic cells contribute to the generation of hepatocytes in the liver. This process is thought mediated by the direct transdifferentiation of bone marrow cells into hepatocytes (Theise et al., 2000), the indirect transdifferentiation of bone marrow cells into oval cells that have the potential to give rise to mature hepatocytes during the cholangiocellular lineage (Oh et al., 2007), and the transfer of genetic materials from bone marrow cells to recipient hepatocytes by cell fusion (Wang et al., 2003). Using the analbuminemic rat model, we investigated whether bone marrow cells gave rise to hepatocytes during post-PH liver regeneration (Arikura et al., 2004). In this process, the original hepatic mass is mainly regenerated by the division of resident hepatocytes.

The livers of one group of F344-alb rats were infused with F344 bone marrow cells via the portal vein immediately after PH (Figure 4A c). The bone marrow of another group of F344-alb rats were hematopoietically reconstituted using whole-body X-irradiation and F344 bone marrow cell transplantation with PH following 4 weeks later (Figure 4A d). Untreated F344-alb rats (Figure 4A a) and those treated with PH alone (Figure 4A b) served as controls. Four weeks after PH, although single cells or two-cell clusters of albumin-positive hepatocytes were seen regardless of bone marrow cell transplantation or PH status in both the control and experimental groups (Figure 4B a), clusters of more than 3 albumin-positive hepatocytes were observed in the livers of recipients that had undergone either bone marrow cell transplantation at the time of PH (Figure 4B b, c) or prior bone marrow reconstitution 4 weeks before PH. Normal albumin mRNA was detected in the RNA that was isolated from the livers of recipient F344-alb rats (Figure 4C a). Normal albumin gene sequences were also detected by PCR in DNA that was isolated from the micro-dissected albumin-positive hepatocyte clusters (Figure 4C b). In a female F344-alb rat that had been transplanted with male F344 bone marrow cells, albumin-positive hepatocyte clusters in the liver were positive for the Y chromosome marker *Sry3* (Figure 4D a, b).

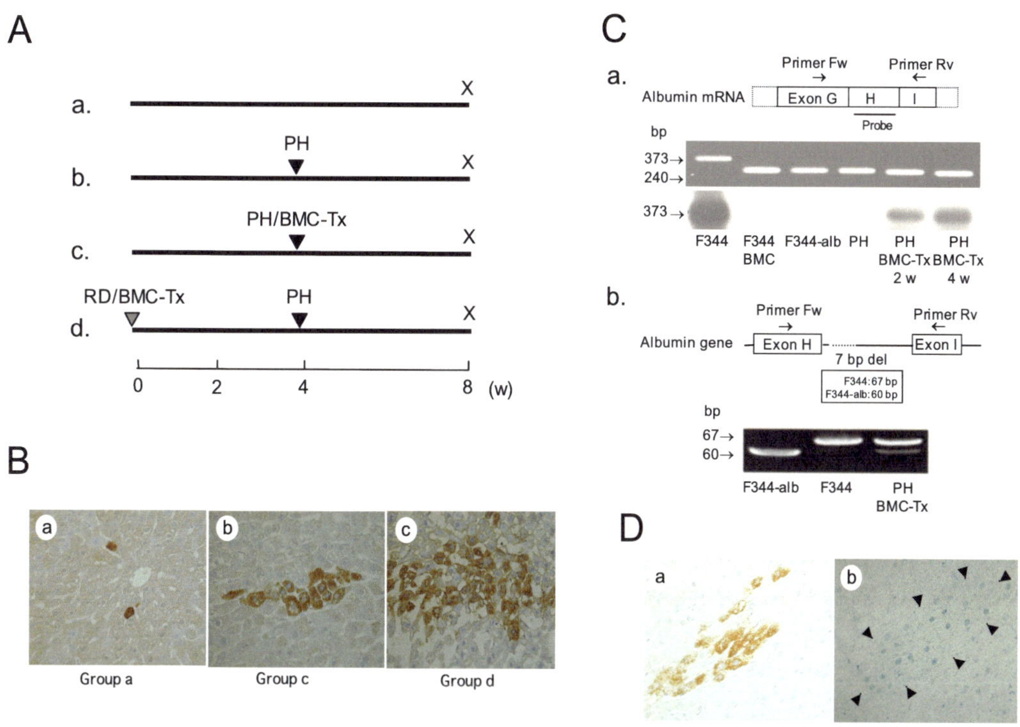

Fig. 4. Detection of bone marrow cell-derived hepatocytes in the F344-alb livers.
A. Experimental groups: a. untreated F344-alb rats; b. F344-alb rats with PH alone; c. F344-alb rats with F344 bone marrow cell transplantation (BMC-Tx) immediately after PH; d. F344-alb rats with bone marrow reconstitution by whole-body X-irradiation (RD) and F344 BMC-Tx followed by PH four weeks later. B. Immunohistochemical staining of the hepatic

tissues for albumin; a. untreated F344-alb rats [Group a in (A)]; b. F344-alb rats with BMC-Tx immediately after PH [Group c in (A)] 2 weeks after PH/BMC-Tx.; c. The same group [Group c in (A)], 4 weeks after PH/BMTx.]. C. a. RT-PCR for albumin mRNA using the primers at exons G and I on RNA isolated from the liver and bone marrow of F344 and livers of untreated F344-alb, F344-alb with PH and F344-alb with PH plus BMC-Tx (2 and 4 weeks, Group c) (upper panel). Although only the exon H skipped albumin mRNA was amplified from the RNAs of control and transplanted F344-alb livers (middle panel), Southern blot analysis of the above PCR products using the exon H probe detected the normal albumin mRNA in the F344-alb livers with BMC-Tx (lower panel).
b. PCR-based detection of normal albumin gene sequence from DNA isolated from an albumin positive hepatocyte cluster in F344-alb liver with RD plus BMC-Tx→PH (Group d) using the primers at exons H (forward) and I (reverse). D. *Sly3 in situ* hybridization. a. Albumin immunostaining of female F344-alb liver with RD/male BMC-Tx→PH (Group d). b. *Sly3 in situ* hybridization in the contiguous section of a.

We also investigated whether hematopoietic stem cells mobilized from the bone marrow into the peripheral blood can give rise to hepatocytes in the regenerating liver after PH (Huiling et al., 2004). The donor F344 rats were repeatedly treated with the recombinant granulocyte colony stimulating factor (G-CSF). Mononuclear cells were isolated from the peripheral blood and infused into the portal veins of F344-alb rats immediately after PH. Clusters of more than 3 albumin-positive hepatocytes and normal albumin mRNA and genomic DNA sequences were detected in the livers of recipient F344-alb rats.

These results demonstrate that F344 bone marrow cells can give rise to albumin-positive hepatocytes during liver regeneration after PH in F344-alb rats. However, the number of albumin-positive hepatocyte clusters was estimated to be 1000-2000 cells in the whole liver, accounting for 0.003-0.004% of the entire hepatic volume. Therefore, bone marrow cells play only a minor role in liver regeneration after PH. Interestingly, we detected large clusters of more than 50 albumin-positive hepatocytes; these clusters may consist of about 360 cells. Because hepatocytes divide one or two times after PH to restore the original hepatic mass, we hypothesized that some of the bone marrow-derived hepatocytes have a high proliferative capacity.

8. Protection from liver injury by bone marrow cell transplantation without liver repopulation

In the liver injury model in which the animals are pretreated with retrorsine (RS) followed by PH, transplanted hepatocytes repopulate the liver to replace a large hepatic mass with donor hepatocytes within 2-4 months (Laconi et al., 1998). However, bone marrow cell transplantation yields few or no donor bone marrow-derived hepatocytes in RS/PH treated animals (Wagers et al., 2002). Therefore, we investigated the effects of hepatocyte transplantation and bone marrow transplantation on RS/PH-treated livers using the analbuminemic rat model (Zhang et al., 2007).

F344-alb rats were given two doses of RS two weeks apart followed by PH in the 4th week (Figure 5A b). F344-alb rats that had received PH alone were used as controls (Figure 5A a). As expected, the RS/PH-treated rats showed a high mortality rate (survival rate 35%), with death occurring between 1 and 11 days after PH, whereas all of the control animals survived (Figure 5B). Serum AST, ALT and bilirubin levels were elevated two days after PH (Figure 5C), and the liver tissue showed marked histological damage. In contrast, when hepatocytes (via the portal vein) or bone marrow cells (via the portal or penile vein) were transplanted immediately after PH in RS-treated F344-alb rats, the survival rate was significantly improved; the improvement was 50% for hepatocyte transplantation and 72.5% for bone marrow cell transplantation (Figure 5B). The liver function test results for these recipient F344-alb rats revealed that the AST, ALT and bilirubin values two days after PH were lower than those of the RS/PH-treated F344-alb rats without cell transplantation, and the levels in the recipient F344-alb rats were almost comparable to those of the untreated F344-alb rats or F344-alb rats with PH alone (Group a) (Figure 5C). Furthermore, the histological damage to the liver tissue was less than that seen in RS/PH-treated F344-alb liver tissue without cell transplantation.

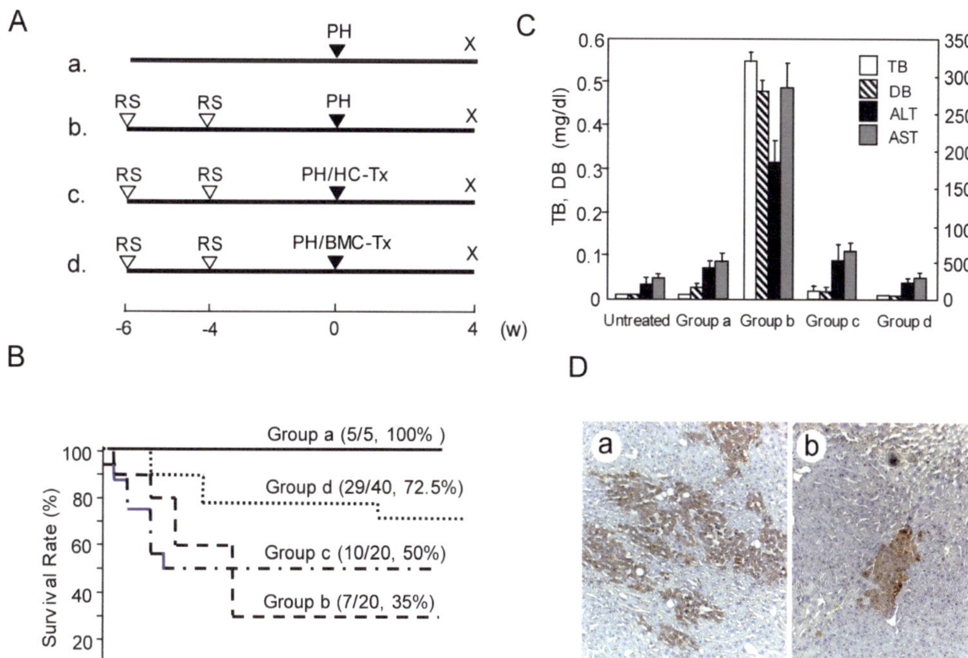

Fig. 5. The effects of hepatocyte transplantation (HC-Tx) and bone marrow cell transplantation (BMC-Tx) in the retrorsine (RS)/PH-induced hepatic injury model. A. Experimental groups: a. F344-alb rats with PH alone; b. F344-alb rats treated with two doses of RS (30 mg/kg of body weight) 2 weeks apart followed by PH two weeks after the second RS treatment; c. F344-alb rats treated with two doses of RS followed by treatment with PH and HC-Tx; d. F344-alb rats treated with two doses of RS followed by treatment with PH and BMC-Tx. B. Survival curve. Most RS/PH-treated F344-alb rats (Group b) died 1-11 days after PH, but the survival rate was significantly higher in the RS/PH-treated F344-alb rats given HC-Tx (Group c) or BMTx (Group d). The numbers in parentheses show survival vs. total numbers of F344-alb and their survival rate 28 days after PH, PH/HC-Tx or PH/BMC-

Tx. C. Liver function tests. TB: Total bilirubin, DB: direct bilirubin, ALT: alanine aminotransferase, AST: aspartate aminotransferase. D. Albumin immunostaining in the F344-alb livers. a. RS+PH/HC-Tx (Group c). b. RS+PH/BMC-Tx (Group d).

In the RS/PH-treated F344-alb rats that received hepatocyte transplantation (Group c), the liver tissue was extensively replaced by albumin-positive F344 hepatocytes (Figure 5D a), the serum albumin level was increased almost to the level of normal rats, and normal albumin mRNA was detected in the liver four weeks after PH. The liver regeneration rate in these rats was almost comparable to the rate in F344-alb rats that underwent PH alone. However, in the RS/PH-treated F344-alb rats treated with bone marrow cell transplantation (Group d), only small clusters of albumin-positive hepatocytes were detected (Figure 5D b). These results are similar to the results seen in rats treated with simultaneous PH and bone marrow transplantation or with bone marrow reconstruction followed by PH as described above (Figure 4B). Furthermore, the liver regeneration rate in the RS/PH-treated animals with bone marrow cell transplantation was as low as that in the RS/PH-treated F344-alb rats without cell transplantation.

Our results indicate that, although the bone marrow-derived hepatocytes do not extensively repopulate the RS/PH-treated liver, they can prevent RS/PH-induced liver injury and increase the potential for recipient survival. Factors derived from the bone marrow cells and hepatocytes may suppress RS/PH-induced liver injury, the mechanism of which remains elusive. Because it takes time for transplanted hepatocytes to repopulate the liver and restore hepatic functions, bone marrow cell transplantation may prevent recipient death from liver injuries that occurs during the period before the transplanted hepatocytes can repopulate the liver.

9. Induction of immunotolerance for allogeneic hepatocytes by donor bone marrow cell transplantation

One of the main challenges that limit the efficiency of hepatocyte transplantation is the large cellular loss after transplantation. The majority of hepatocytes infused into the portal circulation are trapped within the hepatic vessels and cleared by granulocytes, macrophages and Kupffer cells. Therefore, only a small number of infused hepatocytes are integrated into the host hepatic tissue (Han et al., 2009; Okazaki et al., 2008). Moreover, the transplanted allogeneic hepatocytes activate the adaptive immune system of the recipients, thereby activating immunological mechanisms thus contribute to the elimination of these hepatocytes (Han et al., 2009; Okazaki et al., 2008; Bumgardner & Orosz, 2000). However, when host bone marrow is reconstituted using donor bone marrow cells, successful transplantations of allogeneic hepatocytes have been conducted between mouse strains expressing disparate major histocompatibility complex proteins (Vidal et al., 2008). For bone marrow reconstitution, intra-bone marrow bone marrow cell injection is more efficient than intravenous bone marrow cell transfusion (Ikehara, 2005).

We investigated the effect of hematopoietic reconstruction on allogeneic hepatocyte transplantation using Lewis (LEW) rats as donors and F344-alb rats as recipients (Inagaki et al., 2011); these types of rats express disparate major histocompatibility complex protein. As described above, when syngeneic F344 hepatocytes were transplanted into the liver of RS plus PH F344-alb rats (Figure 6A a), the large hepatic mass was replaced by albumin-positive hepatocytes (Figure 6B a). In addition, the serum albumin levels were increased, and no inflammatory changes were observed in the liver four weeks after transplantation. In

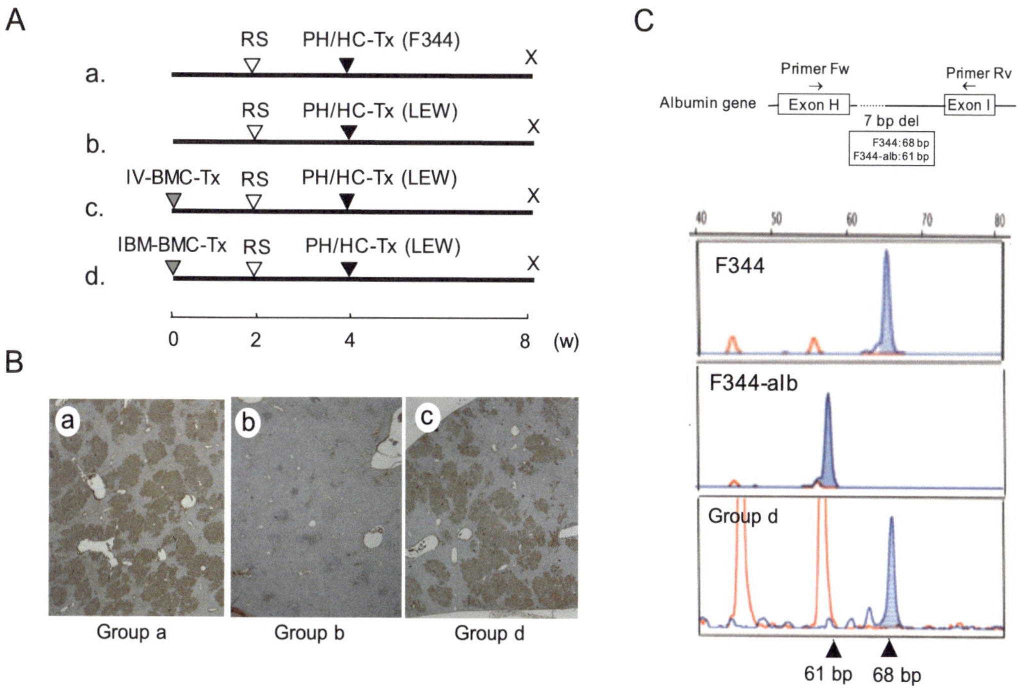

Fig. 6. The tolerance of allogeneic hepatocyte-mediated liver repopulation. A. Experimental groups: a: syngeneic F344 HC-Tx in RS/PH-treated F344-alb rats; b: allogeneic LEW HC-Tx in RS/PH-treated F344-alb rats; c: allogeneic LEW HC-Tx in RS/PH-treated F344-alb rats that received LEW BMC-Tx via intravenous transfusion (IV-BMC-Tx); d: allogeneic LEW HC-Tx in RS/PH-treated F344-alb rats that received LEW BMC-Tx by intra-bone marrow injection (IBM-BMC-Tx). B. Immunohistochemical staining for albumin in recipient F344-alb livers: a: F344-alb rats treated with syngeneic F344 HC-Tx [Group a in (A)]; b: F344-alb rats treated with allogeneic LEW HC-Tx [Group b in (A)]; c: F344-alb rats treated with allogeneic HC-Tx and prior IBM-BMC-Tx [Group d (A)]. C. PCR analysis of bone marrow reconstitution using the primers at albumin gene exons H and I and DNA isolated from the bone marrow cells of untreated F344 and F344-alb rats, as well as F344-alb rats treated with IBM-BMC-Tx [Group d (A)].

contrast, transplantation of allogeneic LEW hepatocytes into F344-alb rats (Figure 6A b) did not induce the repopulation of albumin-positive hepatocytes in recipient livers (Figure 6B b). In addition, serum albumin levels did not increase, and inflammatory cell infiltration was observed in the portal areas of recipient livers. When the bone marrow of the recipient F344-alb rats was reconstituted using LEW bone marrow cells (Figure 6A c, d), albumin-positive hepatocytes repopulated the livers of 1 of 6 F344-alb rats that underwent intravenous bone marrow cell transfusion and 6 of 6 F344-alb rats that underwent intra-bone marrow bone marrow cell injection (Figure 6B c). The latter was associated with increased serum albumin levels. Although slight inflammatory cell infiltration was observed in the portal areas of the recipient livers, no inflammatory changes were detected in the areas repopulated with albumin-positive hepatocytes. When quantitative PCR was used to test bone marrow reconstitution in the recipient F344-alb rats for the normal and analbuminemic albumin gene sequences, 6 of 6 F344-alb rats showed bone marrow reconstitution after intra-bone marrow injection (Figure 6C). In contrast, only 1 of 6 F344-alb rats that underwent intravenous bone marrow cell transfusion showed bone marrow reconstitution. The albumin-positive hepatocyte repopulation and the increase in serum albumin levels depended completely on bone marrow reconstitution by the donor bone marrow cells.

These results indicate that liver repopulation via allogeneic hepatocyte transplantation without the use of immunosuppressants is possible if the recipient bone marrow is efficiently reconstituted using donor bone marrow cells. Intra-bone marrow injection of bone marrow cells induces bone marrow reconstruction in a manner that is more efficient than does intravenous transfusion.

10. Conclusion

Mito et al. (1978) completed the first successful hepatocyte transplantation in the rat spleen 33 years ago. Since then, cosiderable progress has been made in the field of hepatic cell transplantation research. Notably, it has been demonstrated that hepatocytes can be transplanted into the liver and can regenerate the diseased host liver. The use of hepatocyte transplantation is therefore expected to expand as a therapy for human liver disease. For cell transplantation to be used as a human therapy, further investigation is required to address number of problems including efficient ways to suppress the innate and acquired immune responses to the transplanted cells and the short- and long-term risks such as malignant transformation. The F344/F344-alb model provides a useful tool for studies of hepatocyte transplantation in the liver because it can be used to accurately trace the fate and functionality of the transplanted cells. Bone marrow cell transplantation may facilitate cell transplantation therapy via the hepatogenic potential of hematopoietic cells, the putative protective paracrine actions in the liver and the induction of donor-specific immuno-tolerance. A shift from orthotropic liver transplantation, if in part, to hepatocyte transplantation will yield new opportunities to develop therapies for human liver disease.

11. Acknowledgements

This work was supported by the grants from the Japanese Ministry of Science, Education, Sports and Culture.

12. References

Arikura, J., Inagaki, M., Huiling, X., Ozaki, A., Onodera, K., Ogawa, K. & Kasai, S. (2004). Colonization of albumin-producing hepatocytes derived from transplanted F344 rat bone marrow cells in the liver of congenic Nagase's analbuminemic rats. *Journal of Hepatology*, Vol. 41, No. 2, pp. 215-221.

Asamoto, M., Tsuda, H., Kagawa, M., de Caargo, J.L.V., Ito, N. & Nagase, S. (1989). Strain differences in susceptibility to 2-acetylaminofluorene and phenobarbital promotion of rat hepatocarcinogenesis in medium-term assay system: Quantitation of glutathione S-transferase P-positive foci development. *Japanese Journal of Cancer Research,* Vol. 80, No. 10, pp. 939-944.

Block, G.D., Locker, J., Bowen, W.C., Petersen, B.E., Katyal, S., Strom, S.C., Riley, T., Howard, T.A. & Michalopoulos, G.K. (1996). Population expansion, clonal growth, and specific differentiation patterns in primary cultures of hepatocytes induced by HGF/SF, EGF and TGFα in a chemically defined (HGM) medium. *Journal of Cell Biology,* Vol. 132, No. 6, pp.1133-1149.

Bumgardner, G.L. & Orosz, C.G. (2000). Unusual pattern of alloimmunity evoked by allogenic liver parenchymal cells. *Immunological Reviews,* Vol. 174, No. 1, pp. 260-279.

Chiocchetti, A., Toiosano, E., Hisch, E., Silengo, L. & Altruda, F. (1997). Green fluorescent protein as a reporter of gene expression in transgenic mice. *Biochemical Biophys Acta,* Vol. 1352, No. 2, pp. 193-202.

Eckert, J.W., Buerkle, C.J., Major, A.M., Finegold, M.J. and Bandt, M.L. (1995). *In situ* hybridization utilizing a Y chromosome DNA probe. Use as a cell marker for hepatocellular transplantation. *Transplantation,* Vol. 59, No. 1, pp. 109-110.

Esumi, H., Takahashi, Y., Sato, S., Nagase, S. & Sugimura, T. (1983). A seven-base pair deletion in an intron of the albumin gene of analbuminemic rats. *Proceeding of National Academy of Science USA,* Vol. 80, No. 1, pp. 95-99.

Fasset, J., Tobolt, D. & Hansen, L.K. (2006). Type I collagen structure regulates cell morpholoty and EGF signaling in primary rat hepatocytes through cAMP-dependent protein kinase A. *Molecular Biology of The Cell,* Vol. 17, No.1, pp. 345-356.

Gilgenkrantz, H. (2010). Rodent models of liver repopulation. *Methods in Molecular Biology,* Vol. 481, pp. 474-490.

Gupta, S., Rajvanshi, P. & Lee, C,D. (1995). Integration of transplanted hepatocytes into host liver plates demonstrated with dipeptiyl peptidase IV-deficient rat. *Proceeding of National Academy of Science USA,* Vol. 92, No. 13, pp. 5860-5864.

Hamman, K., Clark, H., Montini, E., Al-Dhalimy, M., Gromp, M., Finegold, M. & Harding, C.O. (2005). Low therapeutic threshold for hepatocyte replacement in murine phenylketonuria. *Molecular Therapy,* Vol 12, No. 2, pp. 337-344.

Han, B., Lu, Y., Meng, B. & Qu, B. (2009). Cellular loss after allogenic hepatocytes transplantation. *Transplantation,* Vol. 87, No. 1, pp. 1-5.

Huiling, X., Inagaki, M., Arikura, J., Ozaki, A., Onodera, K., Ogawa, K. & Kasai, S. (2004). Hepatocytes derived from peripheral blood stem cells of granulocyte-colony stimulating factor treated F344 rats in analbuminemic rat livers. *Journal of Surgical Research,* Vol. 122, No. 1, pp.75-82.

Ikehara, S. (2005). Intra-bone marrow-bone marrow transplantation: A new strategy for treatment of stem cell disorders. *Annals of the New York Academy of Science,* Vol. 1051, pp. 626-634.

Inagaki, M., Furukawa, H., Satake, Y., Okada, Y., Chiba, S., Nishikawa, Y. & Ogawa, K. (2011). Replacement of liver parenchyma in analbuminemic rats with allogenic hepatocytes is facilitated by intrabone marrow-bone marrow transplantation. *Cell Transplantation,* Vol. 20, No. 9, pp.1479-1489.

Inoue, M. (1985) Metabolism and transport of amphipathic molecules in analbuminemic ras and human subjects. *Hepatology,* Vol. 5, No. 5, pp.892-898.

Kakizoe, T., Komatsu, H., Honma, Y., Niijima, Y., Kawachi, T., Sugimura, T. & Nagase, S. (1982). High susceptibility of analbuminemic rats to induced bladder cancer. *British Journal of Cancer,* Vol. 45, No. 3, pp. 474-476.

Kaneko, T., Shima, H., Esumi, H., Ochiai, M., Nagase, S., Sugimura, T. & Nagao, M. (1991). Marked increases of two kinds of two-exon-skipped albumin mRNAs with aging and their further increase by treatment with 3'-methyl-4-dimethylaminoazobenzene in Nagase analbuminemic rats. *Proceeding of National Academy of Science USA,* Vol. 88, No. 4, pp. 2807-2811.

Kim, S.H., Kim, J.H. & Akaike, T. (2003). Regulation of cell adhesion signaling by synthetic glycopolymer matrix in primary cultured hepatocyte. *FEBS Letters,* Vol. 553, No. 3, pp. 433-439.

Laconi, E., Oren, R., Mukhopadyay, D.K., Hurston, E., Laconi, S., Pani, P., Dabeva, M.D. & Shafritz, D.A. (1998). Long-term, near total liver replacement by transplantation of isolated hepatocytes in rats treated with retrorsine. *American Journal of Pathology,* Vol. 153, No. 1, pp. 319-329.

Makino, R., Sato, S., Esumi, H., Negishi, C., Takano, M., Sugimura, T., Nagase, S. & Tanaka, H. (1986). Presence of albumin-positive cells in the liver of analbuminemic rats and their increase on treatment with hepatocarcinogens. *Japanese Journal of Cancer Research,* Vol. 77, No. 2, pp. 153-159.

Mao, X., Fujikawa, Y. & Orikin, S.H. (1999). Improved reporter strain for monitoring Cre recombinase-mediated DNA excisions in mice. *Proceeding of National Academy of Science USA,* Vol. 96, No. 9, pp. 5037-5042.

Merion, R.M. (2010). Current status and future of liver transplantation. *Seminar of Liver Disease,* Vol. 30, No. 4, pp. 411-421.

Mito, M., Ebata, H., Kusano, M., Onishi, T., Saito, T. & Sakamoto, S. (1978). Morphology and function of isolated hepatocytes transplanted into the rat spleen. *Transplantation,* Vol. 28, No. 6, pp. 499-505.

Nagase, S., Shimamune, K. & Shumiya, S. (1979). Albumin-deficient rat mutant. *Science,* Vol. 205, No. 4406, pp. 590-591.

Nishikawa, S., Ohta, T., Ogawa, K. & Nagase, S. (1994). Reversion of altered phenotype in primary cultured rat hepatocytes after intrahepatic and intrasplenic transplantation. *Laboratory Investigation,* Vol. 70, No. 6, pp. 925-932.

Ogawa, K., Ohta, T., Inagaki, M. & Nagase, S. (1993). Identification of F344 rat hepatocytes transplanted within the liver of congenic analbuminemic rats by the polymerase chain reaction. *Transplantation,* Vol. 56, No.1, pp. 9-15.

Oh, S.H., Witek, R.P., Bae, S.H., Zheng, D., Jung, Y., Piscaqlia, A.C. & Petersen, B.E. (2007). Bone marrow-derived hepatic oval cells differentiate into hepatocytes in 2-

acetylaminofluorene/partial hepatectomy-induced liver regeneration. *Gastroenterology,* Vol. 132, No. 3, pp. 1077-1087.

Ohta, T., Ogawa, K. & Nagase, S. (1993a). Increase in albumin mRNA by repeated intrahepatic transplantation of F344 rat hepatocytes into the liver of congenic analbuminemic rats. *Biochemical and Biophysical Research Communications,* Vol. 194, No. 2, pp. 601-609.

Ohta, T., Ogawa, K. & Nagase, S. (1993b). Elevation of serum albumin by intrahepatic transplantation of albumin-producing cells does not correct quantitative abnormalities of non-albumin proteins in analbuminemic rats. *Biochemical and Biophysical Research Communications,* Vol. 197, No. 3, pp. 1103-1110.

Ohta, T., Ogawa, K. & Nagase, S. (1994). Analbuminemia does not significantly influence hepatocarcinogenesis on comparing F344 rats and a congenic line carrying the analbuminemic mutation. *Carcinogenesis,* Vol. 15, No. 2, pp. 227-231.

Okazaki, S., Hisha, H., Mizokami, T., Takaki, T., Wang, T., Song, C., Li, Q., Kato, J., Kamiyama, Y. & Ikehara, S. (2008). Successful acceptance of adult liver allografts by intra-bone marrow-bone marrow transplantation. *Stem Cells and Development,* Vol. 17, No. 4, pp. 629-640.

Rothschild, M.A., Oratz, M., & Schreiber, S.S. (1988). Serum albumin. *Hepatology,* Vol. 8, No. 2, pp. 385-401.

Roy-Chowdhury, N. & Roy-Chowdhury, J. Hepatocyte transplantation. *http://www.uptodate.com/contents/*

Sato, K. (1989). Glutathione transferases as markers of preneoplasia and neoplasia. *Advance in Cancer Research,* Vol. 52, pp. 205-255.

Serandour A.L., Loyer, P., Garnier, D., Courselaud, B., Théret, N., Glaise., D, Guguen-Guillouzo, C. & Corlu, A. (2005). TNFα-mediated extracellular matrix remodeling is required for multiple division cycles in rat hepatocytes. *Hepatology,* Vol. 41, No. 3, pp.478-486.

Shalaby, F. & Shafritz, D.A. (1990). Exon skipping during splicing of albumin mRNA precursors in Nagase analbuminemic rats. *Proceeding of National Academy of Science USA,* Vol. 87, No. 4, pp. 2652-2656.

Takahashi, M., Shumiya, S., Maekwa A., Hayashi, Y. & Nagase, S. (1988). High susceptibility of analbuminemic congenic strain of rats with an F344 genetic background to induced bladder cancer and its possible mechanism. *Japanese Journal of Cancer Research,* Vol. 79, No. 6, pp. 705-709.

Theise, N.D., Nimmakayalu, M., Gardner, R., PB. Illei, P.B., Morgan, G., Tepeman, L., Heneqariu, O. & Krause, D.S. (2000). Liver from bone marrow in humans. *Hepatology,* Vol. 32, No. 1, pp. 11-16, 2000.

Vidal, I., Blanchard, N., Alexandre, E., Gandillet, A., Chenard-Neu, M.P., Staedtler, F., Schumacher, M., Bachellier,P., Jaeck, D., Firat, H., Heyd, B. & Richer, L. (2008). Improved xenogenic hepatocyte implantation into nude mouse liver parenchyma with acute liver failure when followed by repeated anti-Fas antibody (Jo2) treatment. *Cell Transplanataion,* Vol. 17, No. 5, pp. 507-524.

Wagers, A.J., Sherwood, R.I., Christensen, J.L. & Weissman, I.L. (2002). Little evidence of developmental plasticity of adult hematopoietic stem cells. *Science,* Vol 297, No. 5590, pp. 2256-2259.

Wang, X., Willenbring, H., Akkari, Y., Torimaru, Y., Foster, M., Al-Dhalimy, M., Lagasse, E., Finegold, M., Olson, M. & Grompe, M. (2003). Cell fusion is the principal source of bone-marow-derived hepatocytes. *Nature,* Vol. 422, No. 6934, pp. 897-901.

Yokota, K. & Ogawa, K. (1978). Application of a genotypic marker for analysis of chemically-induced hepatic carcinogenesis in rats. *Sapporo Medical Journal,* Vol. 56, No. 2, pp. 263-274.

Zhang, B., Inagaki, M., Jiang, B., Miyakoshi, M., Arikura, J., Ogawa, K. & Kasai, S. (2007). Effects of bone marrow and hepatocyte transplantation on liver injury. *Journal of Surgical Research,* Vol. 157, No. 1, pp. 71-80.

8

Rodent Models with Humanized Liver: A Tool to Study Human Pathogens

Ivan Quétier, Nicolas Brezillon and Dina Kremsdorf
INSERM, U845, Pathogenèse des Hépatites Virales B et Immunothérapie, Université Paris Descartes, Sorbonne Paris Cité, Faculté de Médecine René Descartes, CHU Necker, France

1. Introduction

The recent development of small animal models for experimental hepatotropic infection has opened new perspectives for the evaluation of novel therapeutic and/or prophylactic compounds against hepatitis B virus (HBV), hepatitis C virus (HCV) and Plasmodium falciparum, three major hepatic pathogens responsible for millions of deaths each year. Indeed, till now in vitro and in vivo models have their limitations. As example, primary human hepatocytes (PHH) are susceptible to infection by HBV (Gripon et al 1988), HCV (Fournier et al 1998) and by sporozoites (the hepatic stage of Plasmodium falciparum) (Mazier et al 1985), but are hampered by a rapid dedifferentiation of the PHH (the loss of differentiation leads to a loss of susceptibility to infection) and the difficulties of obtaining fresh cells. In vivo, the chimpanzee constitutes the best non-human primate which can be used for studies of HBV, HCV and Plasmodium falciparum (Dandri et al 2005b; Kremsdorf & Brezillon 2007; Moreno et al 2007), but multiple drawbacks, including ethical issues, the inability to produce numerous progeny in a short time (long gestation periods) and exorbitant housing and breeding costs render difficult the accessibility.

For a long time, liver cell transplantation was just a dream; fortunately, experimental biology as led researchers to create new challenging mouse models. Indeed, generation of new mouse models for human hepatocyte transplantation have permitted, for the first time, experimental manipulations of human hepatotropic pathogens of man which are immediate problems of human health, as well as the study of cell transplantation in a regenerative medicine perspective. Here, we will focus on the development of humanized mice models using hepatocyte transplantation to study the three major hepatic pathogens.

2. Transplanted hepatic cells can replace a diseased liver in mice

Few papers laid the foundations for the entire field of liver cell transplantation in mouse. They described and applied a genetic-based animal model for competitive liver regeneration where exogenous transplanted hepatocytes have a selective advantage and can replace the diseased tissue. Two mice models were described: transgenic mice expressing high levels of uPA (urokinase-Plasminogen Activator) (Rhim et al 1994) and mice deficient for the fumaryl acetoacetate hydrolase (FAH) (Grompe et al 1993) (Fig. 1).

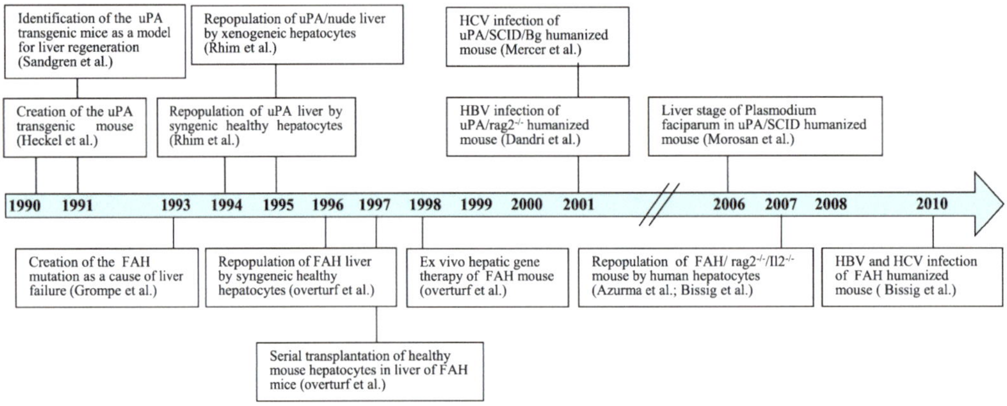

Fig. 1. Steps in the creation mouse models for liver cell transplantation and for infection by liver pathogens. The arrow represents a timeline. Each box represents an independent and initial study describing a mouse model to study liver cell transplantation and the infection of a humanized model by a hepatotropic pathogen. If a group of boxes are connected through the same line to the timeline this means that these studies were published the same year.

2.1 uPA mouse model

The initial observation that opened up the field of liver cell transplantation was serendipitous. With the goal of establishing an *in vivo* system to analyze the coagulation and fibrinolytic systems, Heckel et al. produced transgenic mice expressing high levels of uPA (urokinase-Plasminogen Activator) under control of the albumin enhancer/promoter for liver specific expression (Heckel et al 1990) (Fig.2).

As expected, the transgenic animals showed elevated plasma uPA levels, which often provoked a lethal syndrome of neonatal bleeding, causing the death of numerous transgenic founders. Sandgren et al. observed that some transgenic animals were characterized by a gradual normalization of the liver function over the first weeks (Sandgren et al 1991). The authors concluded that transgene expression was toxic to hepatocytes, and that the surviving animals were viable because deletion of the transgene was occurring, followed by clonal expansion of the rare cells that had lost the deleterious gene. Indeed, even in animals with only a few red spots, the existence of only a few "cured" cells was sufficient to ensure replacement of the diseased liver, providing an in vivo demonstration of the unexpectedly high proliferative potential of adult liver cells.

The same team then demonstrated the ability of a small number of "normal" hepatocytes to repopulate ad-integrum the liver of transgenic uPA mice (Sandgren et al 1991). Indeed, the overexpression of uPA protein in hepatocytes is cytotoxic, giving rise to a continuous liver regeneration process. Under these conditions, hepatocytes which lose the transgene by somatic reversion, as well as healthy transplanted hepatocytes, have a strong survival advantage over resident cells (Rhim et al 1994; Sandgren et al 1991). Throughout the regenerative process, the liver size remained normal, and blood chemistry analyses were used to demonstrate that the engrafted cells were functionally competent. Finally, the authors included an important control to demonstrate that the transplanted liver cells underwent expansion only in the Alb-uPA transgenic and not in normal livers, leading to the critical deduction that a regenerative stimulus, that persists in the transgenic mice from birth until 6 to 8 weeks, when the transgene expressing liver has been replaced by donor cells, or endogenous hepatocytes deleted for the transgene, was necessary to obtain clonal expansion of the transplanted cells. To complete the picture, Rhim et al. introduced the nude gene into the Alb-uPA mice, and used homozygous as well as hemizygous transgenics to demonstrate that xenogenic hepatocytes from rats could reconstitute the diseased livers (Rhim et al 1995).

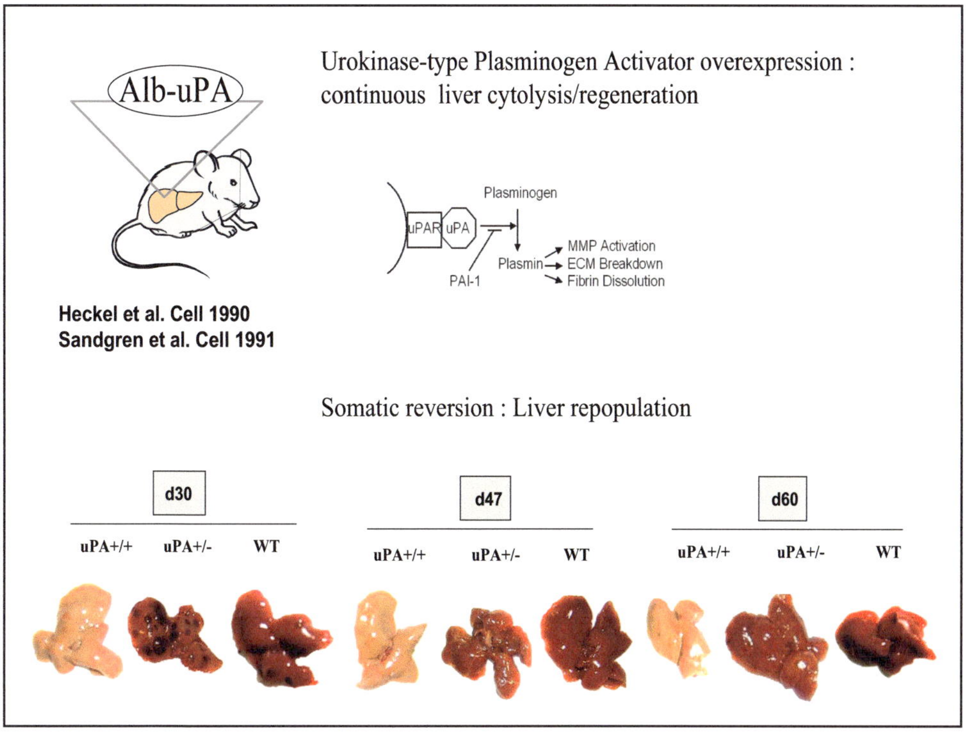

Fig. 2. The Alb-uPA model. Urokinase-type Plasminogen Activator overexpression induces continuous liver cytolysis and regeneration. Gross appearance of liver of Alb-uPA transgenic mice at different time after birth demonstrating somatic reversion of the transgene. Left, homozygous transgenic liver displaying a uniformly white color; center, hemizygous transgenic liver with regeneration nodules, right, non transgenic control.

Based on their proven utility as hosts for liver repopulation, Alb-uPA transgenic mice were backcrossed onto an immunodeficient background (SCID, Rag2$^{-/-}$ or Rag2$^{-/-}$/Pfp$^{-/-}$) to obtain a mouse model which tolerated the xenotransplantation of Human, Woodchuck and Tupaia hepatocytes (Dandri et al 2001a; Dandri et al 2001b; Dandri et al 2005a; Meuleman et al 2005; Petersen et al 1998; Rhim et al 1994; Tateno et al 2004). Because of the reversion process occurring in heterozygous mice for the Alb-uPA transgene, optimum liver repopulation requires intrasplenic transplantation of high quality adult hepatocytes into mice that are homozygous for both the SCID trait and the Alb-uPA transgene, and within one or four weeks of birth. In these conditions, human hepatocytes engrafted and repopulated the mouse parenchyma. Resulting chimeric liver showed satisfactory hepatic architecture and intermingling of the mouse and human subcellular structures, indicating a physiological integration of transplanted cells (Meuleman et al 2005; Tateno et al 2004).

2.2 FAH mouse model

Fumaryl acetoacetate hydrolase (FAH) deficiency causes the human disease hereditary tyrosinaemia type I, an enzyme implicated in the degradation pathway of tyrosine, leading to the accumulation in the liver of toxic metabolites. The inhibitor 2-(2-nitro-4-trifluoro-methylbenzyol)-1,3 cyclohexanedione (NTBC) blocks this pathway at the beginning preventing the generation of these metabolites (Lindstedt et al 1992). Grompe et al. constructed mice with FAH deficiency and described that the NTBC, with treatment begun in utero and maintained thereafter, permitted not only survival of the animal, but also normalized the liver function of the deficient mice (Grompe et al 1993; Grompe et al 1995) (Fig. 3).

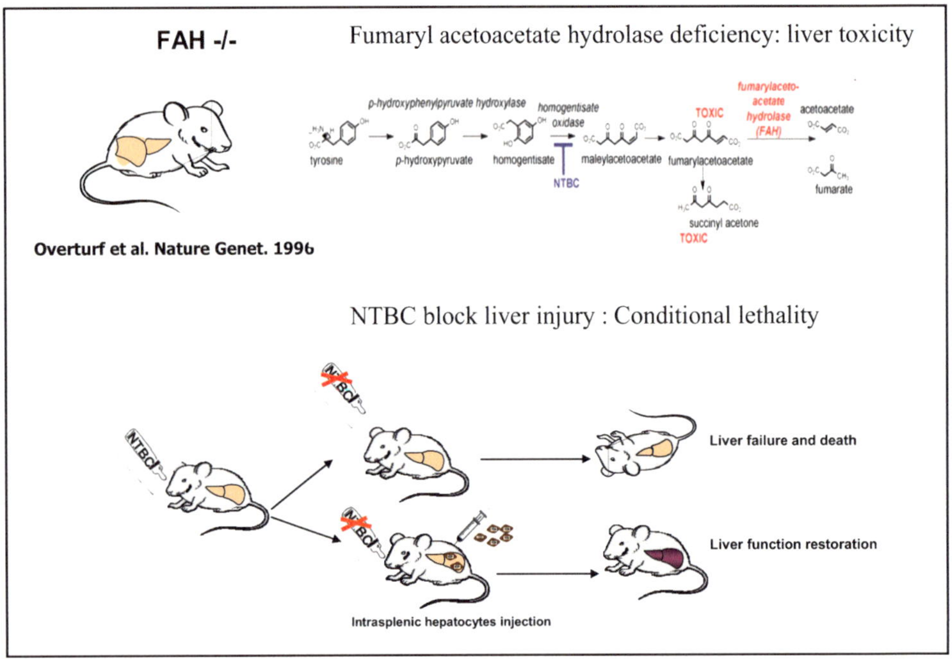

Fig. 3. The FAH model. Fumaryl acetoacetate hydrolase deficiency induces liver toxicity which is blocked by NTBC administration. Mice are submitted to intrasplenic injection of hepatocytes and NTBC drug is withdrawn.

The FAH model was then used for liver transplantation freshly isolated hepatocytes from wild type congenic animals; as in the experiments using the Alb-uPA model, data reported by Overturf et al. demonstrate that a selective advantage of transplanted cells is necessary to obtain repopulation (Overturf et al 1996).

Until recently the transplantation of human hepatocytes into this mouse was not successful: FAH/nude, FAH/NOD/SCID or FAH/Rag1$^{-/-}$ mice appeared to be unable to allow persistence and repopulation by human hepatocytes (Azuma et al 2007). Recently, two different groups have used nearly the same strategy to create a novel mouse model to study human liver cell transplantation. They backcrossed the Rag2-/-/il2-/- mouse with an FAH-/- mouse to give rise to a new mouse model which could be effectively transplanted and repopulated by human hepatocytes, presumably because they lacked not only B and T but also NK cells (Azuma et al 2007; Bissig et al 2007). Differences in methods to obtain satisfactory repopulation have emerged. One group specified the necessity of treating the mice by pre-injection of an adenovirus encoding the uPA protein (Azuma et al 2007) to allow engraftment of human cells, and they found that treatment with an anti human complement agent (Futan) to control the bleeding associated with uPA was not necessary. The second group did not need to use adenovirus encoding the uPA expression, but found that the Futan treatment was necessary (Bissig et al 2007). In both cases, the FAH/Rag2-/-/il2-/- mouse model was successfully transplanted and repopulated (up to 90% of the mouse liver was repopulated by human hepatocytes 3 months post transplantation). Interestingly, it was shown by histological staining that human hepatocytes were interspersed among mouse hepatocytes and did not form individualized clones. Moreover, highly humanized mice permitted long term expansion and maintenance of human cells, and could be used to perform serial transplantation of human hepatocytes from one humanized mouse to a second generation without requiring a new batch of human cells. Finally the humanized livers of the mice expressed a broad range of human markers, including detoxification enzymes (Azuma et al 2007; He et al 2010), and thus, should be useful for pharmacological studies.

3. Mice with chimeric liver: An efficient tool to study hepatotropic pathogens

It was subsequently shown that both humanized mouse models could be infected by HBV and HCV (Fig.1). The uPA/SCID model has already allowed studying HBV or HCV viral life cycle and direct pathogenesis independently to an immune response. In both models, human hepatocytes, maintain their ability to express numerous enzymes implicated in the metabolism and detoxification (p450 family) pathways (Katoh et al 2007; Strom et al 2010) and are suitable to evaluate both the antiviral potential of drugs and the potential toxicity of antiviral compounds. Moreover, the uPA/SCID mice were used to study hepatic stage of Plasmodium falciparum infection.

It is known that distinct HBV genotypes could participate to the severity of liver disease. In order to improve the definition of virological differences among HBV genotypes, Sugiyama and colleagues used the Alb-uPA/SCID mice as a tool to evaluate HBV replication according to viral genotype and confirmed their in vitro previous results, showing a higher replication of genotype C compared to genotype A (Sugiyama et al 2006). Moreover, in a more recent study they have shown that genotype G, which is not detectable in mono-infection, has a higher level of replication in co-infection with HBV genotype H; and that co-infection may cause fibrosis (Tanaka et al 2008).

Another group has taken advantage of immune suppression of the xenograft mouse model to demonstrate that liver disease induced by HBV is not only the result of activation of the immune system, but can be, at least in part, directly mediated by the virus (Meuleman et al 2006). However, it should be noted that the drastic cytopathic effect was observed using a highly pathogenic strain (HBV genotype E) isolated from a patient with fulminant hepatitis. In a recent study, Lutgehetmann et al demonstrated that, as in vitro and in patients, interferon alpha failed to induce ISGs (MxA, OAS, TAP-1) in HBV infected hepatocytes, validating the model for the study of direct interaction between virus and host cells (Lutgehetmann et al 2011)

Petersen et al., using the flexibility of the uPA mouse model, which could be repopulated either by Tupaia belangeri or human hepatocytes and infected by Wooly Monkey HBV or HBV respectively, have tested inhibitors of viral entry (Petersen et al 2008). In both systems, the authors showed that the treatment of repopulated mice with acylated HBV preS-derived lipopeptides prevented viral infection. This alternative approach could benefit patients undergoing liver transplantation to prevent vertical transmission as well as reinfection.

This model can establish long lasting chronic infections and constitute a perfect model to study anti-retroviral treatments. Indeed, it has been validated in different reports showing a good responsiveness to several reverse transcriptase inhibitors (lamivudine, adefovil dipivoxil) (Dandri et al 2005; Tsuge et al 2005). Others steps of viral replication can be targeted, it has been demonstrated that in infected mice HAP BAY 41-4109 (inhibition of capsid formation) was able to diminish HBV viremia (Brezillon et al 2011). Finally, the presence of HBV cccDNA in nucleus of infected human hepatocytes will allow testing new therapeutic approaches to clear hepatocytes or to control transcription from cccDNA (Lutgehetmann et al 2010).

Concerning HCV, infected humanized mice have been used to demonstrate antiviral activity of several molecules, (IFNα2b, BILN2061, Telaprevir, HCV-796) that were already used in clinic, or in pre-clinical trials (Kamiya et al 2010; Kneteman et al 2006; Kneteman et al 2009; Vanwolleghem et al 2007). All these molecules have demonstrated antiviral effect against HCV. Moreover the model has permitted to describe cardio-toxicity of BILN2061, confirming the perfect suitability of humanized mice for antiviral therapy evaluation. This model can also been used to study susceptibility of different viral recombinant strains to actual treatments. This will allow strategies from "bench to bedside" to design specific treatment for each patient (Kurbanov et al 2008).

As for HBV, the design of inhibitors targeting several steps of HCV replication is the key to treat patients. In addition to protease and polymerase inhibitors, some groups tried to target viral entry. Meuleman et al have demonstrated the ability of antibodies directed against cellular surface molecules (CD81 and SR-B1) involved in virus entry to protect human hepatocytes from HCV infection (Meuleman et al 2008; Meuleman et al 2011a; Meuleman et al 2011b). Matsumura et al have shown that amphipatic DNA polymers inhibited HCV post-binding stage and thus blocked de novo infection (Matsumura et al 2009)

It is clear that the immune response to viral infection plays a major role in the outcome of liver disease during HCV infection. To study the involvement of the innate immune system against viral infection, Walters et al., used the immunotolerant Alb-uPA/SCID mouse model to analyze transcriptome profiles of HCV infected versus non infected mice (Walters et al 2006). Globally, in the Alb-uPA/SCID mouse model, HCV infection activates the transcription of interferon-stimulated genes which are in particular implicated in establishing the innate immune response, and thus active in the inhibition of HCV replication. Moreover, and as previously shown in HCV-infected patients and HCV transgenic mice, these authors confirmed in the Alb-uPA/SCID mouse model the relationship between severe HCV infection and perturbation of lipid metabolism (Joyce et al

2009). These observations strongly suggest that liver disease may not be mediated exclusively by an HCV-specific adaptive immune response. Thus, the innate immune response may play a fundamental role in the pathogenesis of HBV and HCV infection.

Infection by the Plasmodium falciparum parasite is restricted to human and closely related species. As for the HBV and HCV viruses, ethical and financial reasons limit the use of non human primates to study this pathogen. Numerous studies have used a humanized mouse that carries human erythrocytes (Moreno et al 2007). Sporozoites, the hepatic stage of the pathogen [for review in the viral cycle of Plasmodium falciparum see (Greenwood et al 2008)] were used to infect chimeric liver of humanized Alb-uPA/SCID mice. The authors demonstrated that the reduction of the innate immune response by anti-macrophage and anti-NK cell treatments both enhance the humanization of Alb-uPA/SCID mice and allowed the infection of human hepatocytes by sporozoites as well as the maturation of the pathogen (Morosan et al 2006). This new model should permit the evaluation of drugs directed specifically against the hepatic stage of the infection. This model has also been used to study biology of P. falciparum, using genetically knock out for Liver-stage antigen-1 parasite, that could show that LSA-1 plays a critical role during late liver-stage schizogony and is thus important in the parasite transition from the liver to blood (Mikolajczak et al 2011). Moreover, this constitutes a starting point to create a future humanized model allowing study of the entire parasite cycle.

4. Concluding remarks

The recent development of small mouse models for experimental HBV, HCV or Plasmodium falciparum infection has opened new perspectives for the evaluation of novel therapeutic and/or prophylactic compounds against these pathogens. These models are physiologically relevant, in that they are based on the transplantation of primary hepatocytes. However, to integrate humanized mouse technology into development process, the technology must be accessible, reproducible and at a reasonable cost. Indeed, both mouse models are relatively complicated to use, but they present the unquestionable advantage of being much less expensive and easier to maintain and breed than primates.

The present challenge is the construction of mice combining human immune and liver cells. Mice with humanized immune systems already represent the model of choice for various lymphotropic pathogens. The addition of human hepatic tissue holds promise for the study of hepatotropic pathogens. Indeed, this will help to understand how hepatotropic pathogens are detected by the immune system, why the majority of individuals fail to mount an effective response, the factors involved in chronic viral persistence versus resolution of infection. A recent report Washburn et al have developed a specific mouse model, humanized with human immune system and liver tissues (Washburn et al 2011). These mice generate a specific immune response against the HCV and seem to develop liver diseases.

5. References

Azuma, H., Paulk, N., Ranade, A., Dorrell, C., Al-Dhalimy, M. , et al. (2007) Robust expansion of human hepatocytes in Fah-/-/Rag2-/-/Il2rg-/- mice. *Nat Biotechnol* 25: pp. 903-910.

Bissig, K.D., Le, T.T., Woods, N.B., Verma, I.M. (2007) Repopulation of adult and neonatal mice with human hepatocytes: a chimeric animal model. *Proc Natl Acad Sci U S A* 104: pp. 20507-20511.

Brezillon, N., Brunelle, M.-N., Massinet, H., Giang, E., Lamant, C., DaSilva, L., Berissi, S., Belghiti, J., Hannoun, L., Puerstinger, G., Wimmer, E., Neyts, J., Hantz, O., Soussan, P., Morosan, S., Kremsdorf, D. (2011) Antiviral Activity of Bay 41-4109 on Hepatitis B Virus in Humanized Alb-uPA/SCID Mice. *PLoS ONE*. Online.

Dandri, M., Burda, M.R., Gocht, A., Torok, E., Pollok, J.M. , et al. (2001a) Woodchuck hepatocytes remain permissive for hepadnavirus infection and mouse liver repopulation after cryopreservation. *Hepatology* 34: pp. 824-833.

Dandri, M., Burda, M.R., Torok, E., Pollok, J.M., Iwanska, A. , et al. (2001b) Repopulation of mouse liver with human hepatocytes and in vivo infection with hepatitis B virus. *Hepatology* 33: pp. 981-988.

Dandri, M., Burda, M.R., Zuckerman, D.M., Wursthorn, K., Matschl, U. , et al. (2005) Chronic infection with hepatitis B viruses and antiviral drug evaluation in uPA mice after liver repopulation with tupaia hepatocytes. *J Hepatol* 42: pp. 54-60.

Dandri, M., Volz, T.K., Lutgehetmann, M., Petersen, J. (2005) Animal models for the study of HBV replication and its variants. *J Clin Virol* 34 Suppl 1: pp. S54-62.

Fournier, C., Sureau, C., Coste, J., Ducos, J., Pageaux, G. , et al. (1998) In vitro infection of adult normal human hepatocytes in primary culture by hepatitis C virus. *J Gen Virol* 79 (Pt 10): pp. 2367-2374.

Greenwood, B.M., Fidock, D.A., Kyle, D.E., Kappe, S.H., Alonso, P.L. , et al. (2008) Malaria: progress, perils, and prospects for eradication. *J Clin Invest* 118: pp. 1266-1276.

Gripon, P., Diot, C., Theze, N., Fourel, I., Loreal, O. , et al. (1988) Hepatitis B virus infection of adult human hepatocytes cultured in the presence of dimethyl sulfoxide. *J Virol* 62: pp. 4136-4143.

Grompe, M., al-Dhalimy, M., Finegold, M., Ou, C.N., Burlingame, T. , et al. (1993) Loss of fumarylacetoacetate hydrolase is responsible for the neonatal hepatic dysfunction phenotype of lethal albino mice. *Genes Dev* 7: pp. 2298-2307.

Grompe, M., Lindstedt, S., al-Dhalimy, M., Kennaway, N.G., Papaconstantinou, J. , et al. (1995) Pharmacological correction of neonatal lethal hepatic dysfunction in a murine model of hereditary tyrosinaemia type I. *Nat Genet* 10: pp. 453-460.

He, Z., Zhang, H., Zhang, X., Xie, D., Chen, Y. , et al. (2010) Liver xeno-repopulation with human hepatocytes in Fah-/-Rag2-/- mice after pharmacological immunosuppression. *Am J Pathol* 177: pp. 1311-1319.

Heckel, J.L., Sandgren, E.P., Degen, J.L., Palmiter, R.D., Brinster, R.L. (1990) Neonatal bleeding in transgenic mice expressing urokinase-type plasminogen activator. *Cell* 62: pp. 447-456.

Joyce, M.A., Walters, K.A., Lamb, S.E., Yeh, M.M., Zhu, L.F. , et al. (2009) HCV induces oxidative and ER stress, and sensitizes infected cells to apoptosis in SCID/Alb-uPA mice. *PLoS Pathog* 5: pp. e1000291.

Kamiya, N., Iwao, E., Hiraga, N., Tsuge, M., Imamura, M. , et al. (2010) Practical evaluation of a mouse with chimeric human liver model for hepatitis C virus infection using an NS3-4A protease inhibitor. *J Gen Virol* 91: pp. 1668-1677.

Katoh, M., Sawada, T., Soeno, Y., Nakajima, M., Tateno, C. , et al. (2007) In vivo drug metabolism model for human cytochrome P450 enzyme using chimeric mice with humanized liver. *J Pharm Sci* 96: pp. 428-437.

Kneteman, N.M., Howe, A.Y., Gao, T., Lewis, J., Pevear, D. , et al. (2009) HCV796: A selective nonstructural protein 5B polymerase inhibitor with potent anti-hepatitis C virus activity in vitro, in mice with chimeric human livers, and in humans infected with hepatitis C virus. *Hepatology* 49: pp. 745-752.
Kneteman, N.M., Weiner, A.J., O'Connell, J., Collett, M., Gao, T. , et al. (2006) Anti-HCV therapies in chimeric scid-Alb/uPA mice parallel outcomes in human clinical application. *Hepatology* 43: pp. 1346-1353.
Kremsdorf, D., Brezillon, N. (2007) New animal models for hepatitis C viral infection and pathogenesis studies. *World J Gastroenterol* 13: pp. 2427-2435.
Kurbanov, F., Tanaka, Y., Chub, E., Maruyama, I., Azlarova, A. , et al. (2008) Molecular epidemiology and interferon susceptibility of the natural recombinant hepatitis C virus strain RF1_2k/1b. *J Infect Dis* 198: pp. 1448-1456.
Lindstedt, S., Holme, E., Lock, E.A., Hjalmarson, O., Strandvik, B. (1992) Treatment of hereditary tyrosinaemia type I by inhibition of 4-hydroxyphenylpyruvate dioxygenase. *Lancet* 340: pp. 813-817.
Lutgehetmann, M., Bornscheuer, T., Volz, T., Allweiss, L., Bockmann, J.H. , et al. (2011) Hepatitis B virus limits response of human hepatocytes to interferon-alpha in chimeric mice. *Gastroenterology* 140: pp. 2074-2083, 2083 e2071-2072.
Lutgehetmann, M., Volz, T., Kopke, A., Broja, T., Tigges, E. , et al. (2010) In vivo proliferation of hepadnavirus-infected hepatocytes induces loss of covalently closed circular DNA in mice. *Hepatology* 52: pp. 16-24.
Matsumura, T., Hu, Z., Kato, T., Dreux, M., Zhang, Y.Y. , et al. (2009) Amphipathic DNA polymers inhibit hepatitis C virus infection by blocking viral entry. *Gastroenterology* 137: pp. 673-681.
Mazier, D., Beaudoin, R.L., Mellouk, S., Druilhe, P., Texier, B. , et al. (1985) Complete development of hepatic stages of Plasmodium falciparum in vitro. *Science* 227: pp. 440-442.
Meuleman, P., Albecka, A., Belouzard, S., Vercauteren, K., Verhoye, L. , et al. (2011a) Griffithsin Has Antiviral Activity against Hepatitis C Virus. *Antimicrob Agents Chemother* 55: pp. 5159-5167.
Meuleman, P., Catanese, M.T., Verhoye, L., Desombere, I., Farhoudi, A. , et al. (2011b) A human monoclonal antibody targeting SR-BI precludes hepatitis C virus infection and viral spread in vitro and in vivo. *Hepatology*.
Meuleman, P., Hesselgesser, J., Paulson, M., Vanwolleghem, T., Desombere, I. , et al. (2008) Anti-CD81 antibodies can prevent a hepatitis C virus infection in vivo. *Hepatology* 48: pp. 1761-1768.
Meuleman, P., Libbrecht, L., De Vos, R., de Hemptinne, B., Gevaert, K. , et al. (2005) Morphological and biochemical characterization of a human liver in a uPA-SCID mouse chimera. *Hepatology* 41: pp. 847-856.
Meuleman, P., Libbrecht, L., Wieland, S., De Vos, R., Habib, N. , et al. (2006) Immune suppression uncovers endogenous cytopathic effects of the hepatitis B virus. *J Virol* 80: pp. 2797-2807.
Mikolajczak, S.A., Sacci, J.B., Jr., De La Vega, P., Camargo, N., VanBuskirk, K. , et al. (2011) Disruption of the Plasmodium falciparum liver-stage antigen-1 locus causes a differentiation defect in late liver-stage parasites. *Cell Microbiol* 13: pp. 1250-1260.
Moreno, A., Perignon, J.L., Morosan, S., Mazier, D., Benito, A. (2007) Plasmodium falciparum-infected mice: more than a tour de force. *Trends Parasitol* 23: pp. 254-259.

Morosan, S., Hez-Deroubaix, S., Lunel, F., Renia, L., Giannini, C. , et al. (2006) Liver-stage development of Plasmodium falciparum, in a humanized mouse model. *J Infect Dis* 193: pp. 996-1004.

Overturf, K., Al-Dhalimy, M., Tanguay, R., Brantly, M., Ou, C.N. , et al. (1996) Hepatocytes corrected by gene therapy are selected in vivo in a murine model of hereditary tyrosinaemia type I. *Nat Genet* 12: pp. 266-273.

Petersen, J., Dandri, M., Gupta, S., Rogler, C.E. (1998) Liver repopulation with xenogenic hepatocytes in B and T cell-deficient mice leads to chronic hepadnavirus infection and clonal growth of hepatocellular carcinoma. *Proc Natl Acad Sci U S A* 95: pp. 310-315.

Petersen, J., Dandri, M., Mier, W., Lutgehetmann, M., Volz, T. , et al. (2008) Prevention of hepatitis B virus infection in vivo by entry inhibitors derived from the large envelope protein. *Nat Biotechnol* 26: pp. 335-341.

Rhim, J.A., Sandgren, E.P., Degen, J.L., Palmiter, R.D., Brinster, R.L. (1994) Replacement of diseased mouse liver by hepatic cell transplantation. *Science* 263: pp. 1149-1152.

Rhim, J.A., Sandgren, E.P., Palmiter, R.D., Brinster, R.L. (1995) Complete reconstitution of mouse liver with xenogeneic hepatocytes. *Proc Natl Acad Sci U S A* 92: pp. 4942-4946.

Sandgren, E.P., Palmiter, R.D., Heckel, J.L., Daugherty, C.C., Brinster, R.L. , et al. (1991) Complete hepatic regeneration after somatic deletion of an albumin-plasminogen activator transgene. *Cell* 66: pp. 245-256.

Strom, S.C., Davila, J., Grompe, M. (2010) Chimeric mice with humanized liver: tools for the study of drug metabolism, excretion, and toxicity. *Methods Mol Biol* 640: pp. 491-509.

Sugiyama, M., Tanaka, Y., Kato, T., Orito, E., Ito, K. , et al. (2006) Influence of hepatitis B virus genotypes on the intra- and extracellular expression of viral DNA and antigens. *Hepatology* 44: pp. 915-924.

Tanaka, Y., Sanchez, L.V., Sugiyama, M., Sakamoto, T., Kurbanov, F. , et al. (2008) Characteristics of hepatitis B virus genotype G coinfected with genotype H in chimeric mice carrying human hepatocytes. *Virology* 376: pp. 408-415.

Tateno, C., Yoshizane, Y., Saito, N., Kataoka, M., Utoh, R. , et al. (2004) Near completely humanized liver in mice shows human-type metabolic responses to drugs. *Am J Pathol* 165: pp. 901-912.

Tsuge, M., Hiraga, N., Takaishi, H., Noguchi, C., Oga, H. , et al. (2005) Infection of human hepatocyte chimeric mouse with genetically engineered hepatitis B virus. *Hepatology* 42: pp. 1046-1054.

Vanwolleghem, T., Meuleman, P., Libbrecht, L., Roskams, T., De Vos, R. , et al. (2007) Ultra-rapid cardiotoxicity of the hepatitis C virus protease inhibitor BILN 2061 in the urokinase-type plasminogen activator mouse. *Gastroenterology* 133: pp. 1144-1155.

Walters, K.A., Joyce, M.A., Thompson, J.C., Proll, S., Wallace, J. , et al. (2006) Application of functional genomics to the chimeric mouse model of HCV infection: optimization of microarray protocols and genomics analysis. *Virol J* 3: pp. 37.

Washburn, M.L., Bility, M.T., Zhang, L., Kovalev, G.I., Buntzman, A. , et al. (2011) A humanized mouse model to study hepatitis C virus infection, immune response, and liver disease. *Gastroenterology* 140: pp. 1334-1344.

9

Liver Parenchyma Regeneration in Connection with Extended Surgical Procedure – Experiment on Large Animal

Vaclav Liska et al.[1,*]
[1]*Department of Surgery*
Teaching Hospital and Medical School Pilsen, Charles University Prague, Czech Republic

1. Introduction

Liver surgery underwent enormous evolution after development and introduction of new technical skills in surgical praxis. Nevertheless many patients with primary or secondary liver malignancies are not indicated to radical surgical therapy that could reach complete remission of malignant disease because the frontiers of liver surgery are limited today by the functional reserves of remnant parenchyma. The main argument to non surgical treatment is increased risk of acute liver failure after extended liver resection, where retained liver parenchyma is to small to sustain the liver functions (Abdalla, 2001). Portal vein embolization (PVE) can multiply the future liver remnant volume (FLRV) in spite of affection of only one of liver lobes by malignant diseases (Makuuchi, 1984, Makuuchi, 1990, Harada, 1997). This procedure was performed firstly in 1984 by Makuuchi (Abdalla, 2001, Makuuchi, 1984, Makuuchi, 1990). PVE of portal branch of with malignancy afflicted liver lobe initiates compensatory hypertrophy of contralateral non-occluded lobe. The occluded lobe underlies atrophy. The compensatory hypertrophy is supposed to be stimulated by increased flow of portal blood, that contains hepatotrophic substances (Kusaka, 2004, Azoulay, 2000). Liver resection after PVE is performed only in 63-96% of patients (Kokudo, 200, Stefano, 2005, Lagasse, 2000). The main reason for this resolution is unsuccessful hypertrophy of FLRV or progression of malignancy. Liver resection after PVE is performed only in 63-96% of patients (Azoulay, 200, Kokudo, 2001, Stefano, 2005). The main reason for this resolution are unsuccessful hypertrophy of FLRV or progression of malignancy.

* Vladislav Treska[1], Hynek Mirka[2], Ondrej Vycital[1], Jan Bruha[1], Pavel Pitule[1], Jana Kopalova[1], Tomas Skalicky[1], Alan Sutnar[1], Jan Benes[3], Jiri Kobr[4], Alena Chlumska[5], Jaroslav Racek[6] and Ladislav Trefil[6]
[1]*Department of Surgery,*
[2]*Department of Radiology,*
[3]*Department of Anaesthesiology and Resuscitation,*
[4]*Department of Pediatrics,*
[5]*Institute of Pathology,*
[6]*Institute of Biochemistry and Haemathology,*
Teaching Hospital and Medical School Pilsen, Charles University Prague, Czech Republic

The proper regeneration of liver parenchyma depends on proliferation of parenchymal and non-parenchymal liver cells. The importance of stem cells or liver oval cells is still under discussion and this mechanism provokes many questions (Lagasse, 2000, Petersen, 1999, Vassilopoulos, 2003). After partial hepatectomy or portal vein ligation increase serum levels of Tumour necrosis factor alpha (TNF-α) and Interleukin-6 (IL-6), which were demonstrated to be involved in priming of hepatocytes and trigger them from G0 to G1 cell cycle phase(Cornell, 1990). These cytokines induce gene activation, which are responsible for G1 phase. Both pleiotrophic cytokines are secreted by non-parenchymal liver cells (mostly Kupffer cells) (Fausto, 2000, Fausto, 2005). TNF-α is superior to IL-6 and stimulates increased secretion of IL-6. In hepatectomized regenerating liver, it is known that this signaling pathway follows the sequence TNF-α → TNFR-1 → NFκB → IL-6 → STAT3 (Michalopoulos, 1997). The proliferation of primed hepatocytes is regulated positively by Hepatocyte growth factor (HGF), Epidermal growth factor (EGF), Insulin-like growth factor (IGF), Transforming growth factor-alpha (TGF-α), etc. The termination of proliferation and stimulation of hepatocyte to differentiation, final remodelation of liver tissue and production of extracellular matrix is controled by Transforming growth factor-beta (TGF-β) (Fukuhara, 2003, Mangnall, 2003, Zimmermann, 2004). TGF-β1 plays the most important role in starting the remodelling of the extracellular matrix and restoration of the original structure of the liver parenchyma. TGF-β1 inhibits DNA synthesis and plays a pivotal role in the down-regulation of liver regeneration as has been demonstrated in toxic models of liver regeneration (Armendariz-Borunda, 1993, Armendariz-Borunda, 1997). TGF-β1 also down-regulates the production of the Hepatocyte growth factor that sustains hepatocyte proliferation (Bustos, 2000). Increased expression of TGF-β1 prevents uncontrolled growth during liver regeneration by the regulation of hepatocyte transition from the G1 to the S phase of the cell cycle (Kusaka, 2006, Oe, 2004). TGF-β1 helps maintain the differentiation of hepatocytes and non-parenchymal liver cells. The proposed HGF/ TGF-β1 ratio could reflect the proliferation/differentiation status of hepatocytes (Lilja, 1999). Increased expression of TGF-β1 was also shown to be a crucial factor for the progression of hepatic fibrosis (Friedman, 2008). This could be explained by increased or prolonged production of the liver extracellular matrix (Viebahn, 2008).

The replication of hepatocytes culminates on 7th days (14per cent of hepatocytes) and the return to quiscent status was observed on 12th day after PVE in swine experimental model (Coelho, 2007, Duncan, 1999). The differencies between PVE and PVL (portal vein ligation) were not proved as statistical significant for reached FLRV (Broering, 2002).

Multipotent mesenchymal stromal cells (MSC) are a fraction of the adult bone marrow stem cell compartment. They contain a subpopulation of mesenchymal stem cells that can differentiate into mesenchymal adult tissues under specified conditions. Differentiation into adipocytes, chondrocytes, osteoblasts and myocytes has been demonstrated in vitro and in vivo (Barry, 2003). Whether and how bone marrow-derived stem cells, and MSC in particular, can contribute to liver regeneration is not entirely clear. Initital reports outlined that bone marrow-derived (stem) cells can transdifferentiate into liver cells or their progenitors after bone marrow transplantation (Jiang, 2002, Lagasse, 2000, Petersen, 1999, Ringe, 2002). Later investigations clarified and complemented these observations, outlining that bone marrow cells can also fuse with resident liver cells (Vassilopoulos, 2003, Wang, 2003). In addition to direct cellular effects, transplanted bone marrow cells can also contribute to liver regeneration by bystander effects, such as the provision of a beneficial proliferative cytokine milieu or antiapoptotic effects (Aldeguer, 2002, Dahlke, 2004, Liska, 2009).

The aim of presented studies was to influence regeneration of liver parenchyma after portal vein embolization/ligation by exogenous cytokines, growth factors and monoclonal antibodies against growth factors with inhibitory funcions or syngeneic Multipotent Mesenchymal Stromal Cells.

2. Methods

2.1 Surgical procedure

All described procedures were prepared and performed after by law of Czech Republic, which is compatible with legislature of European Union. The experimental animal was piglet. The experimental porcine model was established to be as much compatible as possible with PVE/PVL in human medicine.

There were no statistical differencies in weight and age of the piglets undergoing portal vein ligation. The animals were housed under same conditions. In this study there were included 9 piglets in control group, 9 piglets in TNF-α group, 8 piglets in IL-6 group, 6 piglets in MSC group and 7 piglets in MAB TGF-β1 group.

The piglets were premedicated intramuscularly with atropine 1,5 mg and azaperon 1,0 mg/kg. The anesthesia was administered continually through central venous catheter in whole average doses: azaperon 1,0 mg/kg/hour, thiopental 10 mg/kg/hour, ketamin 5-10 mg/kg/hour and fentanyl 1-2 ug/kg/hour. The muscle relaxation was provided by bolus administration of pancuronium 0,1-0,2 mg/kg at the begin of surgery. Animals were intubated and mechanically ventilated during surgical procedure. The monitoring of electrocardiogram, oxygen saturation and central venous pressure was performed. The surgical procedure was performed under aseptic and antiseptic conditions. The antibiotic prophylaxis was administered in total dosis of 1,2 g amoxicillin with clavulanic acid divided into two doses (before surgery and two hours later).

The middle laparotomy was performed. The portal vein branches for caudate, right lateral and right medial lobes (50-60 per cent of supposed liver parenchyma) were prepared and ligated without injury or ligation of hepatic artery branches. The blood flow in hepatic artery branches and occlusion of portal vein branches were controled by Doppler ultrasonography (Medison Sonoace 9900, linear probe with frequency 7,5 MHz, Fig.1). The borders between atrophic and hypertrophic liver lobes were marked by titanium staples to simplify the postoperative ultrasonography.

The recombinant porcine TNF-α in amount 5μg/kilos (rpTNF-α, ProSpec TechnoGene, Israel, 9 piglets from TNF-α group), the recombinant porcine IL-6 in amount 0,5μg/kg (rpIL-6, ProSpec TechnoGene, Israel, 8 piglets from Il-6 group), MSC (autologous bone marrow stem cells cultured in expansion medium (low glucose DMEM with GlutaMAX, without pyruvate, Biochrom) supplemented with 10% fetal bovine serum (FBS, Biochrom), 1% penicillin/streptomycin and 1% glutamine. After 24, 48 and 72 hours, non-adherent cells were removed by changing the culture medium. Adherent cells were then trypsinized (0.5% trypsin-EDTA), harvested and re-plated into new flasks, each time, when cell confluency reached 60% to 80%. To histologically identify transfused MSC within the recipient liver, cultured MSC were labeled with 5-bromo-2-deoxyuridine (BrdU) 60 hours before cell application. 6 piglets, the amount of applied stem cells: 8.75, 14.0, 17.0, 17.5, 43.0 and 61.0 x 10^6 MSC respectively) or physiological solution (9 piglets from control group) were applied into non-occluded portal vein branches (Liska, 2012, Liska, 2009, Liska, 2009, Teoh 2006).

The laparotomy was closed in anatomical layers. At the end of operation the port-a-cath was introduced into the superior caval vein. The animals were extubated and monitored each day for the next fourteen days with a particular emphasis on the clinical examination (attention to wound healing, infection of the port-a-cath and function of the gastrointestinal system) to diagnose possible surgical complications. The postoperative analgezia was provided by intramuscular application of small dosis of Azaperon (10mg).

Administration of MAB TGF-β1 was performed into central venous catheter 24 hours after ligation (7 piglets, 40µg/kg of body weight) (Armendariz-Borunda , 1993, Deneme, 2006, Liska, 2012).

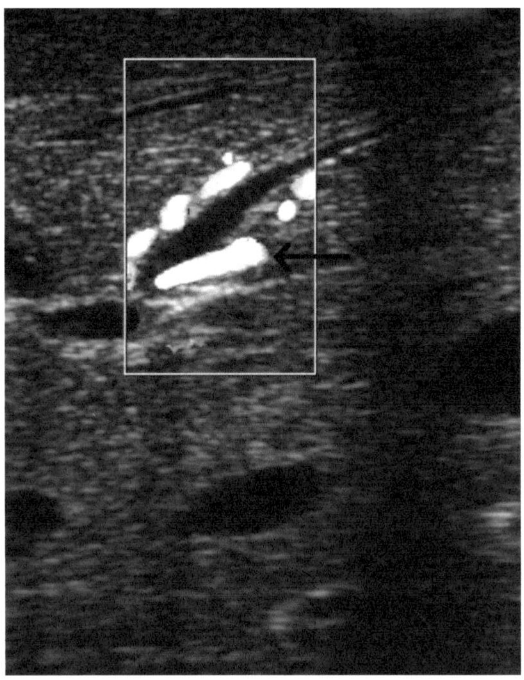

Fig. 1. Doppler ultrasonography of liver parenchyma with occluded portal vein branch and non occluded branch of hepatic artery (arrow).

2.2 Biochemistry and immunoanalysis

The blood samples were collected from central vein catheter at 1. before operation, 2. after ligation of the last portal branch, 3. at the end of operation, 4. 2 hours after operation, 5. 1th postoperative day (p.d.), 6. 3rd p.d., 7. 7th p.d., 8. 10th p.d., 9. 14th p.d.

The following biochemical parametres were estimated by biochemical analysator Olympus 2700: bilirubin, urea, creatinine, alkaline phosphatase (ALP), gamaglutamyltransferase

(GGT), cholinesterase (CHE), aspartylaminotransferase (AST), alanine aminotransferase (ALT) and albumin.

Serum level of C-reactive protein (CRP), studied cytokines (TNF-α and IL-6) and growth factors (TGF-β1, IGF) were meassured by enzyme-linked immunosorbent assay (ELISA) by Auto-EIA II Analyzer (Lasystems Oy Helsinki, Finland).

2.3 Ultrasonography

The ultrasonographic controls were undertaken immediately after operation and on 3rd, 7th, 10th and 14th postoperative day (ultrasound machine Medison Sonoace 9900, convex probe with frequency 3,5 MHz). The diameters of atrophic and hypertrophic (functional liver remnant volume) lobes were meassured in B-modus in all three basic planes (axial, sagittal and coronary). The volume of lobes is counted by using standard ultrasonographic formula, which is used also in human medicine: *axial x sagittal x coronary / 2* (Liska, 2009).

2.4 Histology

The experiment was finished on 14th day by sacrifying of animals under general anaestesia with concentrated solution of potassium chloride administered into central venous catheter.

The histological material from atrophic and hypertrophic parenchyma was examined after staining with hematoxyline-eosine, periodic acid-Schiff (PAS) staining and PAS staining after digestion of preparations with diastase. The proliferation activity was examined using antibody Ki67 (MIB 1 MW, 1:1000 DakoCytomation). We concentrated especially at the measurement of lobulus length, binucleated hepatocyte, length of hepatocytes. The amount of binucleated hepatocytes were measured in 20 microscopic fields by an eyepiece micrometer (Olympus). The size of the hepatocytes and the length of the lobuli were examined twice using the eyepiece micrometer (Olympus).

2.5 Statistical analysis

Statistical analysis was performered by sofware CRAN 2.4.0 and STATISTICA 98 Edition. The meassured parameters (biochemistry, immunoanalysis, ultrasonography) were described by basic statistical variables - mean, median, standard deviation, minimum, maximum and quartile extent. The statistical data were processed graphicaly into Box and Whisker plot diagrams. The comparison of distribution of studied parameters in groups was counted by distribution-free test (Wilcoxon). The Spearman Rank Correlation Coefficient was used because of non-Gausian distribution of parameters values. The whole development of studied parameters in time was compared between groups by parametric test ANOVA.

3. Results

3.1 Functional liver remnant volume

To assess the overall functional effect of MSC infusions, rpTNF-α group, rpIL-6 and MAB TGF-β1 after partial portal vein ligation in the porcine model, the volume of the occluded and non-occluded liver lobes was analyzed by ultrasonography. Repeated ultrasonographic assessments were carried out on p.d. 0, 3, 7, 10 and 14. The volumes of both the ligated right, and the non-ligated left liver lobes were computed by a standard three-dimensional ultrasonographic approach and were confirmed by physical examination during necropsy on day 14. The volume of the hypertrophic lobe increased more rapidly after MSC application in the treatment group when compared to the control group (Figure 2). The control group did not show any change in the volume of the hypertrophic liver lobes within the first three days after surgery. The growth acceleration of the hypertrophic liver lobes in the MSC group was at its maximum between the 3rd and 7th p.d. ($p<0.05$ versus control). However, this stimulating effect slowed down during the second week and there was no statistically significant difference in the size of the hypertrophic liver lobes on day 14. However, the average volume of the non-occluded liver lobe was increased by 30% in the MSC group in contrast to the control group at the end of the experiment (Fig. 2),(Liska, 2009).

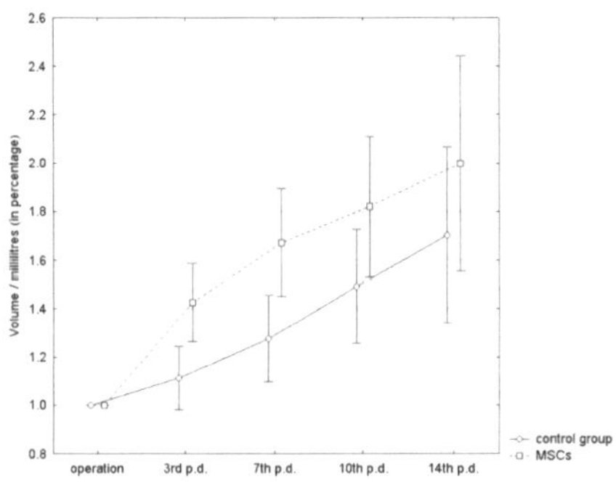

Fig. 2. Volumes of the non-ligated liver lobes are depicted for the control and the MSC-treated group. Liver volume was assessed by 3D ultrasonographic measurements.

In case of TNF-α group the absolute volume of hypertrophic lobes increased after application of TNF-α more rapidly wheareas the control group has no changes in hypertrophic liver lobes volumes in first three days. The acceleration of growth of hypertrophic liver lobes in TNF-α group was maximal on 7th p.d. in comparisson with control group ($p<0.05$), nevertheless this stimulating effect was lost on the 14th p.d. and the differencies were not statistically significant (Fig.3) ,(Liska, 2012).

The absolute volume of hypertrophic lobes grew more rapidly after application of MAB TGF-β1, whereas the control group had a slow but continual growth in the hypertrophic liver lobes during the whole follow-up period. The augmentation of growth of the hypertrophic lobes was maximal between the 3rd and 7th postoperative days in comparison with the control group ($p<0.05$). Nevertheless this growth accelerating effect was lost during the next ultrasonographic controls, and on the 14th postoperative day there were no statistically significant differences (Fig. 4),(Liska, 2012).

IL-6 group demonstrated growth of volume of hypertrophic lobes more rapidly after application of studied cytokine wheareas the control group has slow start of growth in hypertrophic liver lobes in first three days. The acceleration of growth of hypertrophic liver lobes in IL-6 group was maximal on 7th p.d. in comparisson with control group ($p<0.05$), nevertheless this stimulating effect was also lost during the follow up period and on 14th p.d there were no statistically significant differencies as at other experimental groups (Fig.5), (Liska, 2009).

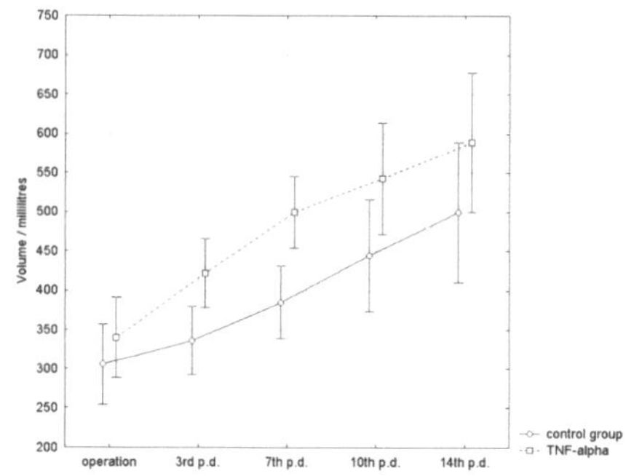

Fig. 3. Comparison of liver hypetrophy between control group and TNF-α group.

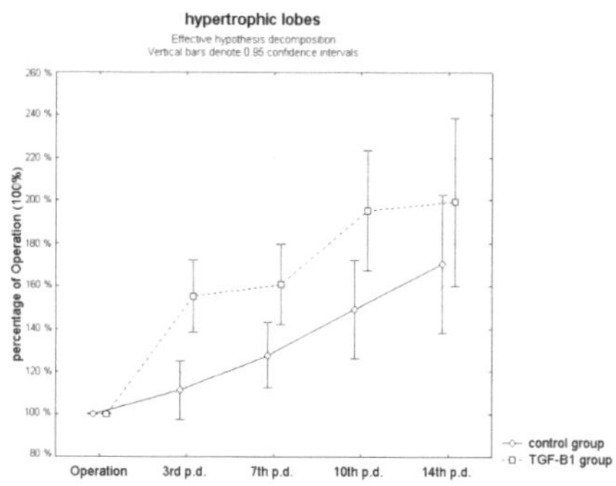

Fig. 4. Comparison of the growth of hypertrophic liver lobes between TGF-β1 and control groups.

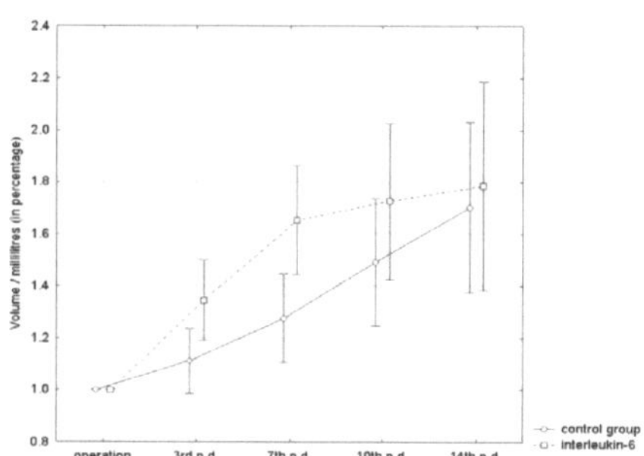

Fig. 5. Comparison of hypertrophic liver lobes growth between IL-6 group and control group.

3.2 Biochemistry

To outline the functional liver capacity after the infusion of MSC and to understand possible systemic side effects, biochemical parameters, as well as cytokine and growth factor levels were analyzed in sera of both animal groups during and after the intervention. All parameters were measured on a clinical analyzer before and during the procedure, as well as on p.d. 1, 3, 7, 10 and 14. Serum levels of AST and ALT increased after portal vein ligation, whereas serum cholestatic markers bilirubin, ALP and GGT remained stable. Synthetic liver function was unchanged with normal levels of CHE (Fig. 6). There was no difference

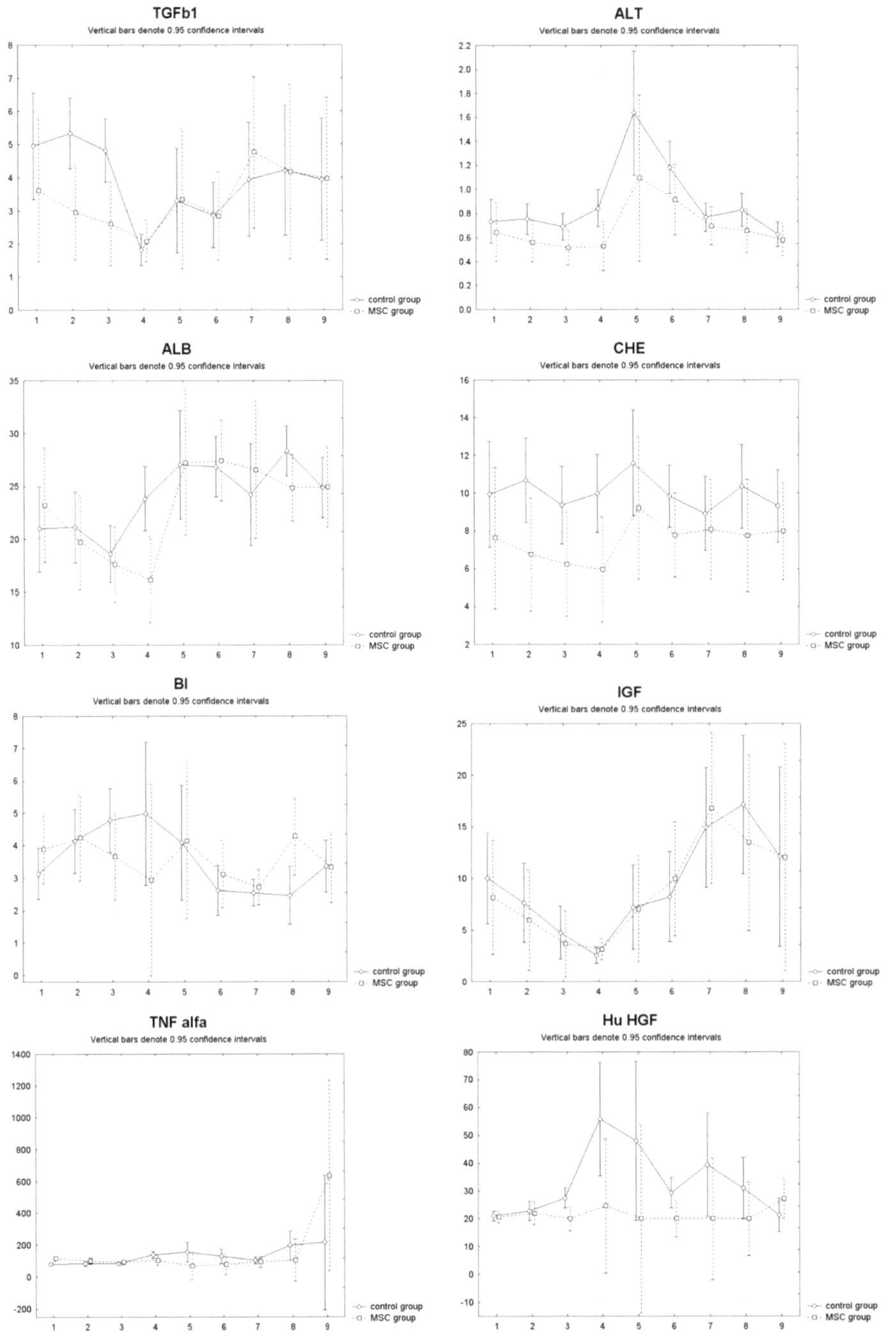

Fig. 6. Comparison of serum biochemistry during the postoperative period between MSC and control group: Blood was drawn from piglets before and during the procedure, as well as on postoperative days 1, 3, 7, 10 and 14. Liver function tests and other biochemical markers were measured on a clinical analyzer and are depicted as mean +/- standard

deviation. (ALB=serum albumin, ALT=alanine aminotransferase, CHE=cholinesterase, BI=serum bilirubin and growth factors TGFb1, IGF and HGF).

between the MSC-treated and the control group regarding these markers. Serum CRP levels were elevated to the same extent in all animals during the postoperative period. Kidney function was stable in all piglets with physiological levels of creatinine and urea (Fig.6). Overall, all treated animals of both groups tolerated the treatment well with neither apparent differences in the clinical course or the biochemistry markers assessed. To describe the regenerative micromilieu within recipient livers, the secretion of growth factors HGF, IGF, TGFβ1 and TNFα was analysed in sera of all animals over time. Both groups expressed the same pattern of TGFß1, IGF and TNFα in the periphery with a tendency towards decrease of TGFß1 and IGF in the early phase and stable TNFα. HGF expression was slightly higher in the control group early after the procedure. Nevertheless, there were no proved statistical differences between studied cytokines and growth factors in particular time points. Overall, the cytokine expression pattern in the periphery indicates that differences in the extent of postoperative liver hypertrophy between the groups are most likely due to micromilieu changes within defined anatomical spaces in the liver (Liska, 2009).

The serum levels of all studied biochemical and immunological parameteres compared between TNF-α and control group are presented in figure 7 and 8. The serum levels of all studied biochemical parameters achieved no statistically signifficant differences. The other studied serum biochemical parameteres were comparable in both experimental groups and the differences did not prove any statistical significancy.

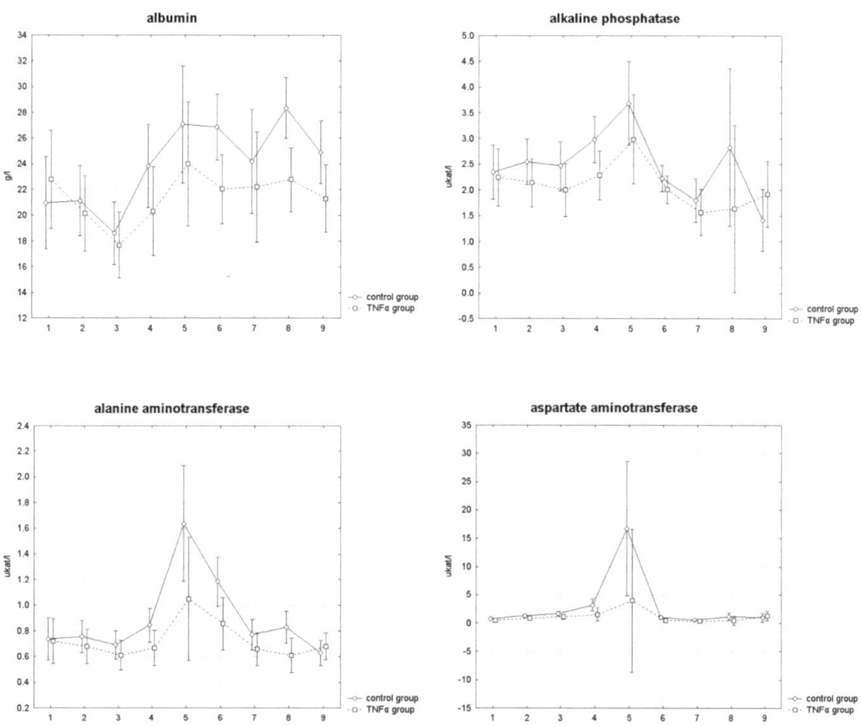

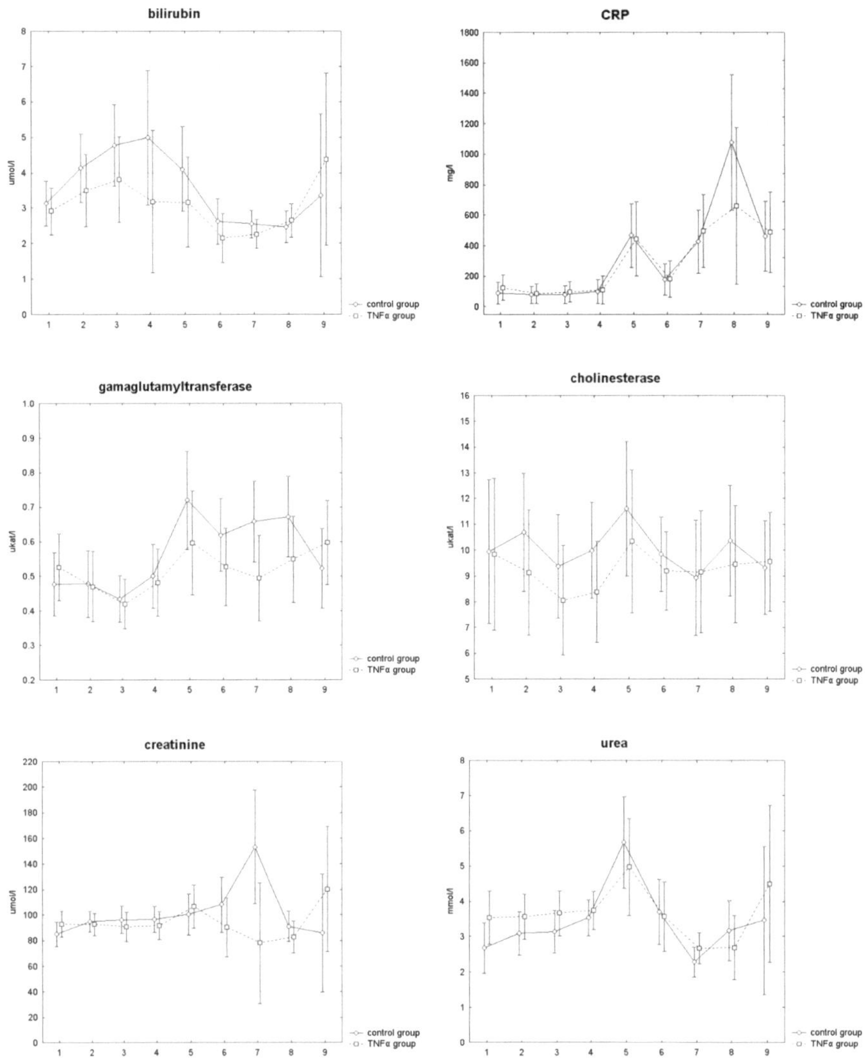

Fig. 7. Comparison of serum levels of studied biochemical parametres (albumin, alkaline phosphatase, alanine aminotransferase, aspartate aminotransferase, bilirubin, C-reactive protein, gamaglutamyltransferase, cholinesterase, creatinine and urea) between TNF-α and control groups during liver hypertrophy. (1) before operation, (2) after ligation of the last portal branch, (3) 30 minutes after partial portal vein ligation - application of TNF-α or physiological solution, (4) 2 hours after partial portal vein ligation, (5) 1st postoperative day p.d., (6) 3rd p.d., (7) 7th p.d., (8) 10th p.d., (9) 14th p.d.

The serum level of TNF-α increased after application of studied cytokine but its serum level got practically normalized on first p.d. (p-value<0.05) (fig.8). The serum levels of all other studied cytokines and growth factors expressed no differences between TNF-α and control group. We did not distingush any serum level changes of CRP (Liska, 2012).

In case of TGF-β1 group the serum levels of all the studied biochemical parameters and the progression of curves are presented in figure 9. All studied serum biochemical parameters were comparable in both experimental groups and the differences shows no statistical significance between the TGF-β1 and the control group at each timepoint (Liska, 2012).

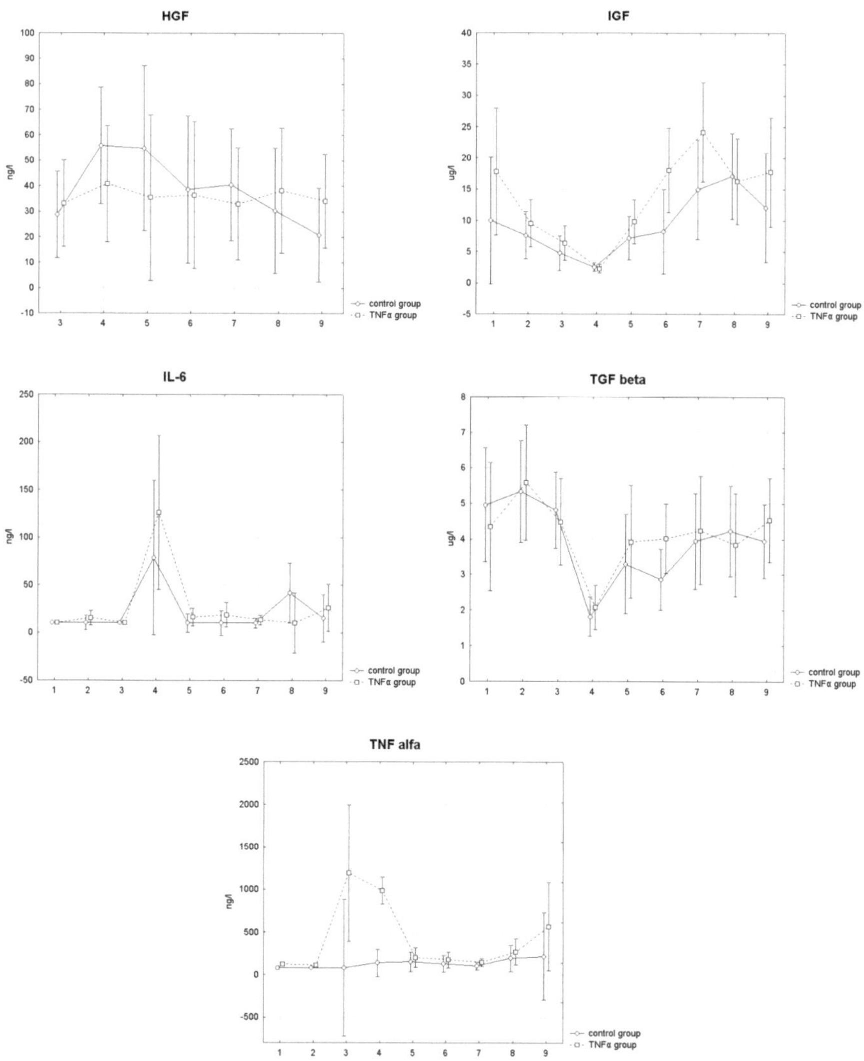

Fig. 8. Comparison of serum levels of studied cytokines and growth factors (HGF, IGF, IL-6, TGFβ-1 and TNF- α) between TNF-α and control groups during liver hypertrophy. (1) before operation, (2) after ligation of the last portal branch, (3) 30 minutes after partial portal vein ligation – application of TNF-α or physiological solution, (4) 2 hours after partial portal vein ligation, (5) 1st postoperative day p.d., (6) 3rd p.d., (7) 7th p.d., (8) 10th p.d., (9) 14th p.d.

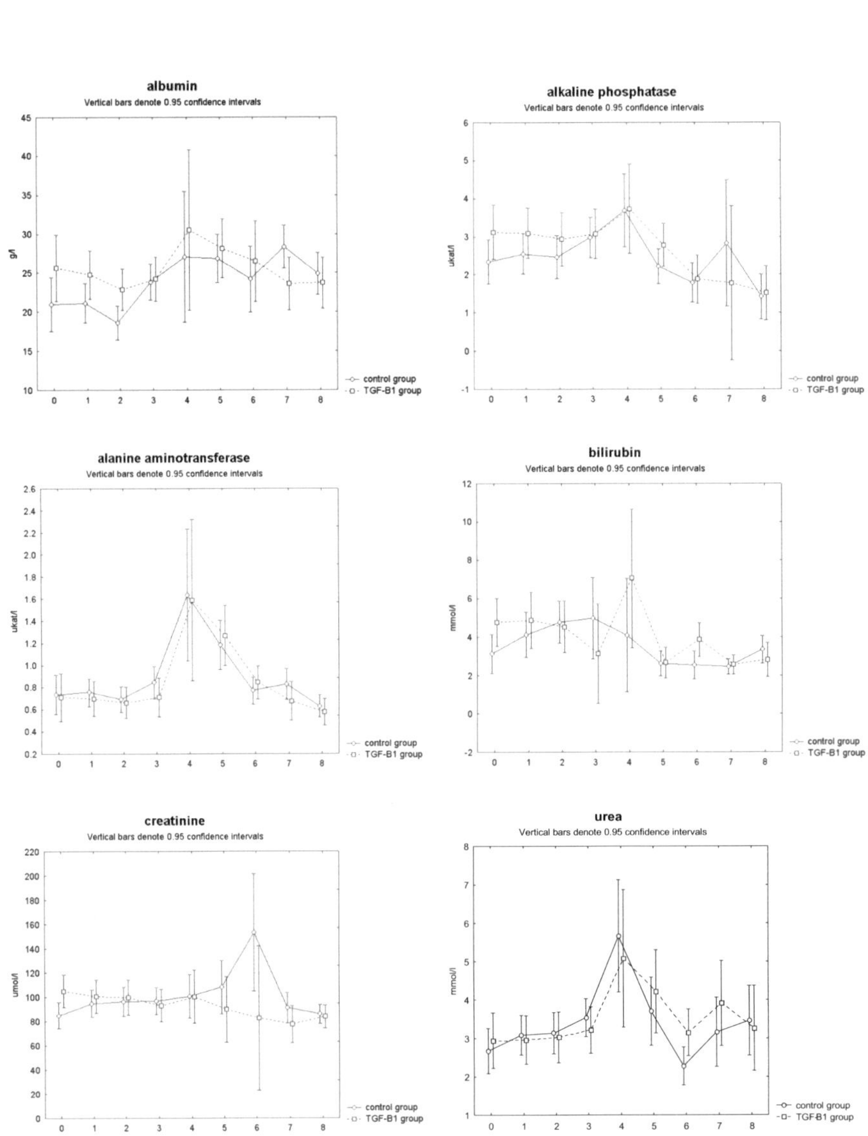

Fig. 9. Comparison of serum levels of some of the studied biochemical parameters between TGF-β1 and control groups. (1) before operation, (2) after ligation of the last portal branch, (3) 30 minutes after partial portal vein ligation, (4) 2 hours after partial portal vein ligation, (5) 1st postoperative day p.d. (before application of MAB TGF-β1 or physiological solution), (6) 3rd p.d., (7) 7th p.d., (8) 10th p.d., (9) 14th p.d.

All studied serum biochemical parameters compared between IL-6 and control groups were comparable and the differences did not prove any statistical significancy in each time point (Fig. 10). The serum levels of all studied cytokines and growth factors expressed no statistically significant differences between IL-6 and control group in each time point (Fig. 11). We did not distinguish any serum level changes of CRP (Liska, 2009).

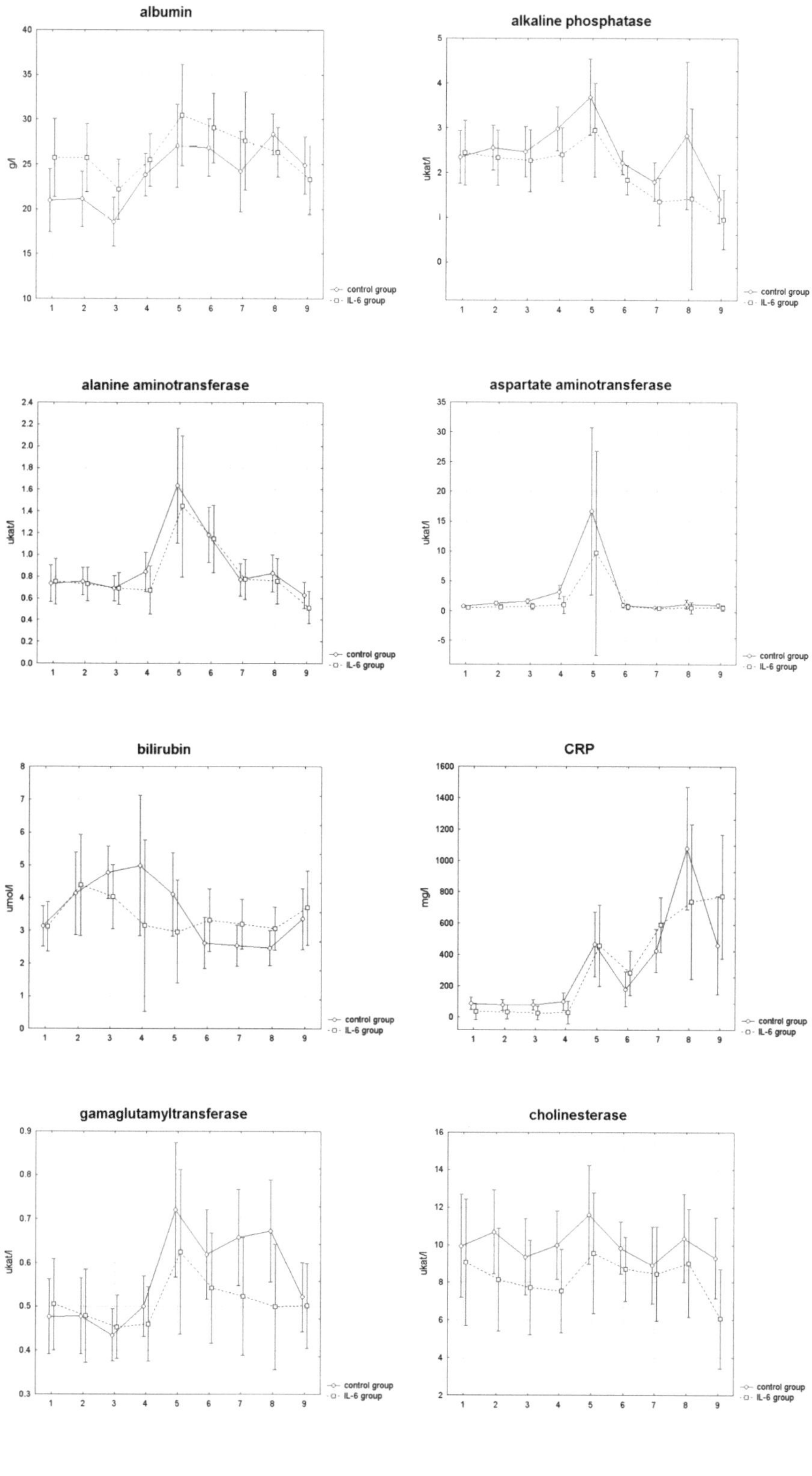
albumin
g/l
alkaline phosphatase
ukat/l
alanine aminotransferase
ukat/l
aspartate aminotransferase
ukat/l
bilirubin
umol/l
CRP
mg/l
gamaglutamyltransferase
ukat/l
cholinesterase
ukat/l
control group
IL-6 group

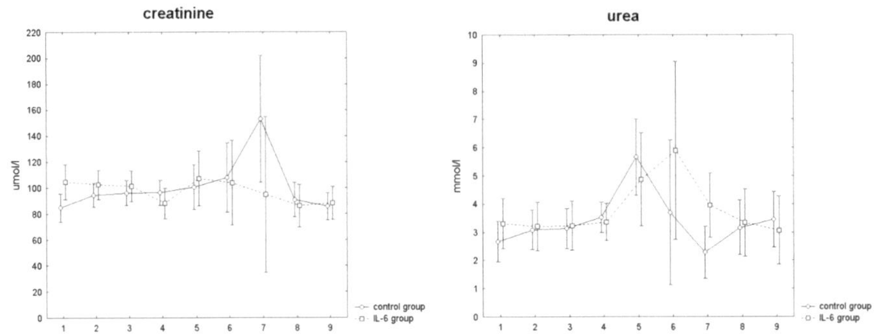

Fig. 10. Comparison of serum levels of studied biochemical parameters between IL-6 group and control group.

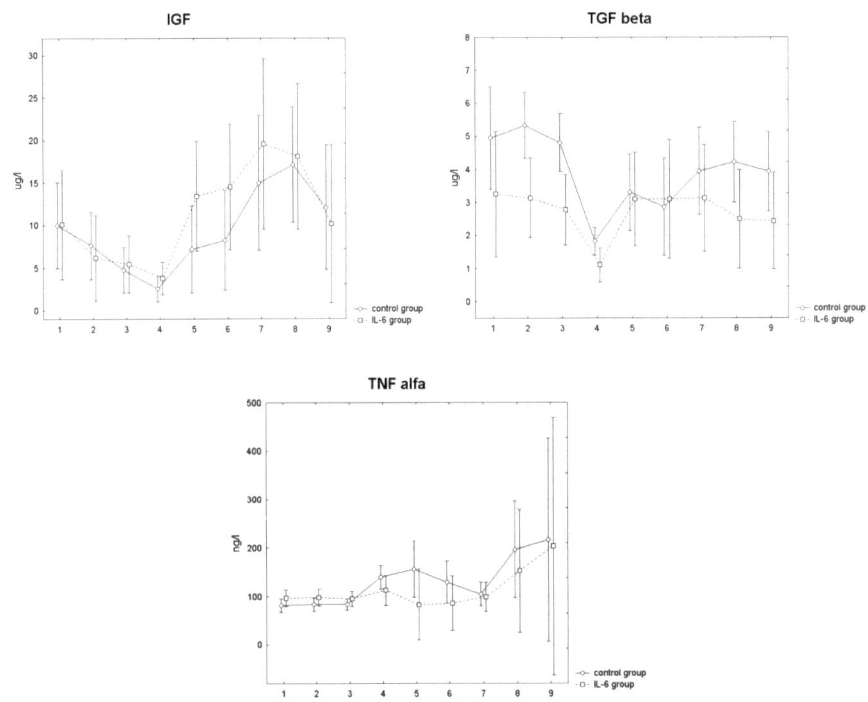

Fig. 11. Comparison of serum levels of studied cytokines and growth factors between IL-6 group and control group.

3.3 Histology and Immunohistology

Histological examination of liver biopsies from MSC and control groups was carried out to outline differences in the proliferative response and the composition and size of liver cells between the groups. Screening for a possible involvement of resident progenitor cells was also performed to try and identify the infused population of MSC. The analysis was performed at the time of necropsy on the 14th p.d. No overall difference in the length of liver lobuli was detected between the groups on H&E staining. However, a tendency towards longer lobuli was demonstrated in the MSC-treated group, potentially indicating a bystander effect of the infused MSC on the size of liver lobuli. Although lobuli length was slightly increased, the size of the individual hepatocytes did not prove to be different between the groups and the number of binucleated hepatocytes was equal in both groups. Two weeks after the intervention, the intrinsic proliferative activity was low in both groups and the Ki67 proliferative index was practically the same as in normal liver tissue.. Only solitary BrdU-stained transplanted MSC were detected within the hypertrophic liver parenchyma (Fig. 12).

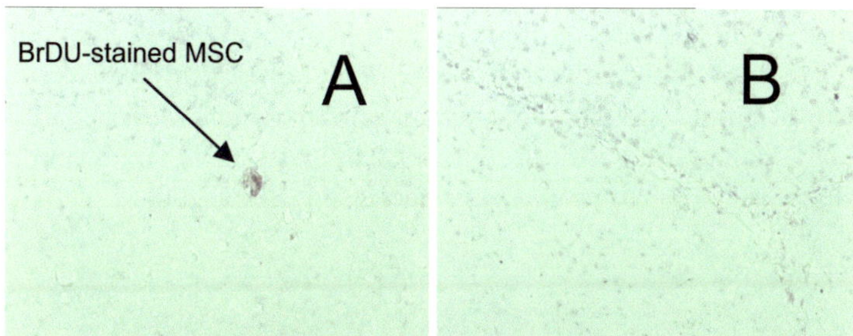

Fig. 12. Identification of infused MSC: Solitary BrdU-stained MSC could be identified in the treatment group on day 14(a). No BrdU-positive cells were identified in control animals (b).

The histological examination of biopsies from liver of TNF- α and control group was burden by the time of collection of specimen, when the proliferative phase of liver regeneration is finished factically. The differences in lobulus length were not statisticaly significant (p-value=0.08), but in the cytokine group there were not only small or normal size lobulus, but also many larger lobulus, which were not present in the control group. The statistical analysis of binucleated hepatocytes did not prove any significant differences in this studied histological parameteres. The length of hepatocytes was not proved also as statistically significant parameter. The proliferative activity in both groups was very decreased (Liska, 2012).

In case of comparison of TGF-β1 and control group the differences in length of lobuli were not statistically significant (see figure 13). Statistical analysis also showed no significant

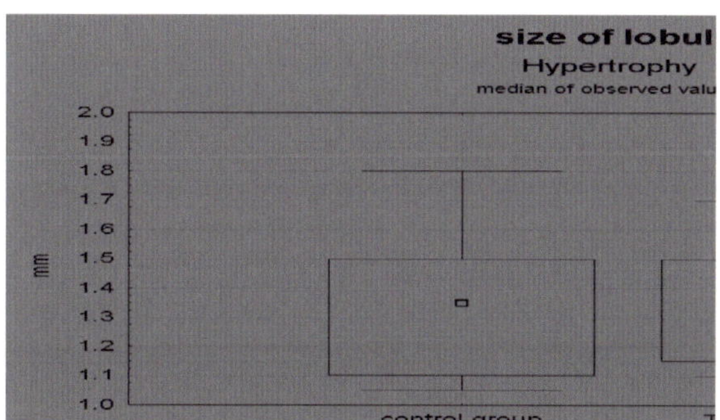

Fig. 13. Comparison of length of lobuli in hypertrophic lobes between TGF-β1 and control groups.

increase in the quantity of binucleated hepatocytes in the hypertrophic liver lobes in the TGF-β1 group (see figure 14). We observed a larger distribution of the number of binucleated hepatocytes per field in the hypertrophic lobe from TGF-β1 group (1-4 binucleated hepatocytes per field) in contrast to the control group (2-3,5 binucleated hepatocytes per field). The test of normality demonstrated a normal distribution of this parameter in each animal in both experimental groups. The size of hepatocytes was also not proved to be a statistically significant parameter for differences between the TGF-β1 group and the control group (see figure 15). Proliferative activity in both groups was greatly reduced.

In case of comparison of histological findings at IL-6 and control groups the differences in lobulus length were not statisticaly significant. The statistical analysis proved significant increased amount of binucleated hepatocytes in hypertrophic liver lobes in IL-6 group (fig. 16). The length of hepatocytes was not proved also as statistically significant parameter. The proliferative activity in both groups was very decreased (Liska, 2009).

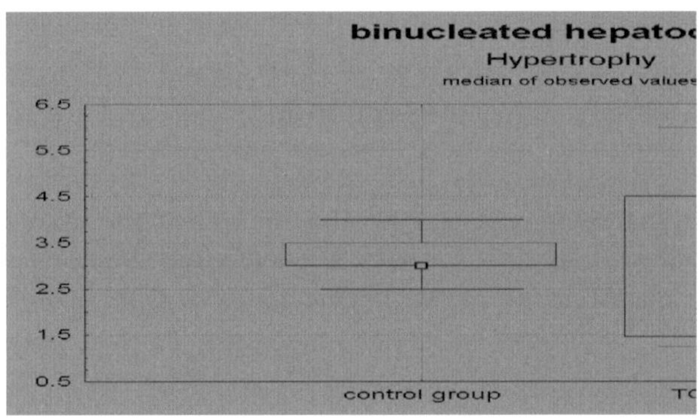

Fig. 14. Comparison of the concentration of binucleated hepatocytes in hypertrophic lobes between TGF-β1 and control groups. The binucleated hepatocytes were detected in 20 microscopic fields.

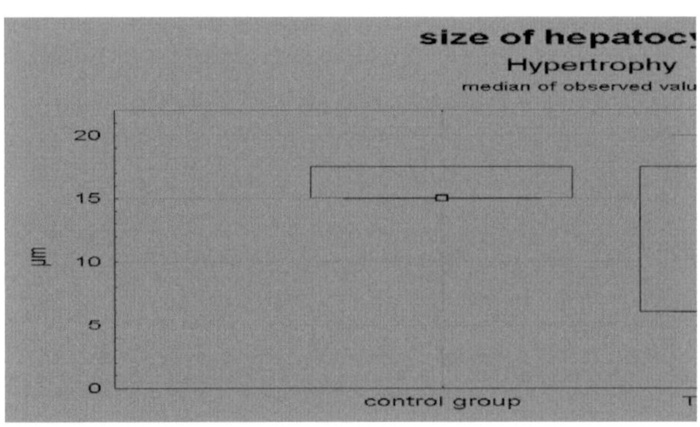

Fig. 15. Comparison of size of hepatocytes in hypertrophic lobes between TGF-β1 and control groups.

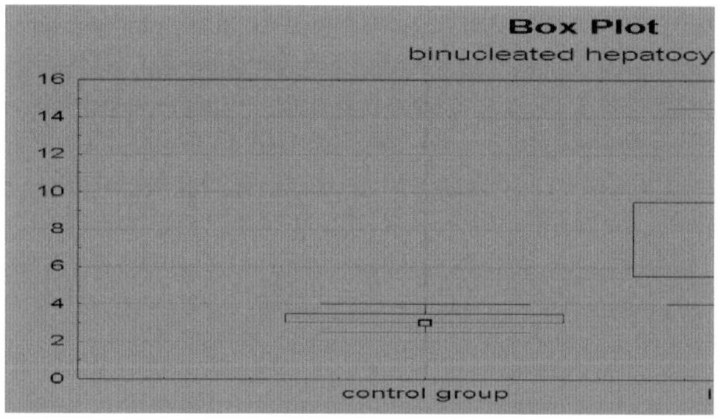

Fig. 16. Comparison of concentration of binucleated hepatocytes in hypertrophic lobes between IL-6 group and control group. The binucleated hepatocyte were detected in 20 microscopical fields.

4. Discussion

4.1 Functional liver remnant volume

In the presented chapter the authors want to review all experiences with influence of experimental portal vein ligation with cytokines, monoclonal antibody against growth factor and mesenchymal stem cells. This model closely mimics the situation of human oncological liver surgery, in which portal vein embolization can be utilized to create a larger future liver remnant volume before resection. The clinical goal of portal vein embolization is to increase the number of patients that can undergo extensive liver resections. This porcine experimental model of liver regeneration is especially applicable to the clinical situation since pigs closely resemble human liver anatomy and fysiology (Liska, 2009, Liska, 2009).

After portal venous infusion of MSC in this model, augmented and accelerated regeneration of the non-occluded liver lobes is outlined. This effect was strongest between the third and seventh postoperative day and slowly weaned off thereafter. However, there was no overall volume difference of statistical significance between both groups at the end of the observation period. This result might have been expected, though, since the body weight to liver weight ratio is tightly controlled in most species (Liska, 2009).

Previous studies have tried to clarify the potential contribution of MSC to liver regeneration in rodent models. Most of these approaches used toxic models, in which the intrinsic self-renewing capacity of the recipient liver was suppressed (Dahlke, 2003, Kang, 2005). Di Bonzo observed only 0,2% human hepatocytes after transplantation of human MSC into immunodeficient mice with chronic liver injury (Di Bonzo, 2008). Overall, whether MSC can significantly contribute to liver regeneration in vivo remains controversial (Dahlke, 2004, Lysy, 2008, Popp, 2006), although MSC can show some features of hepatic differentiation in culture or under immunoprivileged conditions (Najimi, 2007, Ong, 2006). Chamberlain observed differentiation of human MSC into hepatocytes after infusion of MSC into foetal sheep (Chamberlain, 2007) and transplantable hepatocytes were also obtained from heavily growth factor-treated MSC (Banas, 2007, Yamamoto, 2008). Despite the fact that it was not possible to identify transplanted MSC within recipient livers in this model after 14 days, liver hypertrophy was augmented. It is therefore likely that bystander effects of the transplanted

MSC accelerated regeneration in this model. This effect may be well mediated through paracrine effects of the infused MSC on the microenvironment in the periportal regions, where parenchymal regeneration is initiated. Aldeguer et al. suggested that increased production of IL-6 by bone marrow-derived cells, for example, can stimulate intrinsic liver regeneration (Aldeguer, 2002). In addition to being effective in terms of an accelerated regenerative response, the infusion of syngenic MSC into the portal vein also proved to be safe in the preclinical model. Probable side effects of MSC infusions include immunological problems (hypersensitivity, immune complex reactions), metabolic dysregulation and emboli. None of these problems occured in the presented study cohort (Liska, 2009).

The acceleration of growth of hypertrophic liver lobes after application of TNF-α confirmed results gained at in vitro models and in experiments in small laboratory animals (Cornell, 1990). The selected concentration of applied cytokine initiated acceleration of regeneration of liver parenchyma in non-occluded liver lobes (Heinrich, 2006, Teoh, 2006).

The acceleration of growth of hypertrophic liver lobes after application of IL-6 confirmed results gained at in vitro models and in experiments in small laboratory animals (Cornell,1990, Fukuhara, 2003). The selected concentration of applied cytokine initiated acceleration of regeneration of liver parenchyma in non-occluded liver lobes (Baier,2006, Heinrich, 2006).

We presented possibilities of the application of extrinsic MAB TGF-β1 to increase the required future remnant liver volume after partial portal vein ligation. The absolute volume of hypertrophic lobes (left lateral and medial lobes) grew more rapidly after application of MAB TGF-β1, whereas the control group had a slow but continual growth in the hypertrophic liver lobes during the whole follow-up period. The augmentation of growth of the hypertrophic lobes was maximal between the 3rd and 7th postoperative days in comparison with the control group ($p<0.05$). Nevertheless this growth accelerating effect was lost during the next ultrasonographic controls, and on the 14th postoperative day there were no statistically significant differences. The results of the presented study confirmed in large animal experiment previous findings gained using in vitro models and in experiments with small laboratory animals (Armendariz-Borunda, 1993, Armendariz-Borunda, 1997, Deneme, 2006, Delgado-Rizo, 1998). The selected concentration and timing of the application of monoclonal antibody prolonged acceleration of liver parenchyma regeneration in non-occluded liver lobes of animals (Armendariz-Borunda, 1993, Deneme, 2006). The secondary effects that could be hypothesized after application of the monoclonal antibody against key pleiotropic growth factor (changes in immune reactions and homeostasis) were not observed either during application or in the whole postoperative period. In previous described experimental groups (IL-6 and TNF-α group) we have shown the importance and usefulness of cytokines of the first phase of liver regeneration that increase priming of the hepatocytes - Interleukine-6 and Tumor necrosis factor - α (Liska, 2009, Liska, 2012), but these cytokines have pleitropic functions, which could be altered or changed by application of these cytokines from an extrinsic source. The choice of the monoclonal antibody against growth factor that terminates this first phase of liver regeneration and stimulates hepatocytes to differentiation, production of extracellular matrix and remodelation of liver tissue structure, imitates the same effect of these named cytokines (Kusaka, 2004).

The present study describes a new usage of a monoclonal antibody against TGF-β1 in large animal experimental model of partial portal vein ligation, which simulates the situation in human medicine. The achieved acceleration of growth of the hypertrophic liver lobes after application of monoclonal antibody against TGF-β1between 3rd and 7th postoperative days confirmed the key role of the studied cytokine in terminating the regeneration of liver parenchyma after portal vein ligation. The experimental results could be settings for a clinical study in patients with a low response regeneration of future liver remnant volume after portal vein embolization that does not allow surgical treatment (Liska, 2012).

4.2 Biochemistry

All biochemical parameters, including cytokine levels, that were assessed during the observation period, did not outline any significant differences between the MSC-infused and the control group, indicating that both groups were systemically stable throughout the regenerative period. Concerning the role of cytokines it may also be assumed that these act on a more local level within the liver parenchym and differences cannot be observed in the periphery (Liska, 2009).

On the contrary application of TNF-α brought the tendency to differences in serum level of AST, ALT, GGT and creatinine in postoperative period, which were not statistically significant. All these biochemical parameters could be due to operative stress and also to occlusion of portal branches of right liver lobes, which represent significant liver function reserves. The postoperative elevation of these parameters could be hypothesized also to atrophic changes in occluded lobes. This reduction of expected increased serum level of discussed biochemical parameters after application of studied cytokine in non occluded portal vein could present also one of its pleiotropic functions - hepatoprotection agains changes after operative stress and reduction of functional parenchyma (Teoh, 2006, Baier, 2006). This hypothesis should be probably examined on non-resection experimental model (Liska, 2012).

We did not prove any statistically significant differences between serum levels of studied biochemical parameters in particular time points between IL-6 and control groups. It demonstrates also no unsuitable influence of applied cytokine on the liver function. The changes in serum levels of studied cytokines, growth factors were also not observed as different between IL-6 and control group in separate time points (Liska, 2009).

In case of comparison of TGF-β1 and control group no statistically significant differences were shown between serum levels of the studied biochemical parameters at particular points in time. This also demonstrates that there was no unsuitable influence of the applied monoclonal antibody on the liver function. This results support our ideas to choose monoclonal antibody against TGF-β1 for future clinical studies in human liver (Liska, 2012).

4.3 Histology and immunohistochemistry

No histological differences were observed after 14 days between the groups (number of mitotic figures, binucleated hepatocytes, length of lobuli, hepatocyte size), indicating that the initial phase of liver regeneration had been completed in both groups and there was no influence of the infused MSC on liver architecture. Using BrdU staining, only very few of the infused MSC could be detected. Thus, injected MSC do not contribute to liver regeneration through proliferation, but most likely help to establish a micormilieu supportive for intrinsic proliferation of resident hepatocytes. Results acquired by this experimental study confirm the experiences of Furst and Esch, who have previously shown that cell grafts of bone marrow origin (CD133 positive cells) applied after PVE in human surgery can increase the FLRV (Furst, 2007). From the present experimental data, however, it cannot be concluded with certainty that the application of stem cells, including MSC, does not support the growth of liver malignancies to the same extent that it supports liver regeneration. Assuming a bystander effect of MSC on the micromilieu makes this even more likely (Alison, 2006). Thus, further animal investigation is necessary before optimized MSC therapies can be applied in the setting of human medicine (Alison, 2006). In conclusion, it has been hereby shown that the intraportal infusion of syngeneic porcine MSC after PVE in a setting of liver regeneration led to an accelerated and augmented compensatory liver hypertrophy. This effect is most likely due to bystander effects of the transplanted MSC (Liska, 2009).

The increased number of larger lobulus in the hypertrophic parenchyma of TNF-α group in comparison with control group could be explained by incomplete liver regeneration. Because there are practically no mitotic figures or the amount is of same quantity as in the normal liver parenchyma without any surgical procedures or toxic insult, we could hypothesize, that the first phase of liver regeneration is finished and the next phase of regeneration proceeds. It means the remodelation phase and the phase, when the liver microstructure is restored. Next would be objective to future study - the detection of intracellular or extracelular matrix changes during the process of liver regeneration. No differences in the amount of binucleated hepatocytes could be discussed also by end of proliferative phase of liver parenchyma at the time of sampling. The same size of hepatocytes and no atypical hepatocytes in the biopsies could also be explained by the same reason. This hypothesis is supported by restitution of all liver function monitored by biochemical parametres at the moment of sacrifying of experimental animals (Liska, 2012).

The increased number of binucleated hepatocytes in the hypertrophic parenchyma of IL-6 group in comparison with control group could be explained by incomplete liver regeneration at the end of experiment. Because there are practically no mitotic figures or the amount is of same quantity as in the normal liver parenchyma without any surgical procedures or toxic insult, we could hypothesize, that the first phase of liver regeneration is finished and the next phase of regeneration proceeds. It means the remodelation phase and the phase, when the liver microstructure is restored. Next would be objective to future study - the detection of intracellular or extracellular (matrix) changes during the process of liver regeneration. No significant differences in other histological parameters (diameter of lobulus and hepatocytes) were not proved (Liska, 2009).

The larger distribution of the number of binucleated hepatocytes in the hypertrophic parenchyma of TGF-β1 group in comparison with the control group could be explained by incomplete liver regeneration at the end of experiment. Because there are practically no mitotic figures, or the amount is the same as in the normal liver parenchyma without any surgical procedures or toxic insult, it was possible to hypothesize that the first phase of liver regeneration was finished and the next phase of regeneration was proceeding, namely the remodelling phase and the phase when the liver microstructure is restored (Mangnall, 2003). The size of hepatocytes and the length of lobuli were not proved to be statistically different between study and control group. The same size of hepatocytes and length of lobuli in the bioptical samples from the hypertrophic parenchyma could be also explained in the same way. This hypothesis is supported by the restitution of all liver functions monitored by biochemical parameters and completion of the proliferative phase of liver regeneration at the moment of sacrificing of the experimental animals (Liska, 2012).

5. Conclusion

Application of IL-6 and TNF-α augments hypertrophy of FLRV on 7th postoperative day in comparison to control group. In case of application of MAB TGF-β1 we observed maximal increase of FLRV between 3rd and 7th days. Application of MSC augments hypetrophy of FLRV on 3rd day. The biochemical and histological examinations did not prove any important differencies among the groups. The use of TNF-α, IL-6, MAB TGF-β1, MSC could increase the process of liver regeneration after portal vein ligation.These experimental results could be used in clinical practice in patients with risk of acute liver failure after extended liver resection.

6. Summary

The aim of presented studies was to influence regeneration of liver parenchyma after portal vein embolization/ligation. Physiological solution (controls, 9 animals) or TNF-α (9) or interleukin-6 (8) or MSC (6) were applied into non-occluded portal vein branches after ligation of portal vein branches for right lobes. Administration of MAB TGF-β1 was performed 24 hours after ligation (7). Compensatory hypertrophy (FLRV) was followed-up in next 14 days by regular ultrasonography, serum level of biochemical parametres and by histological examinations. Application of IL-6 and TNF-α augments hypertrophy of FLRV on 7th postoperative day in comparison to control group. In case of application of MAB TGF-β1 we observed maximal increase of FLRV between 3rd and 7th days. Application of MSC augments hypetrophy of FLRV on 3rd day. The biochemical and histological examinations did not prove any differences among the groups. The use of TNF-α, IL-6, MAB TGF-β1, MSC could increase the process of liver regeneration after portal vein ligation.

Keywords: Liver surgery, Liver regeneration, Experimental study, Porcine model, Experimental model, Portal vein embolization, TNF-α, Interleukin-6, Mesenchymal stem cells, TGF-β1.

7. Acknowledgement

Supported by grants IGA MZ CR 12025 and Research project MSM 0021620819 (*Replacement of some vital organs*).

8. References

Abdalla, E. K., Hicks M. E. & Vauthey, J. N. (2001). Portal vein embolization: rationale, technique and future prospects. *British Journal of Surgery*, Vol. 88, No. 7, (July 2001), pp. 165-175, ISSN 1365-2168

Alison, M.R. et al.(2006): Stem cell plasticity and tumour formation. *Eur J Cancer*, 42, pp 1247-1256. ISSN: 0959-8049

Aldeguer, X. et al. (2002). Interleukin-6 from intrahepatic cells of bone marrow origin is required for normal murine liver regeneration. *Hepatology*, Vol. 35, No. 1, (January 2002), pp. 40-48, ISSN 1527-3350

Armendariz-Borunda, J. et al. (1993). Transforming growth factor beta gene expression is transiently enhanced at a critical stage during liver regeneration after CCl4 treatment. *Laboratory Investigation*, Vol. 69, No. 3, (1993), pp. 283-294, ISSN 0023-6837

Armendariz-Borunda, J. et al. (1997). Antisense S-oligodeoxynucleotides down-regulate TGFbeta-production by Kupffer cells from CCl4-injured rat livers. *Biochimica et Biophysica Acta*, Vol. 1353, No. 3, (September 1997), pp. 241-252, ISSN 0006-3002

Azoulay, D. et al. (2000). Resection of non-resectable liver metastases from colorectal cancer after percutaneus portal vein embolization. *Annals of Surgery*, Vol. 231, No. 4, (April 2000), pp. 480-486, ISSN 0003-4932

Baier, P. K. et al. (2006). Hepatocyte proliferation and apoptosis in rat liver after liver injury. *Hepato-Gastroenterology*, Vol. 53, No. 71, (September-October 2006), pp. 747-752, ISSN 0172-6390

Banas, A. et al. (2007). Adipose tissue-derived mesenchymal stem cells as a source of human hepatocytes. *Hepatology*, Vol. 46, No. 1, (July 2007), pp. 219-228, ISSN 1527-3350

Barry, F. P. (2003). Biology and clinical applications of mesenchymal stem cells. *Birth Defects Research Part C: Embryo Today*, Vol. 69, No. 3, (August 2003), pp. 250-256, ISSN 1542-975X

Broering, D. C. et al. (2002). Portal vein embolization vs. portal vein ligation for induction of hypertrophy of the future liver remnant. *Journal of Gastrointestinal Surgery*, Vol. 6, No. 6, (November-December 2002), pp. 905-913, ISSN 1091-255X

Bustos, M. et al. (2000). Liver damage using suicide genes: a model for oval cell activation. *The American Journal of Pathology*, Vol. 157, No. 2, (August 2000), pp. 549-559, ISSN 0002-9440

Chamberlain, J. et al. (2007). Efficient generation of human hepatocytes by the intrahepatic delivery of clonal human mesenchymal stem cells in fetal sheep. *Hepatology*, Vol. 46, No. 6, (December 2007), pp. 1935-1945, ISSN 0270-9139

Coelho, M. C. et al. (2007). Expression of interleukin 6 and apoptosis-related genes in suckling and weaning rat model of hepatectomy and liver regeneration. *Journal of Pediatric Surgery*, Vol. 45, No. 4, (April 2007), pp. 613-619, ISSN 0022-3468

Cornell, R. P. (1990). Acute phase responses after acute liver injury by partial hepatectomy in rats as indicators of cytokine release. *Hepatology*, Vol. 11, No. 6, (June 1990), pp. 923-931, ISSN 0270-9139

Dahlke, M. H. et al. (2004). Stem cell therapy of the liver--fusion or fiction? *Liver transplantation*, Vol. 10, No. 4, (April 2004), pp. 471-479, ISSN 1527-6465

Dahlke, M. H. et al. (2003). Liver regeneration in a retrorsine/CCl4-induced acute liver failure model: do bone marrow-derived cells contribute? *Journal of Hepatology*, Vol. 39, No. 3, (September 2003), pp. 365-373, ISSN 0168-8278

Delgado-Rizo, V. et al. (1998). Treatment with anti-transforming growth factor beta antibodies influences an altered pattern of cytokines gene expression in injured rat liver. *Biochimica et Biophysica Acta*, Vol. 1442, No. 1, (October 1998), pp. 20-27, ISSN 0006-3002

Deneme, M. A. et al. (2006). Single dose of anti-transforming growth factor-beta1 monoclonal antibody enhances liver regeneration after partial hepatectomy in biliary-obstructed rats. *The Journal of Surgical Research*, Vol. 136, No. 2, (December 2006), pp. 280-287, ISSN 0022-4804

di Bonzo, L. V. et al. (2008). Human mesenchymal stem cells as a two-edged sword in hepatic regenerative medicine: engraftment and hepatocyte differentiation versus profibrogenic potential. *Gut*, Vol. 57, No. 2, (February 2008), pp. 223-231, ISSN 0017-5749

Duncan, J. R. et al. (1999). Embolization of portal vein branches induces hepatocyte replication in swine: a potential step in gene therapy. *Radiology*, Vol. 210, No. 2, (February 1999), pp. 467-477, ISSN 0033-8419

Fausto, N. & Riehle, K. J. (2005). Mechanisms of liver regeneration and their clinical implications. *Journal of Hepato-Biliary-Pancreatic Surgery*. Vol. 12, No. 3, (2005), pp. 181-189, ISSN 0944-1166

Fausto, N. (2000). Liver regeneration. *Journal of Hepatology*. Vol. 32, Suppl. 1, (2000), pp. 19-31, ISSN 0168-8278

Friedman, S. L. (2008). Mechanisms of hepatic fibrogenesis. *Gastroenterology*, 2008, Vol. 134, No. 6, (May 2008), pp. 1655-1669, ISSN 0016-5085

Fukuhara, Y. et al. (2003). Gene expression in the regenerating rat liver after partial hepatectomy. *Journal of Hepatology*, Vol. 38, No. 6, (June 2003), pp. 784-792, ISSN 0168-8278

Furst, G. Et al. (2007): Portal vein embolization and autologous CD133+ bone marrow stem cells for liver regeneration: initial experience. *Radiology*, 243, pp 171-179. ISSN 0033-8419

Harada, H. et al. (1998). Fate of human liver after hemihepatic portal vein embolization: cell kinetic and morphometric study. *Hepatology*, Vol. 26, No. 5, (November 1997), pp. 1162-1170, ISSN 0270-9139

Heinrich, S. et al. (2006). Portal vein ligation and partial hepatectomy differentially influence growth of intrahepatic metastasis and liver regeneration in mice. *Journal of Hepatology*. 2006, Vol. 45, No. 1, pp. 35-42, ISSN 0168-8278

Jiang, Y. et al. (2002). Pluripotency of mesenchymal stem cells derived from adult marrow. *Nature*, Vol. 418, No. 6893, pp. 41-49, (July 2002), ISSN 0028-0836

Kang, X. et al. (2005). Rat bone marrow mesenchymal stem cells differentiate into hepatocytes in vitro. *World Journal of Gastroenterology*, Vol. 11, No. 22, (2005), pp. 3479-3484, ISSN 1007-9327

Kokudo, N. et al. (2001). Proliferative activity of intrahepatic colorectal metastases after preoperative hemihepatic portal vein embolization. *Hepatology*, Vol. 34, No. 2, (August 2001), pp. 267-272, ISSN 0270-9139

Kusaka, K. et al. (2004). Factors affecting regeneration after portal vein embolization. *Hepatogastroenterology*, Vol. 51, No. 56, (March-April 2004), pp. 532-535, ISSN 0172-6390

Lagasse, E. et al. (2000). Purified hematopoietic stem cells can differentiate into hepatocytes in vivo. *Nature Medicine*, Vol. 6, No. 11, (2000), pp. 1229-1234, ISSN 1078-8956

Lilja, H. et al. (1999). Transforming growth factor beta1 helps maintain differentiated functions in mitogen-treated primary rat hepatocyte cultures. *Molecular Cell Biology Research Communications*, Vol. 1, No. 3, (June 1999), pp. 188-195, ISSN 1522-4724

Liska, V. et al. (2009). Intraportal injection of porcine multipotent mesenchymal stromal cells augments liver regeneration after portal vein embolization. *In Vivo*, Vol. 23, No. 2, (March-April 2009), pp. 229-236, ISSN 0258-851X

Liska, V. et al. (2009). Interleukin-6 augments activation of liver regeneration in porcine model of partial portal vein ligation. *Anticancer Research*. Vol. 29, No. 6, (June 2009), pp. 2371-2377, ISSN 0250-7005

Liska, V. et al.(2009) Cytokines and liver regeneration after partial portal vein ligation in porcine experimental model. *Bratislavske lekarske listy*, Vol. 110, No. 8, (2009), pp. 447-453, ISSN 1336-0345

Liska, V. et al. Tumour Necrosis Factor-Alpha stimulates liver regeneration after parcial portal vein ligation – experimental study on porcine model, *Hepatogastroenterology*, Vol. 59, No 114, (2012), pp. 125-132 , ISSN 0172-6390

Liska V. et al. Inhibition of Transforming Growth Factor Beta-1 augments liver regeneration after partial portal vein ligation in a porcine experimental model. *Hepatogastroenterology*, Vol. 59, No 113, (2012), pp. 264-271, ISSN 0172-6390

Lysy, P. A. at al. (2008). Persistence of a chimerical phenotype after hepatocyte differentiation of human bone marrow mesenchymal stem cells. *Cell Proliferation*. Vol. 41, No. 1, (February 2008), pp. 36-58, ISSN 0960-7722

Makuuchi, M. et al. (1990). Preoperative portal embolization to increase safety of major hepatectomy for hilar bile duct carcinoma: a preliminary report. *Surgery*, Vol. 107, No. 5, (May 1990), pp. 521-527, ISSN 0039-6060

Makuuchi, M., Takayasu, K. & Takayama, T. (1984). Preoperative transcatheter embolization of portal venous branch for patient receiving extended lobectomy du to the bile duct carcinoma. *Journal of Japanese Society for Clinical* Surgery, Vol. 45, No. 1, (1984), pp. 14-20, ISSN 0386-9776

Mangnall, D., Bird, N. C. & Majeed, A. W. (2003). The molecular physiology of liver regeneration following partial hepatectomy. *Liver International*, Vol. 23, No. 2, (April 2003), pp. 124-138, ISSN 1478-3223

Michalopoulos, G. K. & DeFrances, M. C. (1997). Liver regeneration. *Science*, Vol. 237, No. 5309, (April 1997), pp. 60-66, ISSN 0036-8075

Najimi, M. (2007). Adult-derived human liver mesenchymal-like cells as a potential progenitor reservoir of hepatocytes? *Cell Transplantation*, Vol. 16, No. 7, (2007), pp. 717-728, ISSN 0963-6897

Oe, S. et al. (2004). Intact signaling by transforming growth factor beta is not required for termination of liver regeneration in mice. *Hepatology*, Vol. 40, No. 5, (November 2004), pp. 1098-1105, ISSN 0270-9139

Ong, S. Y., Dai, H. & Leong, K. W. (2006) Hepatic differentiation potential of commercially available human mesenchymal stem cells. *Tissue Engineering*, Vol. 12, No. 12, (December 2006), pp. 3477-3485, ISSN 1076-3279

Petersen, B. E. (1999). Bone marrow as a potential source of hepatic oval cells. *Science*, Vol. 284, No. 5417, (May 1999), pp. 1168-1170, ISSN 0036-8075

Popp, F. C. et al. (2006). Therapeutic potential of bone marrow stem cells for liver diseases. *Current Stem Cell Research & Therapy*, Vol. 1, No. 3, (September 2006), pp. 411-418, ISSN 1574-888X

Ringe, J. et al. (2002). Porcine mesenchymal stem cells. Induction of distinct mesenchymal cell lineages. *Cell and Tissue Research*, Vol. 307, No. 3, (March 2002), pp. 321-327, ISSN 0302-766X

Stefano, D. et al. (2005). Preoperative percutaneus portal vein embolization: evaluation of adverse events in 188 patients. *Radiology*, Vol. 234, No. 2, (February 2005), pp. 625-630, ISSN 0033-8419

Teoh, N., Field, J. & Farrell, G. (2006). Interleukin-6 is a key mediator of the hepatoprotective and pro-proliferative effects of ischaemic preconditioning in mice. *Journal of Hepatology*, Vol. 45, No. 1, (July 2006), pp. 20-27, ISSN 0168-8278

Vassilopoulos, G., Wang, P. R. & Russell, D. W. (2003). Transplanted bone marrow regenerates liver by cell fusion. *Nature*, Vol. 422, No. 6934, (April 2003), pp. 901-904, ISSN 0028-0836

Viebahn, C. S. & Yeoh, G. C. (2008). What fires prometheus? The link between inflammation and regeneration following chronic liver injury. *The International Journal of Biochemistry & Cell Biology*, Vol. 40, No. 5, (2008), pp. 855-873, ISSN 1357-2725

Wang, X. et al. (2003). Cell fusion is the principal source of bone-marrow-derived hepatocytes. *Nature*, Vol. 422, No. 6934, (April 2003), pp. 897-901, ISSN 0028-0836

Yamamoto, Y. et al. (2008). A comparative analysis of the transcriptome and signal pathways in hepatic differentiation of human adipose mesenchymal stem cells. *Federation of European Biochemical Societies Journal*, Vol. 275, No. 6, (March 2008), pp. 1260-1273, ISSN 1742-464X

Zimmermann, A. (2004). Regulation of liver regeneration. *Nephrology Dialysis Transplantation*. Vol. 19, Suppl. 4, (2004), pp. iv6-iv10, ISSN 0931-0509

Section 3

Transplantation, Cell Therapies and Liver Bioengineering

10

Liver Transplantation in the Clinic – Progress Made During the Last Three Decades

Marco Carbone[1], Giuseppe Orlando[2,3], Brian Sanders[4], Christopher Booth[2], Tom Soker[2], Quirino Lai[5], Katia Clemente[6], Antonio Famulari[6], Jan P. Lerut[5] and Francesco Pisani[6,*]

1. Introduction

The World Health Organization calculates that over - six hundred and fifty million people worldwide suffer of some form of liver disease, including thirty million Americans. On a worldwide base, approximately one to two million deaths are accounted to liver related diseases annually. Around the globe, China has the world's largest population of Hepatitis B patients (approx. 120 million) with five hundred thousand people dying of liver illnesses every year (CDC, 2007; WHO, 2008). In the US alone, five hundred thousand critical liver problem episodes are reported every year requiring hospitalization with great burden to the patients and a huge cost to the health care system. In the European Union and United States of America alone, over eighty one thousand and twenty six thousand people died of chronic liver disease in 2006, respectively (CDC, 2007; Eurostat, 2007). In these patients, liver transplantation is presently the only proven therapy able to extend survival for end-stage liver disease. It is also the only treatment for severe acute liver failure and to some forms of inborn errors of metabolism.

The road to successful liver grafting in humans has been long and fraught with many obstacles. Experimental attempts at liver transplantation originally took place in the 1950s and 1960s, but human liver transplantation did not become a reality until 1963 (Starzl & Demetris, 1990). Although unsuccessful, Dr. Starzl's attempt at liver transplantation was a milestone in surgery. However, it took nearly 20 years to develop a surgical procedure for orthotopic liver transplantation (OLT) that was safe to apply in humans. In 1983, the National Institutes of Health (NIH) held the Consensus Development Conference on Liver Transplantation. The most important outcome of this conference was OLT became an accepted therapeutic modality for some patients with end-stage liver disease (NIH, 1983). The ideal liver transplant candidate needed to comply with ten conditions (**Table 1**) as well

* [1]*Hepatology Unit, Addenbrooke's Hospital, Cambridge, UK*
[2] *Wake Forest Institute for Regenerative Medicine, Winston Salem, NC, USA*
[3] *Department of General Surgery, Wake Forest Baptist Health, Winston Salem, NC, USA*
[4] *Wake Forest University School of Medicine, Winston Salem, USA*
[5] *Starzl Abdominal Transplant Unit, University Hospitals St.Luc, Université Catholique de Louvain, Brussels, Belgium*
[6]*Renal Failure and Transplant Unit – L'Aquila University, L'Aquila, Italy*

as ten absolute contraindications and five relative contraindications (**Table 2**). Taking into account the multitude of criteria for OLT, few patients were deemed eligible. Furthermore, the University of Pittsburgh was the only liver transplant center in the United States in 1983. Currently, 120 liver transplant centers in the United States are registered with the United Network for Organ Sharing (UNOS), the organization that manages the nation's organ transplant system, and 145 transplant centers from 24 European countries are participating in the European Liver Transplant Registry (ELTR). As reported in the UNOS database, 111,824 liver transplantations have been performed in the United States through December 2010 (UNOS, 2010). Likewise, 100,542 liver transplantations have been performed in Europe with an average of 5,562 transplantations per year in the past decade (ELTR, 2010).

Young patient < 50 years
No viral infection
No alcohol or drug abuse
Normal vessel state
No infection
No (advanced) malignancy
No cardiopulmonary or renal disease
No prior abdominal surgery
Ability to accept the procedure or understand its nature
Ability to accept costs of the procedure

Table 1. The ten conditions to be an ideal liver transplant candidate at the 1984 NIH Consensus Conference

ABSOLUTE
Portal vein thrombosis
Severe hypoxemia due to right to left shunts (HPS)
Sepsis outside the hepatobiliary (HB) system
Primary malignant disease outside the HB system
Metastatic HB malignancy
Active alcoholism
Advanced cardiopulmonary or renal disease
HBsAg and HBeAg positive state
Age > 55 years
Inability to accept the procedure or understand its nature and/or its costs
RELATIVE
Intrahepatic or biliary sepsis
Advanced alcoholic liver disease in the abstinent alcoholic
Age > 50 years
HBsAg positive state
Prior abdominal surgery especially in the right upper quadrant
Portal hypertension surgery

Table 2. The ten absolute and five relative contraindications to liver transplantation at the 1984 NIH Consensus Conference

Since the first OLT was performed, the field has changed dramatically. Improvements in surgery, anaesthesia, immunosuppression, and control of infection have translated into increased access and better patient outcomes. In the pioneering days of OLT, triple-drug therapy (corticosteroids, azathioprine [AZA], and antilymphocyte globulin [ALG]) was used to prevent and treat rejection. The development of a powerful immunosuppressive agent, cyclosporine (CsA), in the late 1970s was one of the most significant events in modern transplantation. By 1984, all transplant centers in the United States used double therapy of corticosteroids and CsA as the maintenance immunosuppressive regimen. During the 1990s, tacrolimus (TAC) emerged as the mainstay maintenance immunosuppressive agent in OLT, with or without corticosteroids. More recently, mycophenolate mofetil (MMF) has replaced the use of AZA in many centers. Before the advent of CsA, 5-year survival after OLT was less than 20%. Current survival rates 1, 3, and 5 years after liver transplantation in the United States are 88%, 80%, and 75%, respectively (UNOS, 2010).

Advancements in surgical techniques and technologies also account for the increased success of OLT. In particular, the standardization of biliary tract reconstruction, advances in retransplantation, and improvements in surgical technology help explain better patient outcomes. Examples of developments in technology include pump-driven veno-venous bypass that does not require recipient heparinization, rapid infusion, and autologous auto-transfusion devices. Additionally, improved procurement and preservation techniques for the donor liver and increased insight into the management of potentially fatal complications have led to improved patient morbidity and mortality.

2. Evolution of liver transplant indications

Nowadays, liver transplantation is indicated for acute or chronic liver failure of any cause (**Table 3**).

Cirrhosis due to chronic hepatitis C infection is one of the most common indications for liver transplantation in the United States and Europe. Despite effective antiviral treatments including Pegylated Interferon, Ribavirin, and direct-acting antiviral agents (DAAs), this indication is likely to remain important for the coming decades given the large prevalence of chronic hepatitis C infection and the propensity of the disease to lead to cirrhosis and hepatocellular carcinoma (HCC) (Merion, 2010).

Chronic hepatitis B has become a less common indication, mostly due to the advent of universal vaccination. Additionally, dramatic improvements in the treatment of hepatitis B, such as the development of nucleoside/nucleotide analogues, has reduced the number of patients with chronic hepatitis B progressing to end-stage liver disease. However, in parts of the world where chronic hepatitis B is endemic, including much of Asia, this remains the most common indication (Perrillo, 2009).

Alcohol-related liver disease is an important indication for OLT in Western countries and is oftentimes associated with concomitant hepatitis C infection. In the past, patients with alcohol-related liver disease and alcohol dependence were often refused access to liver transplantation due to unjust societal allocation of scarce donor organs. However, a careful assessment with the support of a health care professional experienced in the management of patients with addictive behavior is associated with low rates of recidivism after OLT (Lucey,

1992). Nowadays, this indication has become more commonly accepted, as long as patients demonstrate sobriety of at least 6 months duration.

	Year of Transplant									
Primary Diagnosis	**1998**	**1999**	**2000**	**2001**	**2002**	**2003**	**2004**	**2005**	**2006**	**2007**
Total	4,424	4,498	4,595	4,673	4,969	5,351	5,847	6,120	6,362	6,223
Noncholestatic cirrhosis	2,790	2,899	2,956	2,973	3,091	3,151	3,514	3,709	3,732	3,547
Cholestatic liver disease	563	499	475	471	492	513	510	467	552	555
Acute hepatic necrosis	349	405	436	366	369	355	425	448	371	368
Biliary atresia	214	188	161	172	164	175	187	179	143	174
Metabolic liver disease	163	152	165	170	152	183	180	200	215	175
Malignant neoplasm	98	87	84	142	3311	371	481	610	774	783
Other	247	258	318	378	370	603	550	507	575	608

[a] 2008 Annual Report of the U.S. Organ Procurement and Transplantation Network and the Scientific Registry for Transplant Recipients: Transplant Data 1998–2007. Rockville, MD: U.S. Department of Health and Human Services, Health Resources and Services Administration, Healthcare Systems Bureau, Division of Transplantation; 2009

Table 3. Primary Diagnosis of Deceased Donor Liver Transplant Recipients, 1998 to 2007[a]

Cholestatic liver disease is becoming an increasingly uncommon indication for OLT. Data from UNOS show that among cadaveric liver transplants in 1991, 18% were for cholestatic liver disease, compared with 10% in 2000 and only 7.8% in 2008 (UNOS, 2010). The incidence and prevalence of primary biliary cirrhosis (PBC) are steadily increasing whereas the absolute number of OLT performed for PBC is falling (UNOS, 2010; ELTR, 2010). Reasons for this decline in the number of transplants for PBC are not clear but may relate to a changing pattern of disease, increased rates of diagnosis, and more effective treatment. The number of transplants for primary sclerosing cholangitis (PSC) in western countries during the period 1995-2006 has remained stable and represents 8% of all liver transplants. In some areas that have a relatively low prevalence of hepatitis C and alcoholic liver disease (ALD), such as the Scandinavian countries, PSC is the leading indication for OLT, accounting for 16% of the indications (Nordic Liver Transplant Registry, 2010).

Nonalcoholic fatty liver disease (NAFLD) is becoming one of the most common liver disorders in developed countries. Because this disorder can lead to cirrhosis in a number of patients and is associated with an increased risk of hepatocellular carcinoma (HCC), it is an increasingly frequent indication for liver transplantation (Burke, 2004). Considering the current obesity epidemic, NAFLD may become the most common indication for liver transplantation in the coming years.

Primary HCC is a unique and evolving indication for liver transplant. Initially, outcomes in liver transplantations for patients with unresectable HCC were not encouraging. Ninety percent of those transplanted for HCC developed recurrent disease within 2 years. As a result, HCC was considered a contraindication to transplantation for a number of years. Ongoing research on HCC post-transplant elucidated that important predictors of recurrence are tumor characteristics, such as tumor size, stage, and grade, number of nodules, micro- or macrovascular invasion, serum levels of alpha-fetoprotein, and demonstrated absence of extrahepatic spread. In 1996, Mazzaferro and colleagues defined the Milan criteria, which require a single tumor ≤ 5 cm in diameter or no more than three tumor nodules, each ≤ 3 cm in diameter. Liver transplant in patients meeting the Milan criteria have excellent outcomes and low recurrence rates (Mazzaferro, 1996). Augmentations to the Milan criteria are currently being debated in the liver transplant community. The San Francisco criteria, which require either a single lesion ≤ 6.5 cm or up to three lesions none of which is >4.5 cm and with total tumor diameter ≤ 8 cm, have been widely debated, but no consensus has yet emerged (Yao, 2001). Finally, more attention has been given to the role of downstaging by locoregional therapy for otherwise unsuitable candidates. Treatments to shrink nodules, including radiofrequency ablation, transcatheter arterial chemoembolization, and novel thermal and non-thermal techniques for tumor ablation offer strategies for subsequent transplantation in patients with more advanced lesions.

Many inborn errors of metabolism have been successfully treated with liver transplantation (Kayler, 2003).

Acute liver failure has long been considered an appropriate indication for liver transplant. Patients with fulminant hepatic failure should be referred to a transplant center as quickly as possible for critical care management. If given appropriate critical care support, many patients spontaneously recover. Patients predicted to have little chance of spontaneous recovery should undergo transplantation as soon as possible. New technologies using bioartificial liver devices including both a biological component and an artificial scaffold may offer some promises for patients with acute liver failure. However, these technologies have not become widely available yet, and therefore OLT remains an important treatment (Demetriou, 2004; Ellis, 1996).

3. Historical contraindications to OLT

3.1 Absolute contraindications

3.1.1 Splanchnic venous thrombosis

Splanchnic venous thrombosis and portal hypertension surgery are both part of the natural evolution of cirrhosis. Intra-operative mortality during OLT with these conditions was once nearly 50%. Those that survived OLT oftentimes had morbid conditions that diminished the utility of the transplant. In recent times, several technical advances have been made to overcome this problem. Eversion thrombectomy is a good treatment option for a majority of patients with splanchnic venous thrombosis. If eversion thrombectomy is not possible, a free iliac interposition graft between the allograft portal vein and the superior mesenteric vein is indicated. Sometimes it is possible to connect the allograft portal vein to a collateral vein. In cases of extended splanchnic thrombosis, cavo-portal hemi-transposition or combined liver-intestinal transplantation are the last resorts for treatment (Lerut, 1997).

3.1.2 Hepatopulmonary Syndrome (HPS)

HPS is defined as a pO2 < 70 mmHg in an upright position. It is present in up to 20% of cirrhotic patients. Even in cases where oxygen saturation is below 50 mmHg, this situation can be reversed by a successful OLT. The post-transplant recovery is usually more complicated necessitating adapted respiratory care. The pre-transplant baseline macro-aggregated albumin shunt fraction may indicate the limits of correction following OLT (Starzl, 1990).

3.1.3 Sepsis outside the hepatobiliary system

Active extra-hepatic infection compromises outcomes of OLT. Nonetheless, transplantation can be performed successfully if the infection is confined to the lungs or the ascites and does not cause hemodynamic instability (Starzl, 1990). Post-transplant care is usually much more prolonged and expensive if the patient has sepsis outside the hepatobiliary system.

3.1.4 Primary malignant disease outside the hepatobiliary system

Based on the embryological development theory, Starzl advocated the 'en block' or 'cluster transplant' in order to treat extended hepatobiliary malignancies and liver metastases from neuroendocrine tumors. Although initial success was promising, longterm results were disappointing. (Lerut, 2007, 2011; Starzl, 1990).

Pre-OLT malignancies or malignancies discovered incidentally during the OLT-procedure are no longer contraindications to OLT as shown by the Kings' College study. Metastatic hepatobiliary malignancy is an emerging indication for OLT. After successful chemotherapeutic treatment of hepatic and thoracic tumor involvement in children with hepatoblastoma, excellent results have been obtained with OLT. Similarly, excellent longterm results have been obtained for epithelioid hemangio-endothelioma. In the latter group some centers propose sequential or simultaneous hepatopulmonary transplantation (Lerut, 2007).

The role of OLT in the treatment of neuroendocrine tumors with hepatic involvement is continuously evolving. Excellent OLT results can be obtained in select, young (<50 years) patients after R0 resection of the primary tumor for more than six months prior to transplant. Furthermore, if the primary lesion has favorable tumor biology (as expressed by a Ki67 index of < 5-10%) and has a portal vein drainage, results are significantly improved (Bonaccorsi-Riani, 2010).

It is evident that all oncologic patients that undergo OLT will benefit from adapted immunosuppression. Minimization of immunosuppression and use of the m-TOR inhibitor, rapamycin, are of paramount importance. Rapamycin has antitumor activity based on antiangiogenic properties.

The most recent development in the field of OLT is the treatment of metastatic colorectal malignancy. The Oslo-SECA (SEcondary CAncer) study indicates that OLT will have a role in the treatment of colorectal metastases on the condition that adapted chemotherapy and immunosuppression are employed after the transplant procedure. Preliminary results obtained in the Oslo-SECA cases show nearly 50% 5-year survival after OLT (Foss, 2011).

3.1.5 Alcohol abuse

OLT in active alcoholic patients has always been discussed heavily within the medical community (Starzl, 1990). The 'six month abstinence rule' is not generally enforced, and some French groups even advocate OLT in cases of severe acute alcoholic hepatitis (Mathuri, 2005).

3.1.6 Drug abuse

Although also heavily debated, some groups in New York showed that OLT can be successful in recipients on methadone maintenance.

Both alcohol and drug abusers need exceptionally tight follow-up during the pre- and post-transplant period. It is of utmost importance to take familial, professional, and social conditions into consideration in these potential patients (Starzl, 1990).

3.1.7 Advanced cardiopulmonary disease

Two-staged cardiac and hepatic transplantion is becoming more frequent (Starzl, 1990). Several case reports have been published about simultaneous liver and heart transplantation in the context of familial amyloid polyneuropathy (FAP) and hemochromatosis. Simultaneous OLT with coronary, valvular, or arrythmia surgery has also been reported. These surgeries are complex and have stimulated major interest in both cardiovascular and hepatic experts of transplant centers.

3.1.8 Viral infections

Viral cirrhosis was once an absolute contraindication to OLT because of the universal recurrence of the disease in the liver allograft (Lerut, 1998). The landmark paper of Samuel in 1993 showed that adequate antiviral prophylaxis using specific anti-HBs antibodies protects the allograft from reinfection (Samuel, 1993). Further improvement has been achieved by combining nucleoside and nucleotide analogues with immunoglobulins. This prophylactic combination was able to reduce the incidence of allograft reinfection from 100% to 5%. Moreover, those with HBV gain an immunologic advantage from immunoglobulin treatment due to immunoglobulins producing a tolerogenic effect on dendritic cells.

3.1.9 Age

Increasing numbers of transplants are done in patients aged over 65 and 70 years of age (Starzl, 1990). The initial Pittsburgh results have now been confirmed by most transplant centers.

3.1.10 Inability to accept or understand the procedure

OLT is accessible to patients in all levels of society. Successful OLT has been reported in Down Syndrome patients and in drug abusers. Adequate preparation by medical, paramedical, and clinical coordinator teams is of utmost importance in complicated clinical scenarios.

3.2 Relative contraindications

3.2.1 Intra-hepatic or biliary sepsis

Chronic biliary infection is frequent in Caroli disease, primary sclerosing cholangitis, and secondary sclerosing cholangitis. Because the infection is usually confined to the liver, outcome of OLT is excellent as transplantation removes the source of the infection. OLT is especially valuable in these patients as it dramatically improves their quality of life.

3.2.2 Advanced liver disease in abstinent alcoholic patients

The Model for End-Stage Liver Disease (MELD) scoring system aims to reduce liver waitlist mortality by transplanting sicker patients more rapidly. Abstinent alcoholics frequently belong to the sickest patient groups. Nowadays, OLT is a very good indication in such patients if the recipient remains abstinent and is compliant. If the recipient remains abstinent, alcoholic cirrhosis is one of the best indications for OLT as this is the only disease that does not recur in the allograft. Moreover, abstinent alcoholics that receive OLT offer a unique opportunity to study the effect of immunosuppression withdrawal without primary disease involvement in the allograft.

3.2.3 Previous abdominal surgery

Many transplant recipients have undergone previous abdominal and right upper quadrant surgery. These interventions can compromise the transplant procedure. Thus, exploratory or staging laparotomies as well as unnecessary cholecystectomies and cyst fenestrations should be avoided in future OLT patients. Interventional radiology procedures such as the Transjugular Intrahepatic Portosystemic Shunt (TIPSS) are preferred in potential OLT recipients instead of portal hypertension surgery. In cases where portal hypertension surgery cannot be avoided, meso-caval or spleno-renal shunts are the preferred options leaving the hilar region intact.

3.2.4 HIV infection

HIV patients that are well controlled on Highly Active Antiretroviral Therapy (HAART) are generally not contraindicated for OLT. The indication for OLT relates more so to concomitant HCV and/or HBV infection. Co-infected patients are at higher post-OLT infectious risk. Particular attention should be given to the interaction between anti-viral drugs and calcineurin inhibitors.

3.2.5 Positive HBsAg status

See above.

4. Future perspectives

Criteria for placement on the waiting list have become more quantitative. Continuous refinement of the allocation system will improve the management of the waiting list (Metsellaar, 2011).

Alternative techniques such as split liver transplantation (SLT) and living donor liver transplantation (LDLT) will allow for expansion of the allograft pools. The bipartition of the liver is especially important in pediatric patients for whom size-matched whole liver allografts are scarce. Split grafts have been associated with reduction in the risk of death on the pediatric waiting list. However, SLT is much less successful in the adult-adult SLT. Donor selection for splitting and technical and logistical expertise to decrease total ischemia time are important factors in successful transplantation. A better understanding of the liver anatomy and improving surgical skills have allowed living liver donation to become a routine procedure in some centers. Given the major risks of the operation required for the donor, whether this procedure will ever find wide application is unclear.

As there will always be more potential recipients than donors, many researchers are working in the field of artificial tissue engineering and regenerative medicine (RM) (Orlando, 2011a, 2011b, 2011c, 2011d, 2012a, 2012b, 2012c). RM holds the promise of regenerating tissues and organs by either stimulating previously irreparable tissues to heal themselves, by treating liver disease with cell therapies, or by manufacturing tissues *ex vivo* using extracellular matrix (ECM) scaffolds.

This last approach, uses ECM scaffolds that have an intact but decellularized vascular network that is repopulated with autologous or allogeneic stem cells and/or adult cells. Liver ECM scaffolds may be produced from humans or animals. In the latter case, human cells are used to repopulate a scaffold of animal origin, coining a new concept called semi-xenotransplantation (Orlando, 2011a, 2011b). Importantly, preemptive transplantation with regenerated tissues will improve outcomes, especially in cases of metabolic and cystic liver disease.

Future progress in the medical treatment of oncology will enhance outcomes in OLT (Lerut, 2007, 2011; Bonaccorsi-Riani, 2010, Foss, 2011). As treatments of vascular tumors, advanced hepatocellular cancer, cholangiocellular cancer, neuroendocrine tumors, and colorectal liver metastases improve, indications for OLT in the 21st century will become more inclusive of advanced oncologic disease states.

Combined organ transplantation is becoming more frequent as many liver diseases are accompanied by renal function impairment. Nowadays, 15% of all liver recipients have combined liver-kidney transplantation.

Transplant teams should focus more on late morbidity and mortality. Currently, the majority of long-term survivors die from infectious disease, cardiovascular disease, or cancer while having a functional graft. This mortality is directly related to the strength of the maintenance immunosuppression. Minimization or even withdrawal of immunosuppressive protocols must become a priority in organ transplantation (Lerut, 2003, 2008). Unfortunately, tolerance protocols are frequently based on a trial and error approach, as good markers to predict tolerance are not yet available (Lerut, 2006). As the liver is an immunoprivileged organ with relatively high resistance against immune responses, liver recipients should be at the forefront of this research.

5. Conclusion

Since the first transplantation was performed much progress has been made in the field of OLT. Indications for liver transplant have evolved to include previously contraindicated conditions such as those with hepatocellular carcinoma and alcohol-related liver disease. All but one (active sepsis outside the biliary system) contraindication to OLT has been eliminated. As a result, more than 200,000 patients have been transplanted, many with excellent long-term success. With indications to transplantation increasing and contraindications waning, many more patients will be transplanted in the future.

The future of liver transplantation will be no less challenging for its practitioners. The main challenge is the shortage of organs, and many strategies are in place to address this problem. In the near future, immunologic discoveries will allow for an immunosuppression-free state of many recipients. This will guarantee better quality of life and similar survival expectancy as non-transplanted patients. Regenerative medicine technology applied to liver transplantation has the potential to meet the two major needs: namely, the identification of a potentially inexhaustible source of livers and an immunosuppression-free state. In the ideal scenario, livers will be manufactured from autologous cells with no need for anti-rejection therapy.

6. References

Bonaccorsi-Riani E., et al. (2010). Liver transplantation and neuroendocrine tumours: lessons from a single centre experience and from the literature review Transplant Int 23, 2010, 668-678.

Burke A, Lucey MR. (2004). Non-alcoholic fatty liver disease, nonalcoholic steatohepatitis and orthotopic liver transplantation. Am J Transplant, 4, 5, 686–693.

CDC (2007). Centers for Disease Control and Prevention Database.

Demetriou AA, Brown RS, Busuttil RW et al. (2004). Prospective, randomized, multicenter, controlled trial of bioartificial liver in treating acute liver failure. Ann Surg, 239, 2004, 660.

Ellis AJ, Hughes RD, Wendon JA, Dunne J, Langley PG, Kelly JH, Gislason GT, Sussman NL, Williams R. (1996). Pilot-controlled trial of extracorporeal liver assist device in acute liver failure. Hepatology, 24, 1996, 1446-51.

European Liver Transplant Registry. Available at: www.ELTR.org.

Eurostat (2007). Eurostat's Harmonised Regional Statistical Database.

Foss A, Adam R, Dueland S. (2011) Liver transplantation for colorectal metastases: revisiting the concept Transplant Int. 23, 2011, 679-685.

Kayler LK, Rasmussen CS, Dykstra DM, et al. (2003). Liver transplantation in children with metabolic disorders in the United States. Am J Transplant, 3, 2003, 334-339.

Lerut J., Mazza D.,Van Leeuw V.,Laterre P.F.,Donataccio M.,De Ville de Goyet J. et al. (1997). Adult Liver transplantation and abnormalities of splanchnic veins: experience in 53 patients. Transplant Int 10, 1997, 125-132.

Lerut J et al. (1998). Liver transplantation and HBV-related disease: adequate immuno-prophylaxis and delta co-infection as major determinants of long-term prognosis. J Hepatol 30, 1998, 706-714.

Lerut J. (2003). Avoiding steroids in solid organ transplantation. Transplant Int 16, 2003, 213-224.

Lerut J, Sanchez-Fueyo A. (2006).An appraisal of tolerance in liver transplantation. Am J Transplant 6, 2006, 1774-1780.

Lerut, et al. (2008). Tacrolimus monotherapy in liver transplantation:one-year results of a prospective, randomized, double-blind, placebo-controlled study. Ann Surg 248, 2008, 956-967.

Lerut J., Weber M., Orlando G., Dutkowski P. (2007) Vascular and rare liver tumors: A good indication for liver transplantation? J Hepatol 47, 2007, 466-475.

Lerut J., Ciccarelli O., Julliard O, Lannoy V., Gofette P. (2011) Hepatocelluar cancer and liver transplantation: a western experience in : Multidisciplinary Treatment of Hepatocellular Carcinoma. Springer Verlag Eds : Vauthey, Brouquet.

Lucey MR,Merion RM, Henley KS, et al. (1992) Selection for and outcome of liver transplantation in alcoholic liver disease. Gastroenterology, 102,5, 1992, 1736-1741.

Mathuri P. (2005). Is alcoholic hepatitis an indication for transplantation? Current management and outcomes. Liver Transplation, 11, 2005, S21.

Merion R. (2010).Current Status and Future of Liver Transplantation. Semin Liver Dis 30, 2010, 411-421.

Mazzaferro V, Regalia E, Doci R, et al. (1996).Liver transplantation for the treatment of small hepatocellular carcinomas in patients with cirrhosis. N Engl J Med, 334,11,1996,693-699.

Metselaar H.J, Lerut J., Kazemier G. (2011). The true merits of liver allocation according to MELD scores: survival after transplantation tells only one side of the story Transplant Int, 10, 2011.

National Institutes of Health (NIH). (1983) NIH Consensus Development Conference Statement: Liver transplantation. Hepatology. 4(1S), 1983, 107S-110S.

Nordic Liver Transplant Registry. Available at:www.scandiatransplant.org.

Orlando G, Domínguez-Bendala J, Shupe T, Bergman C, Bitar KN, Booth C, Carbone M, Koch K, Lerut JP, Neuberger J, Petersen BE, Ricordi C, Atala A, Stratta RJ, Soker S. (2012). Cell and Organ Bioengineering Technology as Applied to Gastrointestinal Diseases. GUT in press.

Orlando G, Wood KJ, De Coppi P, Baptista PM, Binder KW, Bitar KN, Breuer C, Burnett L, Christ G, Farney A, Figliuzzi M, Holmes JH IV, Koch K, Macchiarini P, Mirmalek-Sani SH, Opara E, Remuzzi A, Rogers J, Saul JM, Seliktar DS, Shapira-Schweitzer K, Smith T, Solomon D, Van Dyke M, Yoo JJ, Zhang Y, Atala A, Stratta RJ, Soker S. (2012). Regenerative medicine as applied to general surgery. Ann Surg, in press.

Orlando G, Wood KJ, Soker S, Stratta RJ. (2011). How regenerative medicine may contribute to the achievement of an immunosuppression-free state. Transplantation, 92, 8, 2011, 36-8.

Orlando G. (2011). Transplantation as a subfield of regenerative medicine. An interview by Lauren Constable. Expert Rev Clin Immunol, 7, 2011, 137-141.

Orlando G, Wood KJ, Stratta RJ, Yoo J, Atala A, Soker S. (2011). Regenerative medicine and organ transplantation: Past, present and future. Transplantation 91, 2011, 1310-7.

Orlando G, Baptista P, Birchall M, Di Coppi P, Farney A, Opara E, Rogers J, Seliktar D, Shapira-Schweitzer K, Stratta RJ, Atala A, Wood KJ, Soker S. (2011). Regenerative medicine as applied to solid organ transplantation: current status and future development. Transpl Int, 24, 2011, 223-232.

Orlando G. (2012). Immunosuppression-free transplantation reconsidered from a regenerative medicine perspective. Exp Rev Clin Immun 2012, in press.

Perrillo R. (2009). Hepatitis B virus prevention strategies for antibody to hepatitis B core antigen-positive liver donation: a survey of North American, European, and Asian-Pacific transplant programs. Liver Transpl, 152, 2009, 223-232.

Starzl T.E., Demetris A.J (1990). Liver Transplantation. Year Book Medical Publishers, Inc. Chicago, USA

United Network for Organ Sharing. Available at: www.UNOS.org

WHO (2008). World Health Organization - Global Burden of Disease: 2004 update (2008). In WHO publications.

11

Potential of Mesenchymal Stem Cells for Liver Regeneration

Melisa Andrea Soland,
Christopher D. Porada and Graça D. Almeida-Porada
*Department of Regenerative Medicine,
Wake Forest Institute for Regenerative Medicine,
USA*

1. Introduction

A wide variety of diseases, including cirrhosis, unresectable hepatic malignancy, ischemia, metabolic and auto-immune disorders, and hepatitis, whether caused by viral agents or drugs/toxins, can trigger hepatic insufficiency and failure, a life-threatening situation for which liver transplantation is the only definitive therapy [1-4]. Over 16,000 patients are currently awaiting the availability of a liver from a compatible donor [5], and many of these patients will die without ever receiving a transplant, due to the current shortage of available donor organs [6]. Furthermore, even when a patient is fortunate enough to find a compatible donor and receive a liver transplant, several factors can still thwart the ultimate success of this procedure. Operative damage, immune rejection towards the new organ, relapse of the pre-existing liver disease, and life-long side effects due to immunosuppression are among the most common complications [7, 8]. Furthermore, after liver transplantation, several long-term morbidities can arise, such as cardiovascular and retinal complications, lymphoproliferative disorders, and chronic renal failure [8-10]. Additionally, it is anticipated that the number of patients in need of liver transplantation will increase in the next decade, due to the obesity epidemic and the higher incidence of Hepatitis C infection. Therefore, new therapeutic approaches that can eliminate the need for partial or complete liver transplantation are urgently needed.

A valuable alternative to entire or partial liver transplantation is the delivery of cells capable of restoring normal organ physiology [11-19]. The use of cell therapy possesses several inherent advantages over organ transplantation: the procedure could be performed in a much less invasive way, the purified cell populations may be less immunogenic [20], and the use of autologous cells could be implemented [21].

Hepatocyte transplantation has been considered one of the most promising alternatives to liver transplantation, as these cells offer the benefit of being fully functional and are therefore able to quickly replace damaged hepatocytes upon delivery [22]. Also, the ability to cryopreserve and store hepatocytes gives the advantage of having a source of cells available when required. However, accessibility of hepatocytes at the required numbers for clinical intervention is still problematic, as human livers are required for their isolation, and the harvesting and storing procedures are difficult and inefficient [23]. Additionally, differentiated hepatocytes have limited proliferation capabilities, and *in vitro* culture may alter their physiological and functional characteristics [24], making it difficult to obtain an adequate number of hepatocytes of sufficient quality for transplantation [25]. Further compounding these difficulties, it has been shown that, following transplantation, only a small percentage of the infused hepatocytes actually survive and durably engraft within the liver [26-30].

A better alternative to the use of adult hepatocytes is exploiting the presence, within the liver, of hepatic stem/progenitor cells (HpSCs), and the intrinsic ability of these cells to extensively expand, differentiate into all mature liver cells, and reconstitute the liver when transplanted, with minimal immunogenicity[31, 32]. HpSCs are found in the Canals of Hering in adult livers, and in ductal plates of the fetal liver. They can be easily isolated by immunoselection using an antibody against the epithelial cell adhesion molecule (EpCAM) and, they comprise approximately 0.5-5% of the liver parenchyma depending on age [33]. Although there is controversy regarding the cell surface markers that define this cell population, positivity for EpCAM, NCAM, or CD133, and negativity for AFP are currently considered to be the most accepted markers. The potential of these cells has been clearly demonstrated in numerous murine studies, and a few studies in humans have confirmed the presence and regenerative properties of HpSCs in the presence of viral hepatitis, cirrhosis or inborn metabolic disorders [34-37].

In the last decades, alternative sources of stem cells have raised great hope for improving the treatment of liver diseases. In particular, the demonstration that cells within the bone marrow contributed, at different levels, to liver parenchymal cells opened the possibility of using autologous cells to treat liver disorders/diseases [38-52]. Amongst these, mesenchymal stem cells (MSC) have been considered an ideal cell source because of their ease of isolation and expansion, their immunomodulatory properties, and their broad differentiation potential [53, 54].

2. Mesenchymal stem cells

Mesenchymal stem cells (MSC), also referred to as marrow stromal cells or stromal precursor cells, were first described in the 1960s by Friedenstein, and shown to belong to the bone marrow stromal microenvironment that supports hematopoietic stem cells and controls the process of hematopoiesis[55]. These cells were also shown to be able to differentiate into multiple lineages of mesenchymal tissues, including bone, cartilage, fat, tendon, and muscle [56-58]. Numerous culture methods and purification procedures such as plastic adherence, Ficoll gradient centrifugation, or cell-sorting using surface markers have all been used to enrich for bone marrow-derived MSC, with each laboratory preferring its own method of isolation. This makes the comparison of results obtained by various laboratories very difficult, since each lab is likely studying somewhat different cell populations, despite the fact that all of these cells have collectively been referred to as MSC. According to the International Society for Cellular Therapy, MSC should have several characteristics in addition to adherence to plastic. They must express CD105, CD73, CD90, but not express CD45, CD34, CD14, CD11b, CD79 or CD19 and HLA-DR surface markers. Furthermore, they must be able to differentiate into osteocytes, chondrocytes, and adipocytes [59]. Although MSC constitute a very small percentage of the nucleated cells present in the BM, between 0.001 and 0.01%, these cells can be expanded exponentially while maintaining their original phenotype and differentiation potential, making it possible to easily obtain adequate numbers for cell-based therapies. These characteristics make MSC ideal agents for cell replacement therapies

2.1 Properties of Mesenchymal stem cells

In keeping with the original findings of Friedenstein, MSC are still most often isolated from the bone marrow. In humans, these "BM-MSC" are usually collected from the superior iliac crest of the pelvis; however, they can also be obtained from the tibial and femoral marrow compartments [60], and the thoracic and lumbar spine [61]. In larger animals, BM-MSC are isolated from the same areas. In contrast, in small animals such as mice, a bone marrow aspiration is not possible, so BM-MSC are harvested by flushing the mid-diaphysis of the tibia or femur [61].

In stark contrast to hematopoietic stem cells, MSC can easily be expanded in culture for many passages without losing their phenotype or pluripotency capability [60]. Indeed, Bruder et al. demonstrated that human BM-MSC can readily be propagated *in vitro* until passage 38±4; after that passage the cells turn flat and very broad, indicating they have reached senescence [62]. Moreover, by plating these cells at a low density and consistently passaging them before they have reached confluence, it is possible to accelerate their growth rate and increase their expansion capacity [63].

Over the past several years, studies have provided compelling evidence that MSC's differentiation capacity far exceeds that originally reported by Friedenstein. Indeed, *in vitro* and *in vivo* transplantation studies have now shown that MSC have the capacity to differentiate not only into mesodermally-derived cell types such as bone [64], cartilage [65], tendon [66], muscle [67], cardiomyocytes [68] and adipose tissue [69], but, even more remarkably, can also give rise to cells that developmental biology classifies as being derived from ectoderm (neurons and astrocytes [70, 71]) and endoderm (pancreatic beta cells [72] and hepatocytes [73]). This extraordinary multipotentiality has generated a great deal of interest in applying MSC to tissue repair/regeneration as well as cell therapy approaches for a variety of diseases/injuries.

In addition to their broad differentiation potential, MSC also appear to possess the intrinsic ability to migrate, or home, to sites of injury following systemic infusion. Importantly, from the standpoint of developing a clinically viable and safe cell-based therapeutic, MSC appear to selectively engraft and differentiate into tissue-specific cells that are missing or defective due to the disease in question, while contributing very little, if at all, to normal/healthy tissue [74-77]. For the past several years, scientists have attempted to elucidate the mechanism by which MSC are selectively attracted to sites of injury. During pathological conditions, several cytokines/chemokines are produced, which will stimulate MSC to express: 1) integrins, by which MSC will bind to endothelial cells, and 2) cytokine/chemokine receptors, by which MSC will migrate towards the inflammatory site. This complex network of signaling allows MSC to establish cell-cell contact and mediate rolling with endothelial cells. Additionally, they also transmigrate into the extracellular matrix by interacting with integrins and fibronectin stimulated by MSC-secreted ligands. Despite these insights, however, more information is required for a complete understanding of this process. This understanding could then be used to develop means of enhancing MSC engraftment after transplantation [78].

The ability of MSC to reprogram to cells specific to other organs/tissues suggested that MSC would have to replace a significant percentage of the damaged cells within a diseased/injured organ to exert a beneficial effect. However, controversy arose in the MSC field when a series of studies were published demonstrating a reproducible therapeutic improvement in the absence of detectable MSC engraftment. These findings sparked additional studies that have now shown that MSC can also mediate tissue repair by acting as "trophic factories", releasing specific cytokines and growth factors that modulate the activity of tissue-specific cells, suppress local inflammation, and inhibit fibrosis and apoptosis, thereby facilitating endogenous tissue regeneration [79]. Adding to the complexity of the functions/effects of MSC, it was recently discovered that MSC can transfer mitochondria or mitochondrial DNA to cells that have been damaged by ischemia and reperfusion. By transferring mitochondria or mitochondrial DNA, MSC can rescue the cells that have non-functional mitochondria, rescuing these cells and enabling regeneration of the tissue [80]. In recent years, it was also shown that MSC express an array of miRNA's, small non-coding RNA's that are involved in gene regulation [81]. It is believed that a single miRNA can regulate several different target genes and a single gene can be regulated by multiple miRNA's. Studies to date have provided evidence that miRNA's are involved with stem cell differentiation, hematopoiesis, immune response, neurogenesis, stress responses, and the development of skeletal and cardiac tissue [82-84]. These regulatory miRNA's have now been shown to be present inside microvesicles that are secreted by MSC, which are then transferred to neighboring cells to regulate their activities. This pathway provides yet another means by which MSC can communicate with injured cells. Following secretion of the microvesicles, the miRNA's contained therein can then enter the injured cell and induce differentiation and/or production of soluble mediators, and stimulate cell-cycle re-entry; the net result of these myriad actions being tissue regeneration [85].

Upon arrival at the site of injury, MSC also fulfill another vital function, which is to modulate the inflammatory microenvironment present within the damaged/diseased tissue. MSC possess an extraordinary ability to modulate immune cells, exerting these effects by releasing soluble factors and by cell-cell contact. MSC are known to inhibit proliferation and maturation of cytotoxic T cells, helper T cells, B cells, dendritic cells, and NK cells, as well as to inhibit NK-mediated cytotoxicity. These broad-ranging actions enable MSC to interfere with each component of the adaptive immune system. MSC can also stimulate the differentiation of Tregs, which can further dampen the immune response. MSC are known to release a host of soluble factors, which have been associated with their immunomodulatory properties including transforming growth factor-β, prostaglandin-E2, inducible NOS, nitric oxide, IL-10, HLA-G, hepatocyte growth factor, and indoleamine 2,3-dioxygenase. By dampening the ongoing inflammation and/or aberrant immune reaction present within the damaged/diseased tissue, MSC facilitate the process of repair/recovery, further adding to the promise of using these cells for regenerative medicine [78].

In addition to the inherent properties of MSC that make them well suited for cellular therapy, it is important to realize that MSC can easily be genetically manipulated *in vitro*, with both viral and non-viral vectors, to enhance their immunosuppressive properties [86, 87], to deliver a protein that is missing/defective in the patient, to induce apoptosis of tumor cells, to promote cell proliferation, to guide their migration to a specific site within the body, and even to direct their differentiation towards a specific cell lineage [88-90], making the range of clinical applications for which MSC could be used almost limitless.

2.2 Sources of Mesenchymal stem cells

In addition to their presence in the bone marrow, these MSC have also been identified in, and isolated from, several other tissues including cord blood (CB-MSC), cord matrix (hWJSCs), amniotic fluid (AF-MSC), placenta, adipose tissue (AT-MSC), brain, liver, lung, and kidney [91-93]. The presence of these cells in several organs/tissues raises the possibility that they could have a crucial function in organ homeostasis, and/or repairing the tissue, and suggests that MSC isolated from these tissues may have a unique transcriptional or proteomic signature that renders these cells biased in terms of homing or differentiation towards the organ of origin. Differences also exist in the cytokine/chemokine molecules produced by MSC from various sources and in their differentiation capabilities [94, 95]. Using fetal liver MSC (FL-MSC) as an example to illustrate these differences, FL-MSC exhibit much more rapid growth kinetics than BM-MSC, due, at least in part, to a greater abundance of transcripts involved in cell cycle regulation, DNA repair and chromatin regulation. In addition, analysis of telomerase activity and telomere length revealed that fetal liver MSC telomeres are longer and these cells possess greater telomerase activity than adult sources of MSC. As a result, these cells are more expandable and they become senescent later in culture [96]. Fetal liver MSC also express more primitive genes, such as Oct-4, Nanog, and SSEA-3 than their adult counterparts, but transcripts involved in differentiation towards more mature cells are reduced relative to MSC from other sources. More importantly from the standpoint of clinical utility, fetal liver MSC also exhibit reduced immunogenicity compared to adult BM-MSC, perhaps due to expression of higher levels of HLA-G1 [97]. In addition to reduced immunogenicity, fetal liver MSC also demonstrated an enhanced immunomodulatory function than BM-MSC when tested for their ability to inhibit T cell proliferation [98]. Despite all the promising characteristics of these cells, very few studies have examined their utility/potential *in vivo*. In one of these studies, rabbit fetal liver MSC were tested for their engraftment, proliferation and differentiation capabilities following *in utero* transplantation. Two routes of administration were analyzed, intrahepatic and intra-amniotic. Both approaches were safe for both the mother and the fetal recipient, but only the intrahepatic route resulted in the formation of donor-derived hepatocytes in the liver. While the levels of hepatocyte production were low, the engraftment persisted for at least 16 weeks after transplantation [99].

Despite their many unique characteristics and promise of offering MSC primed for repair of specific tissues, the inherent difficulty in obtaining organ-specific MSC such as those derived from liver, will likely preclude their widespread use in a clinical setting. Ideally, for cellular therapies, one would like a readily available source of cells that could be used as off-the-shelf therapeutics. MSC are present in significant numbers in discardable tissues such as cord blood, placenta and amniotic fluid, and these MSC have the ability to be expanded and frozen without loss of viability or differentiative potential, making MSC from these tissues an attractive option. Indeed, both AF-MSC [100] and CB-MSC [101] were shown, upon transplantation *in vivo*, to give rise to hepatocytes, suggesting they have definite potential as cellular therapeutic for treating liver diseases. Having considered these two very different

MSC examples, a summary of the properties of the main sources of MSC that are currently being tested for therapeutic purposes appears in **Table I**, to provide a better overall picture of the similarities and differences inherent to MSC isolated from various tissues.

MSC Cells	Purification	Special Characteristics	Advantages	Disadvantages
BM-MSC	Aspiration from the iliac crest, concentration by Ficoll gradient and adhesion to plastic. Number of cells obtained: 1-10 CFU/10^6 MNC (0.01-0.001% of MNC)	- Positive for cell surface markers: CD105, CD73, CD90, CD29, CD44, CD166, STRO-I. - Negative for cell surface markers: CD34, CD45.	- Easy to expand in vitro up to 15 passages with minimal spontaneous differentiation. - Easy to cryopreserve. - Higher capacity for osteogenic differentiation compared to AD-MSC - Cells can be isolated from the patient, modified and injected back into the same patient.	- Invasive and painful extraction. - The number of cells obtained is small. The amount decreases with the patient's age, the presence of osteoporosis, and exposure to chemotherapy or radiation, and varies with gender.
CB-MSC	Collection of umbilical cord blood, purification of MNC by Ficoll gradient and concentration by adhesion to plastic. Number of cells obtained: 0-2.3 clones/10^8 MNC (1.10^3-5.10^3cells/sample)	- Express lower levels of CD90 and CD105 compared to BM-MSC and AT-MSC. - Express hematopoietic genes at higher levels, and consequently higher levels of growth factors than BM-MSC.	- Umbilical cord is discarded at birth; therefore, it represents an untapped source of MSC that can be obtained non-invasively. - Longer survival and expansion in vitro compared to BM-MSC and AT-MSC.	- Low frequency, difficulties and inconsistency in isolation. Only MSC can only be successfully isolated from 10-30% of the collected cord blood samples. - Difficult to differentiate into adipocytes, compared to BM-MSC and AT-MSC. - Slow to differentiate into chondrocytes in vitro, compared to BM-MSC and AT-MSC.
AT-MSC	Lipoaspiration of the fat tissue, collagenase treatment to separate the stromal cell populations and concentration by adhesion to plastic. Number of obtained cells: 1% of adipose cells are MSC.	- Positive cell surface markers: CD49. - Negative cell surface markers: CD106, STRO-I.	- More abundant than BM-MSC or CB-MSC. - Faster proliferation in culture compared to BM-MSC. - More easily differentiable to adipocytes. - Large amounts of fat are discarded during liposuction; therefore, large availability of cells for cell bank. - Similar immunosuppressive ability to BM-MSC. - Cells can be isolated from the patient, modified and injected back into the same patient.	- Do not differentiate well into chondrocytes. - More heterogeneous population that BM-MSC after purification; however, the cells can be sorted to increase homogeneity. - Differentiate less actively to osteoblasts compared to BM-MSC and CB-MSC. - Differentiate more easily to adipocytes compared to CB-MSC.
AFS	Amniocentesis, concentration by adhesion to plastic and selection by c-kit. Number of obtained cells: 0.9-1.5% of cells in the amniotic fluid are MSC.	-Similar surface markers to BM-MSC but lower expression of CD44 and CD105. -Express primitive cell marker: SSEA4. -Express more primitive genes: Oct-4 and Nanog.	- Higher proliferation rate than BM-MSC due to long telomeres. - Even though they express more primitive markers, expansion in vitro and in vivo is not associated with chromosomal instability.	- Less ability to differentiate into chondrocytes than BM-MSC. - Sorting for c-kit should be done to eliminate abundant skin cells derived from fetus.
hWJSCs	Extraction of Wharton's Jelly from umbilical cord, collagenase and hyaluronidase treatment, and final disruption of the tissue by a needle. Number of obtained cells: 4.10^5 cells per sample; 10–15.10^3 cells per centimeter of cord) (usually 15cm/sample).	Express primitive cell marker: SSEA4. -Express more primitive genes: Oct-4 and Nanog.	- Umbilical cord is discarded at birth; therefore, it represents an untapped source of MSC that can be obtained non-invasively. - Higher frequency of MSC than BM or cord blood. - MSC can be isolated from all samples, in contrast to CB-MSC. - Extensive and faster proliferation compared to BM-MSC. - Similar immunosuppressive ability to BM-MSC. - Higher potential to differentiate into chondrocytes than BM-MSC.	- Sorting strategies are encouraged as the population of cell obtained from the Wharton's Jelly is heterogeneous. - Less potential to differentiate into adipocytes than BM-MSC.

Table 1.

2.3 Mesenchymal stem cells for treating liver disease

2.3.1 *In vitro* models to study Mesenchymal stem cell differentiation

One could convincingly argue that the best way to study the differentiation potential of MSC is to perform studies *in vitro*, as studies of this nature allow for creation of a carefully controlled microenvironment which greatly facilitates the delineation of the pathways/mechanisms by which MSC commit to specific lineages and undergo reprogramming. In contrast to *in vivo* studies, where the researcher has little or no control over the myriad local and systemic cues and factors present within the recipient, performing *in vitro* studies enables the researcher to definitively establish the true multipotential capability of MSC at the single cell level, or at the level of a clonally-derived population. Indeed, *in vitro* studies have now revealed that MSC are able to transdifferentiate into cells of the three germ layers, including neuronal and glial cells [102-105], cardiomyocytes [106-110], endothelial cells [111-113], and insulin-producing beta cells [114, 115]. The discovery of this tremendous potential has prompted researchers to perform microarrays studies to understand the molecular mechanisms responsible for the commitment and differentiation of MSC along each of these lineages [116-118]. It is hoped that understanding these pathways will pave the way for the development of methods for efficiently driving MSC differentiation down specific lineage pathways to create the cell type required for therapy. These studies also provided vital information regarding key genes and signaling pathways that are directly involved in maintaining MSC in an undifferentiated state, helping to characterize this cell population and providing clues as to methods for expanding these cells for longer periods of time while maintaining their multipotency.

While these findings were exciting and highlighted the vast potential of MSC for cellular therapy, the most important ability, from the standpoint of therapies for the liver, would be the ability to differentiate into hepatocytes. Accordingly, several protocols have now been developed for the *in vitro* differentiation of both murine and human BM-MSC into hepatocytes [1, 119-125]. These MSC-derived hepatocytes exhibit the same morphology and antigenic profile as native hepatocytes, and they appear to be functional, based upon uptake of low-density lipoprotein, urea production and storage of glycogen. These initial findings with BM-MSC have now been extended to include MSC derived from adipose tissue, amniotic fluid, CB, and Wharton's Jelly, with adipose-derived MSC showing the greatest propensity to differentiate *in vitro* to putative functional hepatocytes. It was initially hypothesized that CB-MSC might harbor an innate capacity to differentiate into hepatocytes, since they constitutively express early as well as more mature hepatic markers and functions [126]. However, this initial assumption was not realized. After several studies, it became clear that CB-MSC differentiate only partially, displaying early and some mature hepatic markers/functions but lacking the expression of other proteins that are critical for liver development [123, 126]. While this discovery initially reduced the enthusiasm for the use of CB-MSC as therapeutics for liver disease, it is important to realize that the immature nature of the hepatocytes they form could still enable them to treat disorders such as metabolic liver disease, in which generation of fully functional mature hepatocytes is not required, as long as the transplanted cells produce adequate levels of the missing/defective enzyme for correction. This limited differentiation capacity does, however, likely preclude their use for treating conditions such as acute hepatic failure.

Only a few groups have analyzed *in vitro* differentiation of hWJSCs. Zhang et al applied a one-step protocol with HGF and FGF-4 and found that, after 21 days, cells expressed hepatocyte markers such as Albumin, AFP, and CK-18 [127]. In other studies, Lin et al. induced differentiation of the cells by co-culturing them with mice liver tissue previously treated with thioacetamide, a chemical used to induce chronic fibrosis of the liver. Two days after induction, hWJSCs expressed hepatic markers, providing evidence that, with the appropriate stimuli, hWJSCs can very rapidly reprogram to adopt a hepatocytic fate. AF-MSC were also tested for their ability to differentiate *in vitro* into hepatocytes. The differentiation process employed by these investigators consisted of two steps: first, the MSC were treated for 1 week with EGF and FGF to commence induction along the hepatic lineage; and second, a maturation step, during which the cells were treated with dexamethasone and oncostatin-M for 2 weeks. The MSC-derived cells obtained at the end of this 2-stage induction protocol expressed several hepatic markers/functions, including albumin production, uptake of low density lipoproteins, glycogen storage, and urea production, promoting the investigators to cautiously refer to them as hepatocyte-like cells [100, 128, 129].

Collectively, the results of these *in vitro* studies provide compelling evidence that MSC derived from a variety of sources all possess the ability to give rise to what appear to be functional hepatocytes, albeit at varying levels. This suggests that MSC could represent viable cellular therapeutics for treating liver disease, and thus provide a much-needed alternative to whole or partial liver transplantation.

2.3.2 *In vivo* models to study Mesenchymal stem cell differentiation

Despite all the knowledge that can be gained from performing *in vitro* studies, they are inherently limited by the need to supply all of the requisite factors to observe the desired differentiation or reprogramming. This becomes problematic when one wishes to discover/investigate novel properties of MSC, since, in most cases, these factors are not yet known. Adding to this problem is the lack of suitable assays to rigorously establish that the "hepatocyte-like cells" generated *in vitro* are, in fact, bona fide hepatocytes that perform all of their required physiologic functions. For this reason, scientists are forced to resort to *in vivo* transplantation studies in the hopes that the required mediators/factors conditions are present within the microenvironment of the target organ, and can coax the transplanted MSC to reprogram towards the desired cellular fate. Performing studies *in vivo* also has the advantage of ensuring that all of the appropriate cues are present to influence migration and homing of MSC to the tissue/organ in question; an essential issue to consider if the ultimate goal is to develop therapies using MSC. Transplantation *in vivo* also provides the opportunity to examine the ability of the MSC-derived cells to seamlessly integrate into the existing cytoarchitecture and adopt appropriate behavioral characteristics. Ideally, studies of this nature would be performed with human MSC and their derivatives, to ensure the clinical translation of the results obtained. Due to ethical and practical issues, however, studies of this nature can clearly not be performed in human subjects. Thus, at the present time, investigators can only test the ability of human stem cells to engraft/differentiate within a xenogeneic setting, using suitable small or large animals as recipients.

2.3.2.1 Mesenchymal stem cells differentiate *in vivo* into hepatocytes

The exciting *in vitro* findings discussed above suggested that MSC could serve as cells for repairing the injured or failing liver. Importantly, MSC can be grown quite readily in culture for extended periods of time without any seeming loss of differentiation capacity. This has two important implications for their use in cellular-based liver therapies. The first of these is that a very small marrow aspirate could be taken from the patient and adequate cells obtained for transplantation, through extensive expansion *in vitro* following isolation. Secondly, by virtue of their ability to be expanded in culture without loss of *in vivo* potential, MSC could be harvested from the patient's own marrow even if the liver disease present was the result of an underlying genetic defect, since MSC are quite amenable to genetic modification/correction using a wide range of viral and non-viral vector systems. Following genetic manipulation, a pure population of genetically corrected autologous MSC could thus be propagated to generate sufficient numbers of cells to achieve meaningful levels of engraftment following transplantation. Based on these promising characteristics, MSC have now been tested in a wide variety of injury/disease model systems for their ability to generate hepatocytes and correct these liver defects. Using MSC isolated from a variety of mouse, rat, and human tissues, investigators have now provided evidence that MSC can mediate varying degrees of correction/repair of the liver following injury due to partial hepatectomy [126, 130-133], treatment with the toxin CCl4 [134-145], injury induced by allyl-alcohol [146, 147], and treatment with 2-acetylaminofluorene [139].

Unfortunately, these studies are confounded by the problem of each group of investigators using MSC defined in different ways, ranging from specific antigen profile to simple plastic adherence. The use of differing definitions for "MSC" can likely explain, at least in part, the differing outcomes, even when using a similar injury model system. One thing that is quite clear from these studies looked at as a whole, however, is that MSC appear to be able to exert beneficial effects in a wide range of injuries and disease states within the liver. Another issue that needs clarification is whether fusion plays a major role in the beneficial effects, since the fusion of donor MSC with host hepatocytes has not yet been addressed in detail in any of these injury/disease models.

Another issue that has complicated interpretation of the data generated from these studies in liver, as well as those conducted looking at the potential of MSC to mediate repair in other organ systems, is the observation that a therapeutic benefit is often observed in the absence of any evidence of engraftment of the transplanted MSC within the damaged organ. Instead, it appears that the transplantation of MSC somehow stimulates the host's liver to repair itself without the donor cells actually having to persist long-term within the recipient. These findings led to a great deal of debate as to whether MSC can actually generate hepatocytes or if, perhaps, all the effects they produce are simply mediated through release of soluble factors. Meticulously executed *in vitro* studies have now provided definitive evidence that MSC can under appropriate conditions be reprogrammed into cells with all of the characteristics of functional hepatocytes that can currently be assessed in culture [37, 135, 148-151]. Thus, it is now presumed that if these hepatocyte induction protocols work well in cultured MSC, *in vivo* organ-specific microenvironment of the recipient liver is likely to be even better suited for inducing the transplanted MSC to differentiate into hepatocytes. Therefore, it seems safe to presume that the beneficial effects of MSC thus far observed in animal injury models have been mediated, at least in part, by MSC differentiating to hepatocytes.

However, the other capabilities of MSC cannot be ignored and may be equally important in the observed therapeutic effects. A variety of evidence from animal studies has now indicated that both MSC's direct differentiation and their indirect effects through secretion of factors which stimulate the regeneration of endogenous cells are likely to play important roles in promoting tissue recovery [79, 152-156]. In support of this conclusion, MSC were shown to provide significant therapeutic benefit during acute hepatic failure by releasing chemotactic cytokines that reduce leukocyte infiltrates and hepatocytes death and increase hepatocyte proliferation [156, 157]. For example, recent studies by Tsai et al. showed that the direct injection of MSC into rats with CCl4-induced liver fibrosis resulted in a significant reduction in the liver fibrosis. However, although MSC engrafted, they did not differentiate into albumin- producing cells, but secreted cytokines that promoted liver regeneration and thereby restored liver function [144].

In addition, other studies have now revealed an additional property of MSC that may indicate that they are ideally suited for treatment of liver diseases involving fibrosis: the ability to enhance fibrous matrix degradation, likely through the induction of metalloproteinases [136, 158-164]. Moreover, other researchers have found that MSC are able to prevent liver fibrosis by suppressing the function of activated hepatic stellate cells, inducing their apoptotic death and diminishing collagen synthesis [155]. Studies like those by Lin et al. have shown that MSC may utilize multiple mechanisms to exert their effects, both engrafting and differentiating into albumin-producing cells, and producing metalloproteinases that significantly reduced the collagen deposits in a rat model of chronic liver fibrosis [165]. However, these promising results must be interpreted carefully and with tempered enthusiasm, because other studies have suggested that under different conditions, transplanted MSC may actually contribute to the myofibroblast pool and thus enhance the fibrotic process within the liver [159, 166-169]. This has led to the current feeling within the field that the effect of MSC will probably vary with the nature of the liver injury/disease that is being treated, the specific experimental model in which the therapy is being tested, and perhaps even the time frame of MSC application, such that MSC could be beneficial if administered at certain stages of disease progression and harmful if administered at other stages. Thus, it appears that the therapeutic potential of MSC may have to be investigated for each specific disease/injury to be treated to delineate the optimal time frame and population to be administered to achieve the desired effect, ensuring they provide benefit rather than harm.

2.3.2.2 The fetal sheep model

All of the afore-mentioned studies exploring the therapeutic properties of MSC in model systems generated by inducing an external stress, such as chemical- or radiation-induced injury or by depleting a specific cell type in the recipient, e.g., partial hepatectomy, have provided compelling evidence that MSC represent valuable cellular therapeutics for liver disease. However, they have also revealed that whether MSC will exert their beneficial or harmful effects is dictated largely by the presence of activated cells and the microenvironment within the injured or diseased organ at the time of transplantation. What is clear is that the microenvironmental conditions that surround MSC play a crucial role in determining the fate adopted by MSC *in vitro* and *in vivo*. The cloning of Dolly the sheep certainly represents the most dramatic example of the power that the microenvironment can exert on cell fate [170]. The microenvironment, in this case the cytoplasm of an enucleated egg, induced the nucleus taken from an adult somatic cell to completely reset its developmental gene expression clock and reveal its true potential. To fully exploit the vast therapeutic potential of MSC, a deep understanding of the mechanisms that control the cell fate and their efficient application to drive differentiation towards the hepatocytic lineage are urgently needed; such understanding will require an adequate model system.

The ideal experimental model would allow transplantation of human MSC, which could then engraft and differentiate/reprogram under normal physiological conditions, in the absence of injury/insult. Additionally, such a model should allow the generation of a broad spectrum of differentiation states of the donor-derived cells in the desired tissue at adequate levels to enable delineation of the mechanisms that participated in their generation. Irrespective of the source of donor cells and mechanisms involved in reprogramming, however, the first key step for proper function is for the cells to reach the target organ. The circulatory system provides an efficient stem cell distribution system throughout life. During fetal life, a series of well-established migratory processes, likely employing the circulatory system, ensure that adequate numbers of appropriate stem/progenitor cells reach the target tissues/organs when needed. This carefully regulated migration is accomplished by the dynamic expression of an array of adhesion molecules and release, by the tissue, of specific chemokines/chemoattractants that alert the circulating stem cells when and where they are most needed. Once the stem cell reaches the target tissue, the permissive milieu induces the entering stem cells to proliferate and differentiate to produce the required type(s) of cells. The existence of this highly permissive milieu is very likely associated with the continuous need for new cells during fetal development.

With these permissive aspects of the developing early gestational-age fetus in mind, we reasoned that it might represent a perfect platform in which to study the properties of human MSC. The transplanted MSC could piggy-back on the naturally occurring migratory pathways, and thus be efficiently disseminated throughout the fetus to the various developing tissues. Once there, they would then be naturally influenced by the host proliferation/differentiation environment to adopt a specific cellular fate, assuming that the transplanted cells harbor the potential in question. By performing the transplant at a point in development when all the organs had begun to differentiate but there was still a need for exponential growth and differentiation, we hypothesized that the fetal milieu would support the possibility of reprogramming of cellular fate through a bombardment of proliferation/differentiation stimuli without forcing the transplanted cells to adopt a specific fate by damaging/inducing regeneration within a particular organ. If the supposition that the appropriate microenvironmental influence can induce a cell with a mature phenotype to regress into an undifferentiated state, directly reprogram a cell to an alternate fate, and/or induce a primitive stem cell to start differentiating into a new lineage, then the fetus should represent an ideal model system in which to examine the full potential of MSC, and other adult stem cells.

In addition to providing a unique signaling environment that can drive migration and differentiation of the transplanted MSC, the fetus also represents a unique recipient from an immunological perspective. In contrast to other model systems routinely used to study stem cell transplantation and the therapeutic potential of MSC, the fetal sheep recipient has an immature, but functioning immune system. In early immunologic development, before thymic processing of mature lymphocytes, the fetus appears to be largely tolerant of foreign antigens [171, 172]. Therefore, if the transplant is performed at the appropriate stage of development, the fetus is able to support the engraftment/differentiation of MSC (and other adult stem cells) in the absence of irradiation or other myeloablative therapies. Furthermore, exposure to foreign antigens during this period often results in sustained tolerance, which can become permanent if the presence of antigen is maintained [100, 173]. By taking advantage of this so-called "window of opportunity" and performing the transplant during the "pre-immune" stage of development, it is possible to reach significant levels of allogeneic sheep cells and xenogeneic human cells within the fetal sheep, in the absence of irradiation or other myeloablative therapies [174-180], to create a lifelong chimera [181], and induce stable, donor-specific immune tolerance.

In addition to the unique characteristics of the fetus itself as a recipient, there are several additional advantages of selecting sheep as an animal model: 1) sheep are fairly close in size to humans during development and throughout life, which should greatly facilitate, or even eliminate the need for, scale-up of the protocol for clinical human therapies once promising results have been obtained in the fetal sheep model, 2) the physiology and developmental processes are similar and therefore, sheep have been for decades the model to study normal fetal growth and fetal abnormalities [182-185], 3) in contrast to mice and rats, sheep are outbred, and thus present a diverse genetic background, just like humans, 4) the development of sheep immune system has been extensively studied and it closely parallels that of humans [186-194], 5) the long lifespan and large size allows the study of cellular fate in the same animal for several years after transplantation, which provides critical answers about long-term efficacy and safety of the therapy in question. Collectively, these properties make the fetal sheep an ideal model in which to test the therapeutic potential of MSC and obtain results that could readily be translated into clinical studies.

2.3.2.2.1 Results obtained in fetal sheep model

In order to investigate the *in vivo* differentiation potential of human MSC in the absence of injury/selective pressure, we isolated several clonal MSC populations from adult BM by magnetic sorting, using an antibody against Stro-I [195]. Although the antigen recognized by this antibody has not yet been identified, we found that by triple-labeling BM cells with antibodies against Stro-I, CD45 and GlyA and selecting for Stro-1+CD45-GlyA- cells, we can reliably obtain a homogenous population that is highly enriched, both phenotypically and functionally, for MSC. This selected population has therefore been used for all of our studies to examine human MSC differentiative potential.

To rigorously test whether MSC could generate significant numbers of hepatocytes *in vivo*, we examined the ability of clonally-derived human MSC from adult BM to generate functional albumin-producing hepatocytes *in vivo* following transplantation into fetal sheep recipients, comparing two routes of administration, intraperitoneal (IP) and intrahepatic (IH) [40]. Human hepatocytes formed after transplantation of BM-MSC into fetal sheep were then identified by HEP-1 staining, coupled with human-specific fluorescence *in situ* hybridization. Our results showed that, although MSC efficiently generated significant numbers of hepatocytes by both routes of administration, the IH injection resulted in a 5-fold increase in the number of hepatocytes generated, when compared to the IP route (12.5% ± 3.5% versus 2.6% ± 0.4%) [196]. In addition to higher levels of hepatocytes, the route of cell administration also exerted a marked effect on the pattern of distribution of the generated hepatocytes. Sheep that received an IP injection exhibited a preferential periportal distribution (in acinar zone 1) of donor-derived hepatocytes that produced high levels of albumin [40], while IH-transplanted animals contained donor-derived (human) hepatocytes dispersed throughout the parenchyma (acinar zone 2) that expressed minimal amounts of albumin. Previous results have demonstrated that localization of the hepatocyte within the liver is strictly associated with the levels of synthesis of certain plasma proteins, such as albumin. Hepatocytes localized in the periportal area of the liver produce higher levels of albumin, compared to hepatocytes situated in other lobular zones [197-199]. These studies thus provided compelling evidence that MSC represent a valuable source of cells for liver repair and regeneration and demonstrate that, by altering the site of injection, generation of hepatocytes occurs in different hepatic zones, and the resultant hepatocytes exhibit differing functionality, just like their naturally-occurring counterparts. These results are highly relevant for designing a potential cellular therapy for liver regeneration, as depending on whether the overall goal of the therapy is to provide hepatocytes to restore the liver architecture or to achieve normal levels of a secreted therapeutic protein into the circulation, different routes of injection would likely be needed. However, if one wishes to achieve functional repopulation of the liver, it is possible that a transplantation approach combining both routes of administration would be the most effective.

In other studies, we evaluated the ability of MSC derived from the fetal kidney to form hepatic cells *in vivo* and *in vitro* [200]. Like their BM counterparts, these cells gave rise to significant numbers of human albumin-producing hepatocyte-like cells upon *in utero* transplantation into fetal sheep. Furthermore, after culture in specific inducing media, cells with hepatocyte-like morphology and phenotype were obtained, suggesting that metanephric-derived MSC could also serve as a source of cells with hepatic repopulating ability. Similar results were also obtained in the fetal sheep model, using a novel, adherent MSC-like cell population isolated from umbilical cord blood, which the authors termed unrestricted somatic stem cells, or "USSC" [201]. This cord blood-derived MSC population was capable of giving rise to albumin-producing human parenchymal hepatic cells at levels of >20% in the recipient liver, in the absence of any injury or genetic defect. Importantly, cell fusion was not required for hepatocyte formation in any of these studies, demonstrating that, at least in this model, human MSC isolated from several different sources all had the ability to directly reprogram to functional hepatocytes.

Another key aspect to assessing the utility of stem cell therapy for regenerative medicine for the liver, and for other organs as well, is the mechanism whereby the transplanted cells replace/repopulate the recipient liver [40]. Indeed, there has been a great deal of controversy about the mechanism by which MSC reprogram and differentiate into other cell lineages, such as hepatocytes. Several researchers have shown that cell fusion could be one of the mechanisms by which MSC appear to give rise to hepatocytes, rather than true reprogramming/transdifferentiation [202]. Furthermore, evidence suggests that the means by which the transplanted MSC contribute to the recipient liver is strictly dependent on the model system employed. For example, an animal model in which proliferation of endogenous hepatocytes has been arrested, such as those using chemical-induced injury, will require replication of the transplanted cells and therefore, favoring transdifferentiation of the transplanted MSC. On the other hand, in an animal model that promotes proliferation of endogenous and MSC-derived hepatocytes, both mechanisms are possible, but fusion seems to be favored.

Using the fetal sheep model made it possible for us to show that MSC could give rise directly to cells within the liver without the need for first forming hematopoietic elements [41]. In more recent studies, we have now shown that the ability to directly contribute to liver repopulation without the need for a hematopoietic intermediate enables the transplanted MSC to rapidly begin contributing to the growing liver, producing cells with hepatic markers within as little as 24 or 48 hours post-transplantation [41]. The findings of these more recent studies confirmed our prior findings regarding the lack of a need for fusion, and furthered our understanding of the mechanism of hepatic repopulation by demonstrating that the generation of hepatocytes occurs independently of the transfer of either mitochondria or membrane-derived vesicles between the transplanted donor cells and the cells of the recipient liver [41]. These findings thus provide strong evidence to support genetic reprogramming and differentiation of the transplanted stem cells. The lack of fusion as a requirement for liver repopulation was in contrast to the results of numerous other studies employing injury models, raising the possibility that the efficacy and mechanism of stem cell repair will likely depend upon not only the stem cell population being transplanted, but also the nature of the injury/defect within the liver, and therefore the conditions within the hepatic microenvironment at the time of stem cell transplantation.

Therefore, we performed studies to begin delineating the mechanism(s) of hepatocyte formation following transplantation of human MSC in the fetal sheep model, which we felt would be ideal for this analysis given the robust generation of human-derived hepatocytes. We labeled human BM-MSC with CFSE, which irreversible stains the plasma membrane [203, 204], or DiD, which labels all cell membranes, membrane-derived vesicles, and intracellular organelles such as mitochondria [205-207]. Consequently, pre-immune fetal sheep were IP injected with either CSFE-positive MSC alone or CFSE-positive MSC in combination with DiD-positive MSC. After transplantation, peripheral blood and peritoneal lavage were assayed for the presence of the cells. At 20h post-transplant, cells were already present in the peripheral blood, and all transplanted cells had exited the peritoneum by 96h. Confocal microscopic analysis for the presence of CFSE+ or DiD+ cells revealed that the transplanted cells initially appeared in the liver at 25h post-transplant, and their numbers then increased, reaching a maximum at 40h post-transplant. The next step was to evaluate if the cells, once in the liver, commenced proliferation before or after initiating differentiation towards tissue-specific cells. At all time points after transplantation, 95% of the CFSE+ or DiD+ cells were also positive for Ki67, indicating that the cells had already begun, or simply continued, to proliferate upon arrival to the liver. These results confirmed that the higher levels of the cells observed at later time points was likely due to the proliferation of the initial MSC that engrafted in the liver and not a result of more cells engrafting in the organ. These studies have important clinical implications, since they suggest that, independently of the low initial percentage of MSC engraftment into a certain tissue, the real contribution of the cell to that tissue does not only depend on the initial engraftment levels but also on the tissue's intrinsic proliferative capacity. Following engraftment of transplanted cells into the liver, hepatoblasts were generated that, due to their intrinsic proliferative capacity [208], continued proliferating and further contributing to the chimeric tissue [196]. In contrast, if one were developing a therapy for which the transplanted cell needed to differentiate into a quiescent cell, such as a terminally differentiated neuron, the contribution of that cell to the tissue would be limited to the initial levels of engraftment.

We next examined the timeline of MSC differentiation into organ-specific cell types in the liver, identifying differentiation of the transplanted cells by their simultaneous positivity for CFSE or DiD and α-Fetoprotein, since during normal fetal liver development, hepatocytes acquire the expression of this protein [209, 210]. At 25h post-transplant, cells that were positive for CFSE or DiD, were already expressing α-fetoprotein, indicating that the transplanted MSC were not only present in the tissue at this first time point of analysis, but they were already differentiated into a hepatocyte-like phenotype. These results thus showed, for the first time, that transplanted MSC engraft within the recipient liver, proliferate, and rapidly commence hepatocytic differentiation. To begin unraveling the mechanism by which MSC seemingly gave rise to hepatocytes in the fetal liver, we performed fluorescence in situ hybridization (FISH) using a human- and a sheep-specific probe, coupled with confocal microscopy for the CFSE or DiD labels. The complete lack of hybridization of the nuclei of CFSE+ or DiD+ cells to the sheep probe conclusively demonstrated that the transplanted human MSC gave rise to hepatocytes independent of fusion or membrane vesicle/organelle transfer, and by true reprogramming/transdifferentiation [211]. In fact, we observed a sequential differentiation program, in which cells gradually expressed markers of differentiation, from the most undifferentiated cell to the mature fully differentiated cell type in the organ in question. Understanding the complete pathway of differentiation could ultimately make it possible to provide a cell driven to a precise point in differentiation to correct of a disease by providing exactly the cell type most needed.

Despite the significance of our findings in the fetal sheep model, it is important to note that, even when using an optimal route of injection, the overall levels of liver engraftment may still be too low to achieve cure in many clinical situations. While the fetus has long been presumed to be immune-naïve, recent studies in mice have suggested that this may not be the case, since syngeneic hematopoietic stem cells engraft at higher levels than allogeneic cells of the same phenotype following *in utero* transplantation. Thus, it is possible that some rudimentary immune surveillance exists within the fetus and limited the levels of engraftment within the liver. MSC are well known for their immune-evading and immunomodulatory properties, but studies in murine and swine models have provided evidence that MSC are not completely invisible to the recipient's immune system, nor immune-inert. Indeed, upon *in vivo* administration, MSC are able to trigger immune responses, resulting in rejection of the transplanted cells [212-216]. Based on these prior studies, we hypothesized that further reducing the immunogenicity of the MSC prior to transplant might enable us to achieve even higher levels of engraftment and hepatocyte generation, both in this "pre-immune" fetal model and, perhaps, even in recipients with a more developed/mature immune system. To test this hypothesis, we genetically modified human MSC to stably express proteins known to exert potent immunomodulatory/immune-evading properties. The proteins we selected were derived from the ubiquitously prevalent human cytomegalovirus (HCMV). This virus is well known to possess multiple immune evasive strategies, which enable it to enter a state of latency in which it is invisible to immune surveillance, only to re-emerge when conditions are favorable, such as during the period of immuno-suppression following bone marrow or solid organ transplant, and wreak havoc on the immuno-compromised patient. HCMV accomplishes its immuno-evasion due largely to its unique short region (US) proteins. We therefore used a retroviral vector to genetically modify MSC to stably express members of the HCMV US protein family that are known to specifically reduce cytotoxic T cell recognition by different mechanisms, and compared the immunogenicity and immunomodulatory properties of these "US-MSC" to unmodified MSC and to MSC transduced with an empty control vector. Our results revealed that MSC expressing US6 (MSC-US6) and US11 (MSC-US11) exhibited the most pronounced reduction in HLA-I expression and accordingly, induced the lowest level of human or sheep PBMNC proliferation in mixed lymphocyte reactions. Moreover, as there are controversial reports regarding whether reduction in HLA-I expression by HCMV US proteins renders infected cells more susceptible to NK killing [217, 218], we next examined whether forced expression of US6 or US11 predisposed MSC to NK lysis. To our surprise, expression of US6 or US11 protein did not increase the ability of NK cells to target MSC. Moreover, expression of US11 actually protected MSC from NK cytotoxic effects [219]. Based on these promising *in vitro* results, we transplanted MSC-US6, MSC-US11 and MSC-E (control cell line transduced with the empty vector) into fetal sheep recipients by IP injection. Tissues were collected at 60 days post-transplant and analyzed for engraftment and hepatocytic differentiation of the transplanted cells. Using both quantitative PCR and immunofluorescence, we determined that expression of either the US6 or the US11 HCMV protein on the transplanted MSC led to significantly enhanced levels of liver engraftment compared to those seen with MSC-E. However, although the increased levels of engrafted cells translated into increased levels of cells expressing HEPAR-I, many of these did not express albumin or Ov-6 [220]. This suggests that the hepatocytes generated by transplantation of these genetically modified cells were of a broad range of differentiation, not immature and not completely mature at the time of tissue collection. These results clearly show that by enhancing the immuno-evasive MSC properties, the levels of engraftment and hepatocyte generation can be significantly increased to provide a more successful regenerative therapy, even in the context of a fetal recipient whose immune system is presumed to be largely immature.

3. Clinical trials

Despite the promising results obtained in animal models, the use of MSC to treat liver diseases is still in its infancy, and very few clinical trials using these cells have been performed. Several concerns still exist over this therapy regarding the best administration route, and the possibility of cellular fusion, with the inherent risks that may accompany the presence of hepatocytes that are potentially genetically unstable within the environment of a diseased liver. In 2007, Mohamadnejad *et al.* reported that infusion of BM-derived MSC via a peripheral vein was found to be well tolerated and to have a definite therapeutic effect, since the quality of life of all 4 transplanted patients was improved by 12 months post-infusion, and the model for end-stage liver disease (MELD) scores for 2 of the 4 patients improved significantly during the course of the trial.

Another 8 patients with end-stage liver disease due to different etiologies received 30-50 million BM-derived MSC injected into a peripheral vein or the portal vein. Treatment was well tolerated by all patients, and liver function improved as verified by MELD scores. However, both of these trials lacked a control arm, and the number of patients was very small. Another study examining the safety and efficacy of umbilical cord-derived MSC (UC-MSC) in 45 patients with decompensated liver cirrhosis demonstrated that both patients that received UC-MSC and those in the control arm that received saline suffered no significant side-effects or complications. However, in patients treated with UC-MSC there was a significant reduction in the volume of ascites when compared with control. Also, UC-MSC therapy significantly improved liver function, as evidenced by the increase of serum albumin levels, decrease in total serum bilirubin levels, and decrease in the sodium model for end-stage liver disease scores [151]. Forty patients with end-stage liver failure due to chronic hepatitis C were selected for a controlled study in which 10 received autologous bone marrow-derived mesenchymal stem cells that were pre-induced to the hepatic lineage in vitro prior to transplant. Three groups were included in this trial: one in which 10 patients received the MSC by an intrasplenic route, another in which the 10 patients received cells by intrahepatic route, and a control group consisting of the remaining 20 patients. Patients in all groups were followed up using clinical and laboratory parameters and evaluated by MELD scores, fatigue scale, and performance status. Both transplanted groups, regardless of administration route, showed significant improvement when compared to the control [172]. Another phase 1 trial, in which four patients with decompensated liver cirrhosis were included, demonstrated that, after infusion of approximately 32 million bone marrow derived MSC through a peripheral vein, MELD scores of 2 patients improved by the end of follow-up as well as the quality of life of all four patients [173]

Collectively, these studies provide hope that BM-derived cells may prove to be a valuable resource for cell-based therapies for liver disease. However, the results of these studies must be interpreted with some trepidation, given the limited number of patients enrolled in each trial and the lack of appropriate controls in some of the studies. Furthermore, since the cells in these trials were autologously-derived, there was no way for the investigators to assess the actual engraftment, persistence, or differentiative potential of the transplanted cells, leaving the mechanism responsible for the observed clinical improvements open to speculation.

4. Conclusion

Presently, chronic liver disease constitutes one of the leading worldwide causes of death. It can be triggered by a wide array of insults, including, but not limited to hepatitis infection, alcohol consumption, exposure to toxic chemicals, and congenital defects. Currently, the only definitive treatment for chronic liver disease is whole or partial liver transplantation. Due to the limited availability of donor livers and the severe morbidity and mortality associated with this treatment, there is an urgent need for new therapeutic approaches. While hepatocyte transplantation represents an option, the limited availability of donor livers and the inability to maintain and expand hepatocytes in culture precludes this option from becoming a clinically viable treatment option. MSC offer several advantages such as: extensive expansion in vitro, multipotent differentiative capacity, the ability to selectively and efficiently migrate to sites of injury following systemic infusion, their potent immunomodulatory and trophic properties, and the ease with which they can be genetically modified, making it possible to use autologous cells, even in the case of underlying genetic disease. MSC can be isolated from a wide range of human tissues and, despite subtle differences, they all share the same beneficial characteristics, making MSC transplantation a promising approach for liver repair/regeneration. However, in order to maximize MSC capabilities for improving/recovering the liver mass and/or function depending on the particular disease/injury, several issues must still be resolved: selection of the most therapeutic MSC source; standardization of the protocols for unequivocally isolating the desired MSC population from each tissue; more complete *in vitro* and *in vivo* characterization of the differentiative potential of the cells; and further optimization of the route, cell dose, timing, and degree of desired MSC differentiation. Once these questions have been answered, the knowledge gained during *in vitro* and *in vivo* studies in animal models could be safely and efficiently translated into humans to develop an appropriate and successful therapy for chronic liver disease.

5. Acknowledgement

This work was supported by National Institutes of Health (Bethesda, MD, USA) grants HL73737 and HL97623.

6. References

[1] Lee, W.M., et al., *Acute liver failure: Summary of a workshop.* Hepatology, 2008. 47(4): p. 1401-15.
[2] Lucey, M.R., et al., *Effect of alcoholic liver disease and hepatitis C infection on waiting list and posttransplant mortality and transplant survival benefit.* Hepatology, 2009. 50(2): p. 400-406.
[3] Sokal, E.M., et al., *End-stage Liver Disease and Liver Transplant: Current Situation and Key Issues.* Journal of Pediatric Gastroenterology and Nutrition, 2008. 47(2): p. 239-246 10.1097/MPG.0b013e318181b21c.
[4] Jalan, R., *Acute liver failure: current management and future prospects.* J Hepatol, 2005. 42 Suppl(1): p. S115-23.
[5] 2012; Available from: http://optn.transplant.hrsa.gov/data/.
[6] Miniño A., S.M., J.X u, and K. Kochanek, *Deaths: final data for 2008.* Natl Vital Stat Rep., 2011. 59(10):1-152.
[7] O'Leary, J.G., R. Lepe, and G.L. Davis, *Indications for Liver Transplantation.* Gastroenterology, 2008. 134(6): p. 1764-1776.
[8] Chung, H., et al., *Retinal Complications in Patients With Solid Organ or Bone Marrow Transplantations.* Transplantation, 2007. 83(6): p. 694-699 10.1097/01.tp.0000259386.59375.8a.
[9] Patel, H., et al., *Posttransplant lymphoproliferative disorder in adult liver transplant recipients: A report of seventeen cases.* Leukemia & Lymphoma, 2007. 48(5): p. 885-891.
[10] Tamsel, S., et al., *Vascular complications after liver transplantation: evaluation with Doppler US.* Abdominal Imaging, 2007. 32(3): p. 339-347.
[11] Alison, M., S. Islam, and S. Lim, *Stem cells in liver regeneration, fibrosis and cancer: the good, the bad and the ugly.* J Pathol, 2008.
[12] Dahlke, M.H., et al., *Stem cell therapy of the liver--fusion or fiction?* Liver Transpl, 2004. 10(4): p. 471-9.
[13] Enns, G.M. and M.T. Millan, *Cell-based therapies for metabolic liver disease.* Mol Genet Metab, 2008. 95(1-2): p. 3-10.
[14] Fausto, N., *Liver regeneration and repair: hepatocytes, progenitor cells, and stem cells.* Hepatology, 2004. 39(6): p. 1477-87.
[15] Kallis, Y.N., M.R. Alison, and S.J. Forbes, *Bone marrow stem cells and liver disease.* Gut, 2007. 56(5): p. 716-24.
[16] Lysy, P.A., et al., *Stem cells for liver tissue repair: current knowledge and perspectives.* World J Gastroenterol, 2008. 14(6): p. 864-75.
[17] Oertel, M. and D.A. Shafritz, *Stem cells, cell transplantation and liver repopulation.* Biochim Biophys Acta, 2008. 1782(2): p. 61-74.
[18] Porada, C.D., E.D. Zanjani, and G. Almeida-Porad, *Adult mesenchymal stem cells: a pluripotent population with multiple applications.* Curr Stem Cell Res Ther, 2006. 1(3): p. 365-9.
[19] Strom, S.C., J.R. Chowdhury, and I.J. Fox, *Hepatocyte transplantation for the treatment of human disease.* Semin Liver Dis, 1999. 19(1): p. 39-48.

[20] Grompe, M., *Principles of therapeutic liver repopulation.* J Inherit Metab Dis, 2006. 29(2-3): p. 421-5.
[21] Almeida-Porada, G., E.D. Zanjani, and C.D. Porada, *Bone marrow stem cells and liver regeneration.* Experimental hematology, 2010. 38(7): p. 574-80.
[22] Nussler, A., et al., *Present status and perspectives of cell-based therapies for liver diseases.* Journal of Hepatology, 2006. 45(1): p. 144-159.
[23] Serralta, A., et al., *Influence of Preservation Solution on the Isolation and Culture of Human Hepatocytes From Liver Grafts.* Cell Transplantation, 2005. 14(10): p. 837-843.
[24] Clayton, D.F. and J.E. Darnell, *Changes in liver-specific compared to common gene transcription during primary culture of mouse hepatocytes.* Molecular and Cellular Biology, 1983. 3(9): p. 1552-1561.
[25] Serralta, A., et al., *Functionality of cultured human hepatocytes from elective samples, cadaveric grafts and hepatectomies.* Toxicology in Vitro. 17(5-6): p. 769-774.
[26] Han, B., et al., *Cellular Loss After Allogenic Hepatocyte Transplantation.* Transplantation, 2009. 87(1): p. 1-5 10.1097/TP.0b013e3181919212.
[27] Grossman, M., et al., *Successful ex vivo gene therapy directed to liver in a patient with familial hypercholesterolaemia.* Nat Genet, 1994. 6(4): p. 335-341.
[28] Fox, I.J., et al., *Treatment of the Crigler Najjar Syndrome Type I with Hepatocyte Transplantation.* New England Journal of Medicine, 1998. 338(20): p. 1422-1427.
[29] Horslen, S.P., et al., *Isolated Hepatocyte Transplantation in an Infant With a Severe Urea Cycle Disorder.* Pediatrics, 2003. 111(6): p. 1262-1267.
[30] Ambrosino, G., et al., *Isolated Hepatocyte Transplantation for Crigler-Najjar Syndrome Type 1.* Cell Transplantation, 2005. 14(2-3): p. 151-157.
[31] Schmelzer, E., E. Wauthier, and L.M. Reid, *The phenotypes of pluripotent human hepatic progenitors.* Stem cells, 2006. 24(8): p. 1852-8.
[32] Susick, R., et al., *Hepatic progenitors and strategies for liver cell therapies.* Annals of the New York Academy of Sciences, 2001. 944: p. 398-419.
[33] Schmelzer, E., et al., *Human hepatic stem cells from fetal and postnatal donors.* The Journal of experimental medicine, 2007. 204(8): p. 1973-87.
[34] Libbrecht, L., et al., *Deep intralobular extension of human hepatic 'progenitor cells' correlates with parenchymal inflammation in chronic viral hepatitis: can 'progenitor cells' migrate?* J Pathol, 2000. 192(3): p. 373-8.
[35] Lowes, K.N., et al., *Oval cell numbers in human chronic liver diseases are directly related to disease severity.* Am J Pathol, 1999. 154(2): p. 537-41.
[36] Xiao, J.C., et al., *Hepatic progenitor cells in human liver cirrhosis: immunohistochemical, electron microscopic and immunofluorencence confocal microscopic findings.* World J Gastroenterol, 2004. 10(8): p. 1208-11.
[37] Khan, A.A., et al., *Human fetal liver-derived stem cell transplantation as supportive modality in the management of end-stage decompensated liver cirrhosis.* Cell Transplantation, 2010. 19(4): p. 409-18.
[38] Almeida-Porada, G., et al., *Differentiative potential of human metanephric mesenchymal cells.* Exp Hematol, 2002. 30(12): p. 1454-62.
[39] Almeida-Porada, G., et al., *Formation of human hepatocytes by human hematopoietic stem cells in sheep.* Blood, 2004. 104(8): p. 2582-90.
[40] Chamberlain, J., et al., *Efficient generation of human hepatocytes by the intrahepatic delivery of clonal human mesenchymal stem cells in fetal sheep.* Hepatology, 2007. 46(6): p. 1935-45.

[41] Colletti, E., Airey, J.A., Liu, W., Simmons, P.J., Zanjani, E.D., Porada, C.D., Almeida-Porada, G., *Generation of tissue-specific cells by MSC does not require fusion or donor to host mitochondrial/membrane transfer.* . Stem Cell Research, 2008. In Press.
[42] Jang, Y.Y., et al., *Hematopoietic stem cells convert into liver cells within days without fusion.* Nat Cell Biol, 2004. 6(6): p. 532-9.
[43] Kollet, O., et al., *HGF, SDF-1, and MMP-9 are involved in stress-induced human CD34+ stem cell recruitment to the liver.* J Clin Invest, 2003. 112(2): p. 160-9.
[44] Krause, D.S., et al., *Multi-organ, multi-lineage engraftment by a single bone marrow-derived stem cell.* Cell, 2001. 105(3): p. 369-77.
[45] Lagasse, E., et al., *Purified hematopoietic stem cells can differentiate into hepatocytes in vivo.* Nat Med, 2000. 6(11): p. 1229-34.
[46] Mackenzie, T.C. and A.W. Flake, *Human mesenchymal stem cells persist, demonstrate site-specific multipotential differentiation, and are present in sites of wound healing and tissue regeneration after transplantation into fetal sheep.* Blood Cells Mol Dis, 2001. 27(3): p. 601-4.
[47] Muraca, M. and A.B. Burlina, *Liver and liver cell transplantation for glycogen storage disease type IA.* Acta Gastroenterol Belg, 2005. 68(4): p. 469-72.
[48] Muraca, M., et al., *Liver repopulation with bone marrow derived cells improves the metabolic disorder in the Gunn rat.* Gut, 2007. 56(12): p. 1725-35.
[49] Nakamura, T., et al., *Significance and therapeutic potential of endothelial progenitor cell transplantation in a cirrhotic liver rat model.* Gastroenterology, 2007. 133(1): p. 91-107 e1.
[50] Newsome, P.N., et al., *Human cord blood-derived cells can differentiate into hepatocytes in the mouse liver with no evidence of cellular fusion.* Gastroenterology, 2003. 124(7): p. 1891-900.
[51] Theise, N.D., et al., *Derivation of hepatocytes from bone marrow cells in mice after radiation-induced myeloablation.* Hepatology, 2000. 31(1): p. 235-40.
[52] Petersen, B.E., et al., *Bone marrow as a potential source of hepatic oval cells.* Science, 1999. 284(5417): p. 1168-70.
[53] Porada, C.D. and G. Almeida-Porada, *Mesenchymal stem cells as therapeutics and vehicles for gene and drug delivery.* Advanced drug delivery reviews, 2010. 62(12): p. 1156-66.
[54] Porada, C.D., E.D. Zanjani, and G. Almeida-Porad, *Adult mesenchymal stem cells: a pluripotent population with multiple applications.* Current stem cell research & therapy, 2006. 1(3): p. 365-9.
[55] Afanasyev B. V, E.E.E., A. R Zander, St Petersburg, S. Pavlov, L. Tolstoy, *A . J . Friedenstein , founder of the mesenchymal stem cell concept.* Cell Ther Transplant, 2010. 1 (3) Pages: 35-38.
[56] Pittenger, M.F., et al., *Multilineage potential of adult human mesenchymal stem cells.* Science, 1999. 284(5411): p. 143-7.
[57] Simmons, P.J. and B. Torok-Storb, *Identification of stromal cell precursors in human bone marrow by a novel monoclonal antibody, STRO-1.* Blood, 1991. 78(1): p. 55-62.
[58] Caplan, A.I., *The mesengenic process.* Clinics in plastic surgery, 1994. 21(3): p. 429-35.
[59] Dominici, M., et al., *Minimal criteria for defining multipotent mesenchymal stromal cells. The International Society for Cellular Therapy position statement.* Cytotherapy, 2006. 8(4): p. 315-7.
[60] Digirolamo, C.M., et al., *Propagation and senescence of human marrow stromal cells in culture: a simple colony-forming assay identifies samples with the greatest potential to propagate and differentiate.* Br J Haematol, 1999. 107(2): p. 275-81.

[61] D'Ippolito, G., et al., *Age-related osteogenic potential of mesenchymal stromal stem cells from human vertebral bone marrow.* J Bone Miner Res, 1999. 14(7): p. 1115-22.
[62] Bruder, S.P., N. Jaiswal, and S.E. Haynesworth, *Growth kinetics, self-renewal, and the osteogenic potential of purified human mesenchymal stem cells during extensive subcultivation and following cryopreservation.* J Cell Biochem, 1997. 64(2): p. 278-94.
[63] Colter, D.C., et al., *Rapid expansion of recycling stem cells in cultures of plastic-adherent cells from human bone marrow.* Proc Natl Acad Sci U S A, 2000. 97(7): p. 3213-8.
[64] Bruder, S.P., et al., *Bone regeneration by implantation of purified, culture-expanded human mesenchymal stem cells.* J Orthop Res, 1998. 16(2): p. 155-62.
[65] Kadiyala, S., et al., *Culture expanded canine mesenchymal stem cells possess osteochondrogenic potential in vivo and in vitro.* Cell Transplant, 1997. 6(2): p. 125-34.
[66] Awad, H.A., et al., *Autologous mesenchymal stem cell-mediated repair of tendon.* Tissue Eng, 1999. 5(3): p. 267-77.
[67] Ferrari, G., et al., *Muscle regeneration by bone marrow-derived myogenic progenitors.* Science, 1998. 279(5356): p. 1528-30.
[68] Toma, C., et al., *Human Mesenchymal Stem Cells Differentiate to a Cardiomyocyte Phenotype in the Adult Murine Heart.* Circulation, 2002. 105(1): p. 93-98.
[69] Prockop, D.J., *Marrow stromal cells as stem cells for nonhematopoietic tissues.* Science, 1997. 276(5309): p. 71-4.
[70] Sanchez-Ramos, J.R., *Neural cells derived from adult bone marrow and umbilical cord blood.* J Neurosci Res, 2002. 69(6): p. 880-93.
[71] Kohyama, J., et al., *Brain from bone: efficient "meta-differentiation" of marrow stroma-derived mature osteoblasts to neurons with Noggin or a demethylating agent.* Differentiation, 2001. 68(4-5): p. 235-44.
[72] Timper, K., et al., *Human adipose tissue-derived mesenchymal stem cells differentiate into insulin, somatostatin, and glucagon expressing cells.* Biochemical and Biophysical Research Communications, 2006. 341(4): p. 1135-1140.
[73] Wong, R.S., *Mesenchymal stem cells: angels or demons?* J Biomed Biotechnol. 2011: p. 459510.
[74] Jiang, W.H., et al., *Migration of intravenously grafted mesenchymal stem cells to injured heart in rats.* Sheng Li Xue Bao, 2005. 57(5): p. 566-72.
[75] Mouiseddine, M., et al., *Human mesenchymal stem cells home specifically to radiation-injured tissues in a non-obese diabetes/severe combined immunodeficiency mouse model.* Br J Radiol, 2007. 80 Spec No 1: p. S49-55.
[76] Fu, X., et al., *Migration of bone marrow-derived mesenchymal stem cells induced by tumor necrosis factor-alpha and its possible role in wound healing.* Wound Repair Regen, 2009. 17(2): p. 185-91.
[77] Spaeth, E., et al., *Inflammation and tumor microenvironments: defining the migratory itinerary of mesenchymal stem cells.* Gene Ther, 2008. 15(10): p. 730-8.
[78] Yagi, H., et al., *Mesenchymal stem cells: Mechanisms of immunomodulation and homing.* Cell Transplant. 19(6): p. 667-79.
[79] Caplan, A.I. and J.E. Dennis, *Mesenchymal stem cells as trophic mediators.* J Cell Biochem, 2006. 98(5): p. 1076-84.
[80] Spees, J.L., et al., *Mitochondrial transfer between cells can rescue aerobic respiration.* Proc Natl Acad Sci U S A, 2006. 103(5): p. 1283-8.
[81] Bartel, D.P., *MicroRNAs: genomics, biogenesis, mechanism, and function.* Cell, 2004. 116(2): p. 281-97.

[82] Krichevsky, A.M., et al., *Specific microRNAs modulate embryonic stem cell-derived neurogenesis.* Stem Cells, 2006. 24(4): p. 857-64.
[83] Chen, J.F., et al., *The role of microRNA-1 and microRNA-133 in skeletal muscle proliferation and differentiation.* Nat Genet, 2006. 38(2): p. 228-33.
[84] Pedersen, I.M., et al., *Interferon modulation of cellular microRNAs as an antiviral mechanism.* Nature, 2007. 449(7164): p. 919-22.
[85] Guo, L., R.C. Zhao, and Y. Wu, *The role of microRNAs in self-renewal and differentiation of mesenchymal stem cells.* Exp Hematol. 39(6): p. 608-16.
[86] Soland M., M.B., C. D Porada, E.. Zanjani, S. St Jeor and G. Almeida-Porada, *Modulation of Mesenchymal Stem Cell Immunogenicity through Forced Expression of Human Cytomegalovirus Proteins* Blood, 1468:2416a, 2008.
[87] Yamagami, T., Almeida-Porada, G. , *Exploiting molecules involved in fetal-maternal tolerance to overcome immunologic barriers.* ProQuest, 2008: p. 1-145.
[88] Phillips, M.I. and Y.L. Tang, *Genetic modification of stem cells for transplantation.* Adv Drug Deliv Rev, 2008. 60(2): p. 160-72.
[89] Hodgkinson, C.P., et al., *Genetic engineering of mesenchymal stem cells and its application in human disease therapy.* Hum Gene Ther. 21(11): p. 1513-26.
[90] Porada, C.D., et al., *Phenotypic correction of hemophilia A in sheep by postnatal intraperitoneal transplantation of FVIII-expressing MSC.* Exp Hematol. 39(12): p. 1124-1135 e4.
[91] Almeida-Porada, G.a., et al., *Differentiative potential of human metanephric mesenchymal cells.* Experimental hematology, 2002. 30(12): p. 1454-1462.
[92] Fan, C.G., et al., *Characterization and Neural Differentiation of Fetal Lung Mesenchymal Stem Cells.* Cell Transplantation. 14(5): p. 311-321.
[93] in 't Anker, P.S., et al., *Mesenchymal stem cells in human second-trimester bone marrow, liver, lung, and spleen exhibit a similar immunophenotype but a heterogeneous multilineage differentiation potential.* Haematologica, 2003. 88(8): p. 845-852.
[94] Wagner, W., et al., *Comparative characteristics of mesenchymal stem cells from human bone marrow, adipose tissue, and umbilical cord blood.* Exp Hematol, 2005. 33(11): p. 1402-16.
[95] Wang, T.H., Y.S. Lee, and S.M. Hwang, *Transcriptome analysis of common gene expression in human mesenchymal stem cells derived from four different origins.* Methods Mol Biol. 698: p. 405-17.
[96] Guillot, P.V., et al., *Human first-trimester fetal MSC express pluripotency markers and grow faster and have longer telomeres than adult MSC.* Stem Cells, 2007. 25(3): p. 646-54.
[97] Giuliani, M., et al., *Long-lasting inhibitory effects of fetal liver mesenchymal stem cells on T-lymphocyte proliferation.* PLoS One. 6(5): p. e19988.
[98] Chen, P.M., et al., *Immunomodulatory properties of human adult and fetal multipotent mesenchymal stem cells.* J Biomed Sci. 18: p. 49.
[99] Moreno, R., et al., *Fetal liver-derived mesenchymal stem cell engraftment after allogeneic in utero transplantation into rabbits.* Stem Cells Dev. 21(2): p. 284-95.
[100] De Coppi, P., et al., *Isolation of amniotic stem cell lines with potential for therapy.* Nature biotechnology, 2007. 25(1): p. 100-6.
[101] Kogler, G., et al., *A new human somatic stem cell from placental cord blood with intrinsic pluripotent differentiation potential.* The Journal of experimental medicine, 2004. 200(2): p. 123-35.
[102] Choong, P.F., et al., *Generating neuron-like cells from BM-derived mesenchymal stromal cells in vitro.* Cytotherapy, 2007. 9(2): p. 170-83.

[103] Franco Lambert, A.P., et al., *Differentiation of human adipose-derived adult stem cells into neuronal tissue: does it work?* Differentiation, 2009. 77(3): p. 221-8.
[104] Kennea, N.L., et al., *Differentiation of human fetal mesenchymal stem cells into cells with an oligodendrocyte phenotype.* Cell Cycle, 2009. 8(7): p. 1069-79.
[105] Kim, S., et al., *Neural differentiation potential of peripheral blood- and bone-marrow-derived precursor cells.* Brain Res, 2006. 1123(1): p. 27-33.
[106] Moscoso, I., et al., *Differentiation "in vitro" of primary and immortalized porcine mesenchymal stem cells into cardiomyocytes for cell transplantation.* Transplant Proc, 2005. 37(1): p. 481-2.
[107] Tokcaer-Keskin, Z., et al., *Timing of induction of cardiomyocyte differentiation for in vitro cultured mesenchymal stem cells: a perspective for emergencies.* Can J Physiol Pharmacol, 2009. 87(2): p. 143-50.
[108] Wang, T., et al., *Cell-to-cell contact induces mesenchymal stem cell to differentiate into cardiomyocyte and smooth muscle cell.* Int J Cardiol, 2006. 109(1): p. 74-81.
[109] Xie, X.J., et al., *Differentiation of bone marrow mesenchymal stem cells induced by myocardial medium under hypoxic conditions.* Acta Pharmacol Sin, 2006. 27(9): p. 1153-8.
[110] Xu, W., et al., *Mesenchymal stem cells from adult human bone marrow differentiate into a cardiomyocyte phenotype in vitro.* Exp Biol Med (Maywood), 2004. 229(7): p. 623-31.
[111] Gang, E.J., et al., *In vitro endothelial potential of human UC blood-derived mesenchymal stem cells.* Cytotherapy, 2006. 8(3): p. 215-27.
[112] Oskowitz, A., et al., *Serum-deprived human multipotent mesenchymal stromal cells (MSCs) are highly angiogenic.* Stem Cell Res. 6(3): p. 215-25.
[113] Vater, C., P. Kasten, and M. Stiehler, *Culture media for the differentiation of mesenchymal stromal cells.* Acta Biomater. 7(2): p. 463-77.
[114] Chen, L.B., X.B. Jiang, and L. Yang, *Differentiation of rat marrow mesenchymal stem cells into pancreatic islet beta-cells.* World J Gastroenterol, 2004. 10(20): p. 3016-20.
[115] Choi, K.S., et al., *In vitro trans-differentiation of rat mesenchymal cells into insulin-producing cells by rat pancreatic extract.* Biochem Biophys Res Commun, 2005. 330(4): p. 1299-305.
[116] Hishikawa, K., et al., *Gene expression profile of human mesenchymal stem cells during osteogenesis in three-dimensional thermoreversible gelation polymer.* Biochem Biophys Res Commun, 2004. 317(4): p. 1103-7.
[117] Lee, R.H., et al., *Characterization and expression analysis of mesenchymal stem cells from human bone marrow and adipose tissue.* Cell Physiol Biochem, 2004. 14(4-6): p. 311-24.
[118] Nakamura, T., et al., *Temporal gene expression changes during adipogenesis in human mesenchymal stem cells.* Biochem Biophys Res Commun, 2003. 303(1): p. 306-12.
[119] Lange, C., et al., *Liver-specific gene expression in mesenchymal stem cells is induced by liver cells.* World J Gastroenterol, 2005. 11(29): p. 4497-504.
[120] Piryaei, A., et al., *Differentiation of bone marrow-derived mesenchymal stem cells into hepatocyte-like cells on nanofibers and their transplantation into a carbon tetrachloride-induced liver fibrosis model.* Stem Cell Rev. 7(1): p. 103-18.
[121] Sgodda, M., et al., *Hepatocyte differentiation of mesenchymal stem cells from rat peritoneal adipose tissue in vitro and in vivo.* Exp Cell Res, 2007. 313(13): p. 2875-86.
[122] Snykers, S., et al., *Hepatic differentiation of mesenchymal stem cells: in vitro strategies.* Methods Mol Biol. 698: p. 305-14.
[123] Zhao, Q., et al., *Differentiation of human umbilical cord mesenchymal stromal cells into low immunogenic hepatocyte-like cells.* Cytotherapy, 2009. 11(4): p. 414-26.

[124] Lee, K.D., et al., *In vitro hepatic differentiation of human mesenchymal stem cells.* Hepatology, 2004. 40(6): p. 1275-84.

[125] Stock, P., et al., *The generation of hepatocytes from mesenchymal stem cells and engraftment into murine liver.* Nat Protoc. 5(4): p. 617-27.

[126] Campard, D., et al., *Native umbilical cord matrix stem cells express hepatic markers and differentiate into hepatocyte-like cells.* Gastroenterology, 2008. 134(3): p. 833-48.

[127] Zhang, Y.N., P.C. Lie, and X. Wei, *Differentiation of mesenchymal stromal cells derived from umbilical cord Wharton's jelly into hepatocyte-like cells.* Cytotherapy, 2009. 11(5): p. 548-58.

[128] Roubelakis, M.G., et al., *Molecular and proteomic characterization of human mesenchymal stem cells derived from amniotic fluid: comparison to bone marrow mesenchymal stem cells.* Stem Cells Dev, 2007. 16(6): p. 931-52.

[129] Zagoura, D.S., et al., *Therapeutic potential of a distinct population of human amniotic fluid mesenchymal stem cells and their secreted molecules in mice with acute hepatic failure.* Gut.

[130] Kim, D.H., et al., *Effect of partial hepatectomy on in vivo engraftment after intravenous administration of human adipose tissue stromal cells in mouse.* Microsurgery, 2003. 23(5): p. 424-31.

[131] Lysy, P.A., et al., *Persistence of a chimerical phenotype after hepatocyte differentiation of human bone marrow mesenchymal stem cells.* Cell Prolif, 2008. 41(1): p. 36-58.

[132] Hipp, J. and A. Atala, *Sources of stem cells for regenerative medicine.* Stem Cell Rev, 2008. 4(1): p. 3-11.

[133] Miyazaki, M., et al., *Isolation of a bone marrow-derived stem cell line with high proliferation potential and its application for preventing acute fatal liver failure.* Stem Cells, 2007. 25(11): p. 2855-63.

[134] Banas, A., et al., *Rapid hepatic fate specification of adipose-derived stem cells and their therapeutic potential for liver failure.* J Gastroenterol Hepatol, 2008.

[135] Banas, A., et al., *Adipose tissue-derived mesenchymal stem cells as a source of human hepatocytes.* Hepatology, 2007. 46(1): p. 219-28.

[136] Fang, B., et al., *Systemic infusion of FLK1(+) mesenchymal stem cells ameliorate carbon tetrachloride-induced liver fibrosis in mice.* Transplantation, 2004. 78(1): p. 83-8.

[137] Ishikawa, T., et al., *Fibroblast growth factor 2 facilitates the differentiation of transplanted bone marrow cells into hepatocytes.* Cell Tissue Res, 2006. 323(2): p. 221-31.

[138] Luk, J.M., et al., *Hepatic potential of bone marrow stromal cells: development of in vitro co-culture and intra-portal transplantation models.* J Immunol Methods, 2005. 305(1): p. 39-47.

[139] Okumoto, K., et al., *Characteristics of rat bone marrow cells differentiated into a liver cell lineage and dynamics of the transplanted cells in the injured liver.* J Gastroenterol, 2006. 41(1): p. 62-9.

[140] Oyagi, S., et al., *Therapeutic effect of transplanting HGF-treated bone marrow mesenchymal cells into CCl4-injured rats.* J Hepatol, 2006. 44(4): p. 742-8.

[141] Seo, M.J., et al., *Differentiation of human adipose stromal cells into hepatic lineage in vitro and in vivo.* Biochem Biophys Res Commun, 2005. 328(1): p. 258-64.

[142] Zheng, J.F. and L.J. Liang, *Intra-portal transplantation of bone marrow stromal cells ameliorates liver fibrosis in mice.* Hepatobiliary Pancreat Dis Int, 2008. 7(3): p. 264-70.

[143] Jung, K.H., et al., *Effect of human umbilical cord blood-derived mesenchymal stem cells in a cirrhotic rat model.* Liver International, 2009. 29(6): p. 898-909.

[144] Tsai, P.C., et al., *The therapeutic potential of human umbilical mesenchymal stem cells from Wharton's jelly in the treatment of rat liver fibrosis.* Liver Transpl, 2009. 15(5): p. 484-95.
[145] Yan, Y., et al., *Mesenchymal stem cells from human umbilical cords ameliorate mouse hepatic injury in vivo.* Liver International, 2009. 29(3): p. 356-365.
[146] Popp, F.C., et al., *No contribution of multipotent mesenchymal stromal cells to liver regeneration in a rat model of prolonged hepatic injury.* Stem Cells, 2007. 25(3): p. 639-45.
[147] Grossman, M., S.E. Raper, and J.M. Wilson, *Towards liver-directed gene therapy: retrovirus-mediated gene transfer into human hepatocytes.* Somat Cell Mol Genet, 1991. 17(6): p. 601-7.
[148] Aurich, H., et al., *Hepatocyte differentiation of mesenchymal stem cells from human adipose tissue in vitro promotes hepatic integration in vivo.* Gut, 2008.
[149] Pan, R.L., et al., *Fetal liver-conditioned medium induces hepatic specification from mouse bone marrow mesenchymal stromal cells: a novel strategy for hepatic transdifferentiation.* Cytotherapy, 2008. 10(7): p. 668-75.
[150] Stock, P., et al., *Hepatocytes derived from adult stem cells.* Transplant Proc, 2008. 40(2): p. 620-3.
[151] Zhang, Z., et al., *Human umbilical cord mesenchymal stem cells improve liver function and ascites in decompensated liver cirrhosis patients.* J Gastroenterol Hepatol, 2012. 27 Suppl 2: p. 112-20.
[152] Banas, A., et al., *IFATS collection: in vivo therapeutic potential of human adipose tissue mesenchymal stem cells after transplantation into mice with liver injury.* Stem Cells, 2008. 26(10): p. 2705-12.
[153] Haynesworth, S.E., M.A. Baber, and A.I. Caplan, *Cytokine expression by human marrow-derived mesenchymal progenitor cells in vitro: effects of dexamethasone and IL-1 alpha.* J Cell Physiol, 1996. 166(3): p. 585-92.
[154] Khurana, S. and A. Mukhopadhyay, *In vitro transdifferentiation of adult hematopoietic stem cells: An alternative source of engraftable hepatocytes.* J Hepatol, 2008. 49(6): p. 998-1007.
[155] Parekkadan, B., et al., *Immunomodulation of activated hepatic stellate cells by mesenchymal stem cells.* Biochem Biophys Res Commun, 2007. 363(2): p. 247-52.
[156] Parekkadan, B., et al., *Mesenchymal stem cell-derived molecules reverse fulminant hepatic failure.* PLoS One, 2007. 2(9): p. e941.
[157] van Poll, D., et al., *Mesenchymal stem cell-derived molecules directly modulate hepatocellular death and regeneration in vitro and in vivo.* Hepatology, 2008. 47(5): p. 1634-43.
[158] Abdel Aziz, M.T., et al., *Therapeutic potential of bone marrow-derived mesenchymal stem cells on experimental liver fibrosis.* Clin Biochem, 2007. 40(12): p. 893-9.
[159] Taniguchi, E., et al., *Endothelial progenitor cell transplantation improves the survival following liver injury in mice.* Gastroenterology, 2006. 130(2): p. 521-31.
[160] Higashiyama, R., et al., *Bone marrow-derived cells express matrix metalloproteinases and contribute to regression of liver fibrosis in mice.* Hepatology, 2007. 45(1): p. 213-22.
[161] Li, J.T., et al., *Molecular mechanism of hepatic stellate cell activation and antifibrotic therapeutic strategies.* J Gastroenterol, 2008. 43(6): p. 419-28.
[162] Sakaida, I., et al., *Transplantation of bone marrow cells reduces CCl4-induced liver fibrosis in mice.* Hepatology, 2004. 40(6): p. 1304-11.

[163] Zhao, D.C., et al., *Bone marrow-derived mesenchymal stem cells protect against experimental liver fibrosis in rats.* World J Gastroenterol, 2005. 11(22): p. 3431-40.
[164] Zhao, Z.H., et al., *[Dynamic expression of matrix metalloproteinase-2, membrane type-matrix metalloproteinase-2 in experimental hepatic fibrosis and its reversal in rat].* Zhonghua Shi Yan He Lin Chuang Bing Du Xue Za Zhi, 2004. 18(4): p. 328-31.
[165] Lin, S.Z., et al., *Transplantation of human Wharton's Jelly-derived stem cells alleviates chemically induced liver fibrosis in rats.* Cell Transplant. 19(11): p. 1451-63.
[166] Asawa, S., et al., *Participation of bone marrow cells in biliary fibrosis after bile duct ligation.* J Gastroenterol Hepatol, 2007. 22(11): p. 2001-8.
[167] Baba, S., et al., *Commitment of bone marrow cells to hepatic stellate cells in mouse.* J Hepatol, 2004. 40(2): p. 255-60.
[168] Kisseleva, T., et al., *Bone marrow-derived fibrocytes participate in pathogenesis of liver fibrosis.* J Hepatol, 2006. 45(3): p. 429-38.
[169] Russo, F.P., et al., *The bone marrow functionally contributes to liver fibrosis.* Gastroenterology, 2006. 130(6): p. 1807-21.
[170] Campbell, K.H., et al., *Sheep cloned by nuclear transfer from a cultured cell line.* Nature, 1996. 380(6569): p. 64-6.
[171] Kharaziha, P., et al., *Improvement of liver function in liver cirrhosis patients after autologous mesenchymal stem cell injection: a phase I-II clinical trial.* European journal of gastroenterology & hepatology, 2009. 21(10): p. 1199-205.
[172] Amer, M.E., et al., *Clinical and laboratory evaluation of patients with end-stage liver cell failure injected with bone marrow-derived hepatocyte-like cells.* European journal of gastroenterology & hepatology, 2011. 23(10): p. 936-41.
[173] Mohamadnejad, M., et al., *Phase 1 trial of autologous bone marrow mesenchymal stem cell transplantation in patients with decompensated liver cirrhosis.* Arch Iran Med, 2007. 10(4): p. 459-66.
[174] Almeida-Porada, G., C. Porada, and E.D. Zanjani, *Adult stem cell plasticity and methods of detection.* Rev Clin Exp Hematol, 2001. 5(1): p. 26-41.
[175] Almeida-Porada, G., C. Porada, and E.D. Zanjani, *Plasticity of human stem cells in the fetal sheep model of human stem cell transplantation.* Int J Hematol, 2004. 79(1): p. 1-6.
[176] Almeida-Porada, G. and E.D. Zanjani, *A large animal noninjury model for study of human stem cell plasticity.* Blood Cells Mol Dis, 2004. 32(1): p. 77-81.
[177] Almeida-Porada M.G., P.C., ElShabrawy D., Simmons P.J., Zanjani E.D. , *Human marrow stromal cells (MSC) represent a latent pool of stem cells capable of generating long-term hematopoietic cells.* . Blood, 2001. 98(1): p. 713a.
[178] Flake, A.W. and E.D. Zanjani, *In utero transplantation of hematopoietic stem cells.* Crit Rev Oncol Hematol, 1993. 15(1): p. 35-48.
[179] Zanjani, E.D., G. Almeida-Porada, and A.W. Flake, *Retention and multilineage expression of human hematopoietic stem cells in human-sheep chimeras.* Stem Cells, 1995. 13(2): p. 101-11.
[180] Zanjani, E.D., G. Almeida-Porada, and A.W. Flake, *The human/sheep xenograft model: a large animal model of human hematopoiesis.* Int J Hematol, 1996. 63(3): p. 179-92.
[181] Zanjani, E.D., et al., *Transplantation of hematopoietic stem cells in utero.* Stem Cells, 1997. 15 Suppl 1: p. 79-92; discussion 93.
[182] Barbera, A., et al., *Early ultrasonographic detection of fetal growth retardation in an ovine model of placental insufficiency.* Am J Obstet Gynecol, 1995. 173(4): p. 1071-4.

[183] Beierle, E.A., M.R. Langham, Jr., and S. Cassin, *In utero lung growth of fetal sheep with diaphragmatic hernia and tracheal stenosis.* J Pediatr Surg, 1996. 31(1): p. 141-6; discussion 146-7.
[184] Morrison, J.L., *Sheep models of intrauterine growth restriction: fetal adaptations and consequences.* Clin Exp Pharmacol Physiol, 2008. 35(7): p. 730-43.
[185] Stelnicki, E.J., et al., *A new in utero model for lateral facial clefts.* J Craniofac Surg, 1997. 8(6): p. 460-5.
[186] Cahill, R.N., et al., *The ontogeny of T cell recirculation during foetal life.* Semin Immunol, 1999. 11(2): p. 105-14.
[187] Jennings, R.W., et al., *Ontogeny of fetal sheep polymorphonuclear leukocyte phagocytosis.* J Pediatr Surg, 1991. 26(7): p. 853-5.
[188] Miyasaka, M. and B. Morris, *The ontogeny of the lymphoid system and immune responsiveness in sheep.* Prog Vet Microbiol Immunol, 1988. 4: p. 21-55.
[189] Osburn, B.I., *The ontogeny of the ruminant immune system and its significance in the understanding of maternal-fetal-neonatal relationships.* Adv Exp Med Biol, 1981. 137: p. 91-103.
[190] Raghunathan, R., et al., *Ontogeny of the immune system: fetal lamb as a model.* Pediatr Res, 1984. 18(5): p. 451-6.
[191] Sawyer, M., J. Moe, and B.I. Osburn, *Ontogeny of immunity and leukocytes in the ovine fetus and elevation of immunoglobulins related to congenital infection.* Am J Vet Res, 1978. 39(4): p. 643-8.
[192] Silverstein, A.M., R.A. Prendergast, and K.L. Kraner, *Fetal Response to Antigenic Stimulus. Iv. Rejection of Skin Homografts by the Fetal Lamb.* J Exp Med, 1964. 119: p. 955-64.
[193] Silverstein, A.M., et al., *Fetal response to antigenic stimulus. II. Antibody production by the fetal lamb.* J Exp Med, 1963. 117: p. 799-812.
[194] Skopal-Chase, J.L., et al., *Immune ontogeny and engraftment receptivity in the sheep fetus.* Fetal Diagn Ther, 2009. 25(1): p. 102-10.
[195] Simmons, P.J., et al., *Isolation, characterization and functional activity of human marrow stromal progenitors in hemopoiesis.* Prog Clin Biol Res, 1994. 389: p. 271-80.
[196] Chamberlain, J., et al., *Efficient generation of human hepatocytes by the intrahepatic delivery of clonal human mesenchymal stem cells in fetal sheep.* Hepatology, 2007.
[197] Feldmann, G., et al., *Functional hepatocellular heterogeneity for the production of plasma proteins.* Enzyme, 1992. 46(1-3): p. 139-54.
[198] Krishna, M., R.V. Lloyd, and K.P. Batts, *Detection of albumin messenger RNA in hepatic and extrahepatic neoplasms. A marker of hepatocellular differentiation.* Am J Surg Pathol, 1997. 21(2): p. 147-52.
[199] Racine, L., et al., *Distribution of albumin, alpha 1-inhibitor 3 and their respective mRNAs in periportal and perivenous rat hepatocytes isolated by the digitonin-collagenase technique.* Biochem J, 1995. 305 (Pt 1): p. 263-8.
[200] De Ugarte, D.A., et al., *Comparison of multi-lineage cells from human adipose tissue and bone marrow.* Cells Tissues Organs, 2003. 174(3): p. 101-9.
[201] Kogler, G., et al., *A new human somatic stem cell from placental cord blood with intrinsic pluripotent differentiation potential.* J Exp Med, 2004. 200(2): p. 123-35.
[202] Wang, X., et al., *Cell fusion is the principal source of bone-marrow-derived hepatocytes.* Nature, 2003. 422(6934): p. 897-901.

[203] Quah, B.J., H.S. Warren, and C.R. Parish, *Monitoring lymphocyte proliferation in vitro and in vivo with the intracellular fluorescent dye carboxyfluorescein diacetate succinimidyl ester.* Nat Protoc, 2007. 2(9): p. 2049-56.

[204] Slavik, J.M., et al., *Rapamycin-resistant proliferation of CD8+ T cells correlates with p27kip1 down-regulation and bcl-xL induction, and is prevented by an inhibitor of phosphoinositide 3-kinase activity.* J Biol Chem, 2004. 279(2): p. 910-9.

[205] Anderson, W.M. and D. Trgovcich-Zacok, *Carbocyanine dyes with long alkyl side-chains: broad spectrum inhibitors of mitochondrial electron transport chain activity.* Biochem Pharmacol, 1995. 49(9): p. 1303-11.

[206] Onfelt, B., et al., *Structurally distinct membrane nanotubes between human macrophages support long-distance vesicular traffic or surfing of bacteria.* J Immunol, 2006. 177(12): p. 8476-83.

[207] Zorov, D.B., et al., *Examining intracellular organelle function using fluorescent probes: from animalcules to quantum dots.* Circ Res, 2004. 95(3): p. 239-52.

[208] Mahieu-Caputo, D., et al., *Repopulation of athymic mouse liver by cryopreserved early human fetal hepatoblasts.* Hum Gene Ther, 2004. 15(12): p. 1219-28.

[209] Gouon-Evans, V., et al., *BMP-4 is required for hepatic specification of mouse embryonic stem cell-derived definitive endoderm.* Nat Biotechnol, 2006. 24(11): p. 1402-11.

[210] Nava, S., et al., *Characterization of cells in the developing human liver.* Differentiation, 2005. 73(5): p. 249-60.

[211] Colletti, E.J., et al., *Generation of tissue-specific cells from MSC does not require fusion or donor-to-host mitochondrial/membrane transfer.* Stem Cell Res, 2009. 2(2): p. 125-38.

[212] Nauta, A.J., et al., *Donor-derived mesenchymal stem cells are immunogenic in an allogeneic host and stimulate donor graft rejection in a nonmyeloablative setting,* 2006, *Transplantation*. p. 2114-2120.

[213] Eliopoulos, N., et al., *Allogeneic marrow stromal cells are immune rejected by MHC class I- and class II-mismatched recipient mice,* 2005, *Blood*. p. 106: 4057-4065.

[214] Badillo, A.T., et al., *Murine bone marrow stromal progenitor cells elicit an in vivo cellular and humoral alloimmune response.* Biol Blood Marrow Transplant, 2007. 13(4): p. 412-22.

[215] Poncelet, A.J., et al., *Although pig allogeneic mesenchymal stem cells are not immunogenic in vitro, intracardiac injection elicits an immune response in vivo.* Transplantation, 2007. 83(6): p. 783-90.

[216] Camp, D.M., et al., *Cellular immune response to intrastriatally implanted allogeneic bone marrow stromal cells in a rat model of Parkinson's disease.* J Neuroinflammation, 2009. 6: p. 17.

[217] Fletcher, J.M., H.G. Prentice, and J.E. Grundy, *Natural killer cell lysis of cytomegalovirus (CMV)-infected cells correlates with virally induced changes in cell surface lymphocyte function-associated antigen-3 (LFA-3) expression and not with the CMV-induced down-regulation of cell surface class I HLA.* J Immunol, 1998. 161(5): p. 2365-74.

[218] Brutkiewicz, R.R. and R.M. Welsh, *Major histocompatibility complex class I antigens and the control of viral infections by natural killer cells.* J Virol, 1995. 69(7): p. 3967-71.

[219] Soland, M., et al., *Modulation of Mesenchymal Stem Cell Immunogenicity through Forced Expression of Human Cytomegalovirus Proteins* Blood, 1468:2416a, 2008.

[220] Soland, M., E. J Colletti, M. Bego, C. Sanada, C. D Porada, E D Zanjani, S St. Jeor and G. Almeida-Porada, *Modulation of Mesenchymal Stem Cell MHC-I Complex Increases Engraftment In Vivo* Blood, 2010. 2811, 1457a.

12

Cell Based Therapy for Chronic Liver Disease: Role of Fetal Liver Cells in Restoration of the Liver Cell Functions

Chaturvedula Tripura[1], Aleem Khan[2] and Gopal Pande[1,*]
[1]*CSIR- Centre for Cellular and Molecular Biology, Hyderabad,*
[2]*Centre for Liver Research and Diagnostics, Deccan Medical College, Owaisi Hospital, Kanchan Bagh, Hyderabad, India*

1. Introduction

Liver is the largest organ in the human body and it functions like a metabolic factory. Disruption of its anatomical structure, which is often caused by excessive fibrosis of the extracellular matrix (ECM), when left unattended, can lead to cirrhosis of the liver and cause permanent and irreversible damage to its organization and function, with fatal consequences. Liver cirrhosis, which is generally the end result of chronic liver disease (CLD), can be caused due to many etiological reasons including (a) long term infections with hepatitis B and C viruses, (b) uncontrolled alcohol abuse, (c) excessive exposure to metabolic products of metals like iron and copper (d) autoimmune inflammation of the liver (e) nonalcoholic fatty liver disease (NAFLD) and (f) nonalcoholic steatohepatitis (NASH) (reviewed in [1]). Histo-pathologically, a hallmark of liver cirrhosis is the abnormal production and storage of collagen molecules in the ECM, formation of a scar tissue that replaces normal parenchyma and blockage of the portal flow of blood through the organ, thus affecting normal hepatocellular activity and ultimately total loss of liver functions [2, 3]. Cirrhosis of the liver in early stages is largely asymptomatic, therefore remains undetected by physical examination and other available tests. Diagnosis of fibrosis at early stages and prevention of its progression to cirrhosis is a very important factor in the management of the disease. Among the different options available for treatment, this review would focus on cell based therapy for liver cirrhosis with a special attention on the challenges and procedures of using human fetal liver cells. Methods to improve the clinical application of cell and tissue imaging of liver in the management of cirrhosis would also be discussed, briefly.

2. Treatment options for liver cirrhosis

The established choices of treatment for cirrhosis are very limited and in most cases withdrawl of the underlying causative agent is used as the first line of treatment. Anti-viral

* Corresponding Author

therapy and biochemical modulation of liver metabolism are some of the classical treatment strategies but more recently several cell therapy based options have been used albeit many are still experimental and limited to preclinical studies. Some of the approaches for treatment of cirrhosis are discussed below.

2.1 Liver transplantation

Liver transplant is considered to be one of the best curative treatment solutions available for advanced liver cirrhosis. However, the treatment procedure carries operative risk, is expensive, requires life long immunosuppression, and there is also a risk of graft rejection that requires a re-transplantation of the organ. One of the major limitations of liver transplantation is the availability of donor liver. There is a constant rise in number of patients on the waiting list for donor liver and an acute shortage on the availability, which is the leading cause for the increase in morbidity and mortality. This gap in the demand and supply of a donor liver tissue is partially filled by obtaining liver tissues from living donors and doing auxiliary liver transplantation. However, while opting for this procedure the possible risk to the donor and the benefits of the recipient must be considered. In this scenario, there is a clear need to look for alternative strategies for the therapy of end stage liver diseases that would be more effective and safe in reducing the tissue scarring of fibrotic livers and in redemption of normal liver function.

2.2 Cell based therapy of liver diseases

Cell based approaches for treatment of liver diseases offer novel but challenging alternatives to liver transplantation. Several types of stem and progenitor cells have been explored for their possible use in this field. Cell based therapies could be initiated by either (a) the activation or mobilization of autologous stem cells to the site of injury or (b) by the infusion of heterologous (or autologous) stem and progenitor cells from different sources.

2.2.1 Activation of autologous regenerative cells

The therapeutic role of autologous regeneration by resident cells in the liver or by mobilized cells from the bone marrow has a significant role to play in the auto regeneration of the normal tissue following liver injury. The failure of these mechanisms leads to the activation of degenerative cascades in the liver such as necrosis, cell death and abnormal accumulation of collagen. The autologous cell based therapies utilise the reactivation mechanisms where either the non responding resident cells in the liver are activated or fresh cells from the bone marrow are mobilised to the site of injury.

2.2.1.1 Activation of resident regenerative cells in the liver by injectible growth factors

In a cirrhotic liver the normal architecture is completely disrupted by abnormal accumulation of ECM components that block the vascular supply leading to the death of liver parenchyma which contributes to the decrease in liver function. The liver has a spontaneous regeneration potential to compensate for this loss of liver parenchyma by division of the existing hepatocytes and hepatic progenitor cells (HPC). However due to extensive scarring of tissue the regenerating cells are prevented from regaining their normal function. It is proposed that ECM remodelling can lead to resolution of this blockage and reactivate the resident regenerative cells in the affected area. Earlier work on ECM remodelling by using matrix modifying factors such as hepatocyte growth factor (HGF) have shown prevention and/or regression of fibrosis in animal models of liver and pulmonary injury [4]. Administration of human rHGF or gene transfer of human HGF to rats with hepatic fibrosis/ cirrhosis caused by di-methyl-nitrosamine prevented the onset and progression of hepatic fibrosis/cirrhosis [5, 6]. Some of the studies have revealed that HGF mediates this process by directly antagonizing the pro-fibrotic actions of Transforming growth factor (TGF)-β1 [7, 8]. In addition to these molecules, studies on basic fibroblast growth factor (b-FGF, FGF-2), in animal models, has shown their participation in tissue regeneration, angiogenesis and in wound healing processes [9]. A recent report has shown that activation of HPC might be linked up with ECM remodelling. Degradation of collagen I and subsequent laminin deposition seem to be important prerequisites for HPC activation and expansion [10]. Based upon these results it appears that a better understanding of the factors that govern HPC proliferation and the resultant ECM changes would provide clues to improve regeneration of liver cells in chronic liver disease.

2.2.1.2 *In situ* mobilization of bone marrow cells to the liver

Liver cirrhosis is associated with an intermittent mobilization of different types of bone marrow cells that are committed to differentiate to hepatocytes [11]. This process could be triggered either spontaneously upon liver injury or by administration of growth factors that could stimulate the migration of stem cells from the bone marrow into peripheral blood. Granulocyte-colony stimulating factor (G-CSF) is one such mobilizing agent that is receiving considerable attention recently in the field of liver therapy. Several studies have indicated that G-CSF may be effective in mobilizing bone marrow cells into the peripheral blood that contribute to liver repair [12, 13]. In rats G-CSF was shown to contribute to liver repair in a double mechanism by increasing the bone marrow derived liver repopulation, and also by activating the endogenous oval cells, that express G-CSF receptor (G-CSFR) [14].

In a clinical trial, 8 patients affected by severe liver cirrhosis were administered G-CSF and the treatment was well tolerated in all the patients during a follow-up of 8 months, and mobilization of bone marrow stem cells co-expressing epithelial and stem markers was noted [15, 16] . Two independent studies on a group of 24 patients and 18 patients [17, 18] with severe liver cirrhosis resulted in a dose-dependent mobilization of good CD34+/CD133+ bone marrow stem cells and proved that the procedure was safe, but did not achieve any significant clinical improvement. Treatment with G-CSF was associated with the induction of HPC proliferation within 7 days of administration [19]. In a recent report G-CSF based mobilization of bone marrow cells was used successfully to treat patients even with acute on chronic-liver failure (ACLF), and a significant recovery in the clinical condition was noted [20]. Use of G-CSF could thus promote the *in situ* mobilization and regeneration of the liver tissue without excessive intervention and is slowly gaining its importance in the field of liver therapy.

2.2.2 Infusion of therapeutic cells

The discovery of stem cells has revolutionized the field of medicine offering potential options for the management of various chronic disorders. Different types of stem and progenitor cells from various sources are available with a broad potential for differentiation and application in tissue regeneration and newer sources are being explored. Several clinical studies have been reported and reviewed where regenerative cells had been infused for the purpose of liver therapy and many other studies are currently in progress.

Based on the donor tissue source and the differentiation potential of cells the results on the therapeutic efficacy of several cell types has been described below.

2.2.2.1 Infusion of adult hepatocytes and Bio-artificial Livers (BALs)

Adult hepatocytes are the fully mature functional cells of the liver that are highly specialized with the ability to divide and are an ideal source for transplantation. Though hepatocyte transplantation has been recognized as an attractive option for the management of metabolic liver disease some 35 years ago [21], lack of availability of livers for cell isolation, difficulty in expansion of hepatocytes *in vitro* and their sensitivity to freeze-thaw are major limitations for their routine use in cell therapy.

Allogeneic primary hepatocytes isolated from cadaver livers and infused via the splenic artery or the portal vein showed an improvement in the clinical condition [22, 23, 24]. Many preclinical studies and clinical applications of this technique have been made to cure metabolic liver disorders and end-stage liver diseases [25]. In most instances, hepatocyte transplantation has been able to grant a clinical improvement for up to 12 months [26].

Implantable hepatocyte-based devices and extra corporeal liver assist devices represent another alternative for the treatment of end-stage liver disease [27]. BAL devices are intended to support the failing functions of the organ, and include both a biological component (parenchymal cells) and an artificial scaffold serving as an interface with patient blood or plasma. Two main clinical trials have evaluated the efficacy of liver assist devices in patients with fulminant liver failure. Owing to the difficulty in supply of human hepatocytes, the devices included either purified pig hepatocytes (HepatAssist) or cell lines derived from liver cancer cells (ELAD) [28, 29]. Both trials failed to demonstrate a beneficial effect on survival. In the future, such devices are likely to re-emerge as a result of new technologies that allow growth and differentiation of large amounts of liver cells *in vitro* from stem/precursor cells. However, BALs will have to challenge cost-effectiveness in comparison with more convenient artificial devices.

2.2.2.2 HPC's and fetal liver cells (hepatoblasts)

Isolation, expansion and differentiation of adult HPCs to functional hepatocytes has been described by several workers [30, 31, 32], however identification of a specific marker of HPCs still remains a challenge. Clinical studies on the repopulation of the human liver by HPCs are still awaited.

Compared to the adult HPCs, fetal liver progenitor cells are highly proliferative, less immunogenic and more resistant to cryopreservation. The human fetal liver, between 10 and 18 weeks of gestation, contains a large number of actively dividing hepatic stem and progenitor cells that are termed as hepatoblasts. These are bi-potent cells , that can give rise to both hepatocytes and bile duct cells; they co-express hepatocyte markers (such as albumin, α-fetoprotein (AFP), α-1 microglobulin, glycogen, glucose-6-phosphatase (G-6-P) and Hep-Par-1) and biliary markers, for example, gamma glutamyl transpeptidase (GGT), dipeptidyl peptidase IV (DPPIV), cytokeratin (CK)-19 and Das-1-monoclonal antibody-reactive antigen, [33, 34]. These markers are expressed throughout the second trimester and thus offer opportunities to isolate and study large numbers of progenitor cells from abortuses.

2.2.2.3 Haematopoietic Stem Cells (HSCs) and Mesenchymal stem cells (MSCs) from the bone marrow

HSCs are committed progenitor cells of the bone marrow and can be extensively expanded without loss of pluripotency. The bone marrow is an important source of autologous HSCs, and MSCs [35, 36]. Infusion of unsorted autologous bone marrow stem cells through the portal vein and hepatic artery in patients with cirrhosis, showed an improvement in Child Pugh score and albumin levels [16, 37]. A significant increase of liver function in cirrhotic and hepatocellular carcinoma patients was observed following autologous bone marrow stem cell transplantation prior to surgery [38]. However, use of unsorted bone marrow stem cells or MSCs must be treated with caution as these cells can also differentiate into myofibroblasts which are the scar forming cells of the liver [39]. Recent data also provides evidence to this in a rodent model where use of whole bone marrow as cell therapy led to the worsening of liver fibrosis [40]. Macrophages which are the cells of haematopoietic origin, are known to play a critical role in regulating liver fibrosis in murine models [41] and a single intra-portal administration of macrophages has recently been shown to reduce fibrosis in a murine model of liver injury and increase regeneration [40].

MSCs and HSCs can also be isolated from other tissues such as peripheral blood, adipose tissue, umbilical cord blood (UCB) and placenta [42]. Adipose tissue derived MSCs have a good proliferative capacity *in vitro* and *in vivo* and can differentiate into hepatic cells [43-46]. The placenta and UCB are also important sources of young MSCs and HSCs that can be obtained without invasiveness or harm to donor and provide no ethical barriers for basic and clinical applications [47, 48]. Recently UCB cells were used on a diverse group of end stage cirrhotic patients with an improvement in the life span in these patients [49] raising a hope on the use of these cells for liver cell therapy.

MSCs and HSCs from all sources display a high degree of plasticity giving rise to a wide range of phenotypes, including hepatocyte-like cells. The low immunogenic property of these cells have shown promising results with the use of these cells in *in vitro* studies and clinical trials [50-55].

2.2.2.4 Embryonic and Induced pluripotent stem cells (iPSCs)

Embryonic stem cells (ESCs) derived from the inner cell mass of 5-6 day old embryos have the advantage of being able to proliferate in an unlimited fashion and might constitute an easily available source to obtain a large number of transplantable cells for regenerative treatments. By manipulating the factors responsible for maintaining the undifferentiated state of these cells and by exposure to appropriate growth factors *in vitro*, ESCs can be directed towards the hepatic lineage. ESC-derived hepatocyte-like cells were able to colonize the injured liver and function as mature hepatocytes without teratoma formation in several animal models of liver disease [26, 56, 57]. Due to the propensity of these cells to form both malignant and non-malignant tumors, caution has to be still exerted for their use in transplantation. The ethical issues regarding the use of human embryonic stem cells will also have implications for their clinical use.

Induced pluripotent stem cells offer a solution to this ethical concern and the risk of rejection related to embryonic stem cells. iPSCs are generated *in vitro* by genetic reprogramming of adult somatic cells with certain factors, to form pluripotent stem cells with embryonic-like differentiation potential [58]. Hepatocytes derived from iPS cells have a reasonable synthetic and metabolic capacity [59], and seem to be similar to cells derived from ES cells [60-62]. However, the same concerns of tumor formation or reversion to more primitive state with uncontrolled expansion within the recipient still remain. In spite of these limitations iPS-derived hepatocytes are a very promising population for cell therapies in hepatology.

3. Management of end stage liver disease with fetal liver cells

Hepatic progenitors derived from the livers of electively aborted fetuses of 5 to 20 weeks gestation are generally designated as "multipotent." Fetal cells seem to have an edge over embryonic stem cells in that, being less versatile, they may not form teratomas and exhibit an important property of being less immunogenic by the little to no expression of the Class II HLA marker on their surface, which otherwise can trigger a rejection reaction [63]. Thus, tissue matching that is a prime requirement in blood transfusions, organ transplants, and allogenic adult stem cell transplantation is not necessary when transplanting these cells.

3.1 Hepatic progenitors in developing liver

The developing human (and also murine) liver during early to mid gestation has been shown to comprise stem and progenitor cells that can give rise to the different adult cell types. The developmental plasticity of fetal liver parenchymal cells, at this stage of organ development, makes them a very good source for cell therapy. During the first trimester of fetal development, the liver serves as the site for hematopoiesis and later during the mid-gestation the hematopoietic cells migrate to the bone marrow and the liver starts functioning primarily as a hepatic organ (Figure 1). In the first trimester, human fetal liver is the site of synthesis of progenitor and stem cells of many lineages but during the second trimester, markers for hepatoblasts continue to be expressed but the expression of markers for other lineages is reduced [63, 64]. This offers an opportunity to isolate and study large numbers of hepatic progenitor cells without haematopoietic potential during the second trimester of gestation. It is interesting that hepatic progenitor cells isolated from human fetal liver proliferate for several months and retained their normal karyotypes, thereby indicating a strong telomerase activity in them [65].

3.1.1 Cell surface and intracellular markers for progenitor cells

Many different markers have been used for the identification of progenitor cells. Bi-potential progenitors have been isolated from fetal mouse liver in a number of studies. Petersen et al have established that a Thy-1+ve cell population within the rat fetal liver also expressed CK-18, a hepatocyte marker [66]. Hepatic progenitor cells have also been reported to express c-kit, a hepatic stem cell marker, along with CD34 and Thy-1 [67]. In a study from our lab, CD34 cells from human fetal liver were found to be expressing hepatic and biliary markers as well [64]. In another study with human fetal liver cells, Thy-1^{+} (a haematopoietic marker) cell populations were isolated that were found to be positive for progenitor (CD34, c-kit, CK14, M2PK, OV6), biliary (CK19) and hepatic (HepPar1) markers, revealing their progenitor as well as hepatic and biliary nature [68]. Expression of these specific markers by each cell type is used for the isolation of progenitor cells.

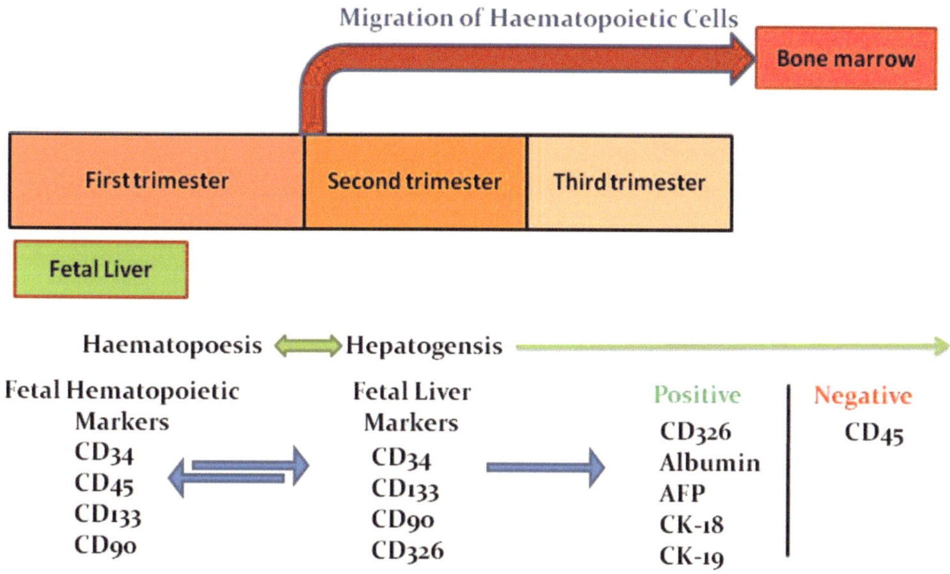

Fig. 1. Transition of cells during liver development

Obtaining good quality and quantity of fetal progenitor cells is very important to ensure repopulation of the liver upon transfusion of these cells. Care must be ensured at every step of tissue collection, isolation of progenitor cells and transfusion of the cells in order to study the efficiency of these cells in treating liver diseases. Some of the important challenges and procedures in these approaches are discussed below.

3.2 Ethical and logistic challenges in obtaining human fetal tissues

The use of human fetal tissues for preclinical or clinical work is always controversial due to religious and ethical issues. The main challenge is the procurement of fetal tissue, therefore the first important step is to identify a maternity/obstetric hospital/department willing to donate aborted fetuses of known gestation period and acceptable medical history for research in good clinical grade condition. The following measures need to be ensured for collection and use of fetal material:

a. **Ethical clearance:** Clearances to work with human fetal material must be obtained from the hospital and the research institute's ethical committee where the fetal cells would be isolated and processed, prior to beginning of the work.
b. **Informed consent:** The mother must be properly informed of the procedure and the written consent must be obtained for the donation prior to the abortion. The process of abortion and donation must be maintained as separate procedures to ensure that the fetus is donated voluntarily and no financial gain is involved in the process.
c. **Screening procedures:** All donors of the fetus must be serologically screened for syphilis, toxoplasmosis, rubella, hepatitis B and C, human immunodeficiency virus 1, cytomegalovirus, parvovirus, and herpes simplex types 1 and 2.
d. **Gestational age:** The gestational age of the fetus is very important and should not cross the second trimester of gestation. The sample should preferably be chosen from an

elective abortion in order to ensure that the tissue is intact and must be collected immediately after the procedure and stored in cold conditions.

e. **Transportation of fetal material:** Once the tissue is collected it must be stored in a clean sterile container with controlled temperature to ensure that the cells are not affected with the temperature changes during the transport. The tissue must be transported to the processing lab as quickly as possible maintaining proper sterile procedures. The cells must be isolated in clean sterile conditions within a maximum of 4 hours of the abortion. If the cells are to be used immediately for transfusion then the processed cells must be enumerated and transported to the hospital in sterile vials ensuring proper temperature conditions.
f. **Storage of cells:** The cells that would not be used immediately should be enumerated and stored with a safe cryo preservative such as DMSO in liquid nitrogen until further use. The viability of the cells during the freeze thaw must be checked before the frozen cells are used for transfusion.

3.3 Preparation of cells from human fetal tissue for therapeutic use:

Once a fetus of the appropriate gestation age is obtained in good medical condition, cell isolation should be done within 4 hours in order to obtain good viability of cells. Two main steps are involved in this process: (a) Isolation of total viable cells from the liver, and (b) enrichment of progenitor cells suitable for transfusion.

3.3.1 Isolation of viable liver cells

The cells in the liver tissue are connected by intercellular connections and tight junctions embedded in an ECM that needs to be disrupted or dissolved for single cell isolation. Three different approaches for isolating viable liver cells with some modifications are principally used:

1. Mechanical dispersion using partial homogenization and forcing the liver tissue through steel meshes [69].
2. The second method of hepatocyte isolation includes removal of calcium and potassium from liver by perfusion and reverse perfusion with calcium binding agent like citrate, Ethylene diamine tetraacetic acid (EDTA), ethylene glycol tetraacetic acid (EGTA), tetra phenyl borate (TBP), pyrophosphate, versine and ATP in calcium free Locke's solution as a sole means of separating intracellular spaces [70, 71].
3. The third method involves the dissolution of intracellular junctions by using proteolytic and matrix dissolving enzymes like trypsin, papain, lysozyme, neuraminidase and pepsin [72, 73].
4. Recently a new protocol for efficient and quick isolation of fetal liver cells from 18-24 weeks of gestation was developed. The protocol involved a 5 step portal vein *in situ* liver perfusion technique using both EGTA and limited exposure to collagenase that resulted in a greater and efficient cell yield [74].

The human fetal liver cells thus isolated can be used as such for treatment of liver failure in model animals and even in experimental clinical trials or can be enriched further for specific progenitor cells.

3.3.2 Enrichment of progenitor cells

The total liver cells that are isolated are a mixture of different population of cells from which the progenitor cells are isolated based on their specific properties by using any of the following enrichment methods. However the method of using the surface markers for enrichment is more efficient in specifically selecting progenitor cells.

a. **Enrichment based on size and density:** Cells of a particular lineage or function have a specific size and mass and this property has been used to separate out desired population of cells from a mixture of cells. Several investigators have used physical methods like density gradient centrifugation and counter flow centrifugation elutriation methods for the enrichment of the hepatic progenitors. Yaswen et al., demonstrated the enrichment of hepatic oval cells (hepatic progenitor) using Centrifugal elutriation (Beckman Instruments, Palo Alto, Calif.) of the cell suspension [75]. Enriched fraction was characterized histochemically for gamma-glutamyl transpeptidase, peroxidase, alkaline phosphatase, glucose-6-phosphatase activities, albumin and alpha-fetoprotein by immunocytochemical methods [75].
b. **Enrichment based on surface markers:** The enrichment of cells can be achieved by using surface markers specifically expressed by the progenitor cells. The process uses antibody specific for the surface marker and separation is achieved by either magnetically tagged (MACS) or fluorescently tagged (FACS) antibodies. In our experience, though FACS is more sensitive for enriching a specific cell type based on the marker expression, the cell yield is not very efficient. Hence enrichment using MACS is more efficient in giving a good yield of the desired population of cells.
   Studies with human livers have shown that Epithelial cell adhesion molecule (EpCAM) is expressed by hepatic stem cells, hepatoblasts and committed progenitors but not expressed in mature hepatocytes [76, 77]. Thus, sorting for EpCAM results in only progenitor cells but in distinct ratios of hepatic stem cells to hepatoblasts depending upon whether the tissue is fetal, neonatal, or adult. In our recent study we have also enriched the hepatic progenitors using EpCAM and further characterized using liver specific and stem cell markers such as CD29, CD90, CD49f, CD34 and we found that EpCAM +ve cells expressed intermediate levels HLA class I but no HLA class II. Our study demonstrated the usefulness of EpCAM as a novel surface marker for enrichment of hepatic progenitors [63]. Earlier we have also used CD34+/CD45- as one the marker for the hepatic progenitors [64].
c. **Separation based on Functional markers:** Certain types of cells express functional receptors on their surface, which transport molecules out of the cells. One such transport protein is the ATP-binding cassette sub-family G member 2 (ABCG2) transporter protein that specifically excludes the Hoechst dye. Separation of progenitor cells can be achieved using this functional marker where the cells specifically exclude the Hoechst dye [78, 79].

3.3.3 Characterization of liver progenitors

The enriched progenitor cells must be further characterized for the expression of hepatic and biliary markers by using methods such as flow cytometry, immune histochemistry and RT-PCR. Hepatic progenitor cells express many markers that are similar to hepatocytes or bile duct cells, and also share some of the haematopoietic markers, AFP, certain keratin markers [e.g., cytokeratin 19 (CK 19)], and Gamma glutamyl transpeptidase (GGT) [80]. In our study, fetal progenitor cells enriched with EpCAM marker were further characterized for the expression of liver epithelial markers (CK18), biliary specific marker (CK19) and hepatic markers (albumin, AFP) by RT-PCR. FACS indicated that the cells were positive for CD29, CD49f, CD90, CD34, albumin and AFP and negative for HLA class II and CD45. Immunocytochemical staining confirmed the expression of CK18 and albumin [63]. In a study by Lie et al., epithelial progenitor cells from human fetal livers were isolated by cell culture and characterized for their expression of liver epithelial markers (cytokeratin [CK8 and CK18) and biliary-specific markers (CK7 and CK19) by real time PCR. FACS analysis was used to confirm the expression of hepatic markers such as CD117, CD147, CD90, CD44, and absence of hematopoietic markers such as CD34 and CD45. Hepatic differentiation potential of these cells was also confirmed both *in vitro* and *in vivo* [81].

3.4 Treatment of patients with human fetal liver derived cells

Treatment of patients with fetal cells is slowly gaining its importance in the field of cell therapy with encouraging outcomes and improvements in the clinical conditions. However, the success of such procedures is largely dependent on several factors such as age of the patient, type and stage of the disease progression and the route and site of cell injection chosen. Each of these factors are variable for every patient and must be chosen carefully before administering the cells. Some of these factors and the success of the treatment procedures are discussed below:

3.4.1 Categories of patients selected for treatment with fetal cells

Several end stage liver diseases have been treated by liver cell based therapy with different degrees of success; a summary of these findings so far, is given below

1. **Acute liver failure** is characterized by loss of liver cells leading to encephalopathy. Initial study by Habibullah et al., showed that fetal hepatocyte transplantation may be beneficial in patients with ALF in grade III or IV encephalopathy [82]. Also, the transplanted hepatocytes may proliferate under the influence of hepatotropic factors thereby increasing their total metabolic and detoxifying capacity. In another study five patients comatose with acute liver failure received transplantation of 1.3×10^{9} to 3.9×10^{10} cryopreserved hepatocytes through intrasplenic and intraportal infusion and three patients showed improved result in encephalopathy score and some liver functions [83]. At our centre we have also performed intra-peritoneal transplantation of hepatocytes in a 26 yr old acute fatty liver of a pregnant patient who recovered within two days of transplantation [84].
2. **Metabolic Liver Disease** Hepatocyte transplantation into the liver corrected deficiency. In a landmark trial a child with Crigler-Najjar type I, suffering from dangerous hyper bilirubinaemia, was given $7.5\text{x}10^{9}$ allogenic donor hepatocytes by infusion via portal vein catheter [85]. This procedure resulted in reduction of serum bilirubin levels. Recently, in our study on Crigler-Najjar type I syndrome the patient treated with fetal liver derived hepatic progenitors showed a decrease in the total bilirubin and increase in the conjugated bilirubin [86]. Hepatocyte transplantation in a 4 year old patient with infantile Refsum disease led to partial clearance of abnormal bile acids with pipecholic acid being reduced to 60 per cent of pre-transplantation levels. The child was able to stand and walk 6 months after hepatocyte transplantation [87].

3. **Cirrhosis of liver** This category of liver disease is more appropriate for cell therapy and efficacy of various sources of cells has been demonstrated. Late stage progressive hepatic fibrosis characterized by distortion of hepatic architecture, necrosis of hepatocyte and formation of regenerative nodules leads to cirrhosis. Cell based therapy is emerging as an efficient method of treatment in the management of decompensated liver cirrhosis [88]. Treatment of liver cirrhosis using different sources of cells has been used by several investigators. Procedure and results of these studies has been discussed in section 3.4.3.

3.4.2 Infusion of cells into patients

The survival, proliferation and engraftment of transplanted/infused therapeutic fetal cells are critically dependent on the route of cell delivery and the physiological condition of site of transplantation. The most appropriate site of transplantation is directly into the liver, however because of some limitations of the clinical conditions, several ectopic sites such as the peritoneum, spleen also have been attempted.

3.4.2.1 Sites and routes for cell infusion/transplantation

Several routes to deliver the cells into the liver have been used with varying degrees of success. Cells for transplantation have usually been delivered to liver through hepatic artery [89] or through portal vein [90]. (The different routes for cell infusion are schematically depicted in figure 2).

a. **Hepatic artery**: Hepatic artery can be accessed through trans femoral, trans radial, trans brachial. In our experience hepatic artery route is more convenient compared to hepatic portal vein [91]. Though highly efficient, direct deliveries into liver might pose the risk of occlusion and in certain cases fibrosis, due to portal hypertension and embolism of cells.
b. **Portal Vein**: Portal vein can be accessed by percutaneous trans hepatic approach or trans jugular approach. The hepatic portal vein drains blood from the gastrointestinal tract and spleen into the liver. It is more often used compared to the hepatic artery because multiple vascular accesses are more practical through the vein. However the limitations of being a major procedure and the risk of bleeding must be considered for using the trans-jugular approach for delivery of cells through portal vein
   Accessing portal vein through percutaneous trans-hepatic approach is difficult in the presence of ascites which is a common clinical problem in the end stage of liver disease.
c. **Splenic Approach**: The most appropriate ectopic site is the splenic pulp within the spleen. Spleen has a rich blood supply which is accessible to hepatic portal circulation, leading to the translocation of the transplanted cells to the hepatic sinusoids. Direct intrasplenic injection was suggested as a better method to transplant hepatocytes compared to the splenic artery infusion, since the latter led to vascular occlusion with hepatocytes, gastric erosion, and large areas of splenic necrosis [92]. In a recent study fetal cells were transplanted in a patient with end stage cirrhosis via the intrasplenic infusion through the splenic artery leading to improvement in the clinical condition [74]. Studies have demonstrated that after injection into the splenic pulp, most hepatocytes immediately translocate to splenic veins and then to hepatic sinusoids, although hepatocytes trapped in splenic sinusoids may engraft in the spleen itself [93].

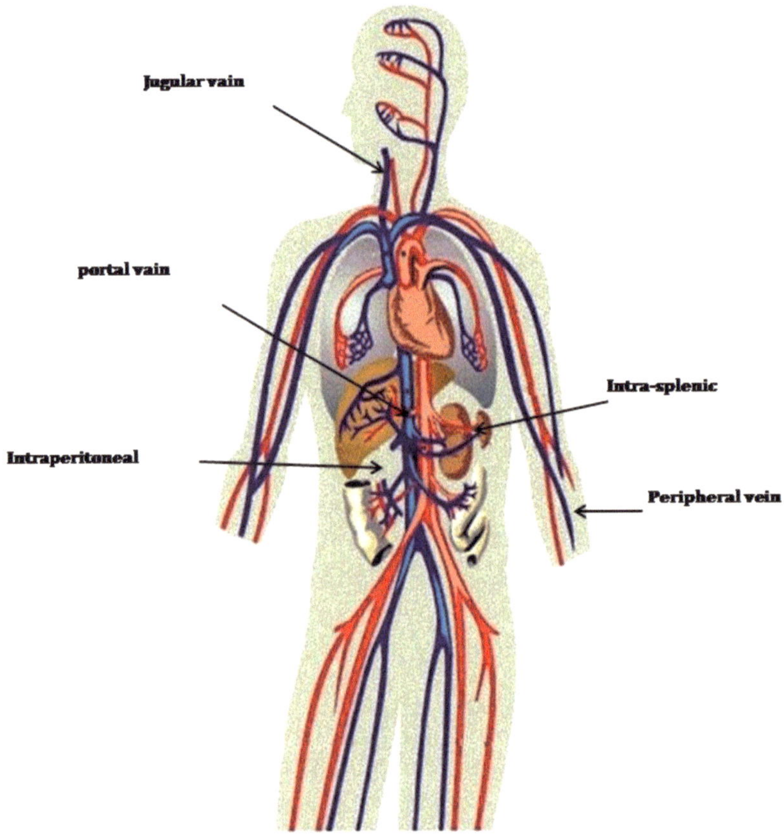

Fig. 2. Different routes for cell infusion

3.4.3 Efficacy and safety evaluation

The safety and efficacy of the fetal and other cell based treatment of end stage liver disease need to be constantly evaluated and revised as per the new protocols as they become available. Comparison of the different cell types and the routes of cell delivery has provided different alternatives for treatment of this disease. For example, our group demonstrated the efficacy of fetal hepatic progenitor cells, delivered through the hepatic artery in 25 cases of de-compensated liver cirrhosis, resulting in significant clinical improvement in more than 80 % cases. The hepatic angiogram showed no sign of thrombosis/narrowing/ischemia in the hepatic artery when analyzed after successive time intervals (Figure 3) [91].

Further data on immune status of allogenic transplantation of human fetal derived progenitors showed no significant changes in the immune status (T cells, NK cells, and Cytokines). The availability of long-term (> 3 years) follow-up data further confirms the safety and efficacy of this therapy. None of the patients recruited in this study developed any procedure/therapy related complications as demonstrated by liver angiograms that were taken pre- and post-transplantation of cells. In another study from our lab [86], infusion of bone marrow derived CD34+ cells in 5 cases by the same route also showed good outcomes in a 6 month follow up.

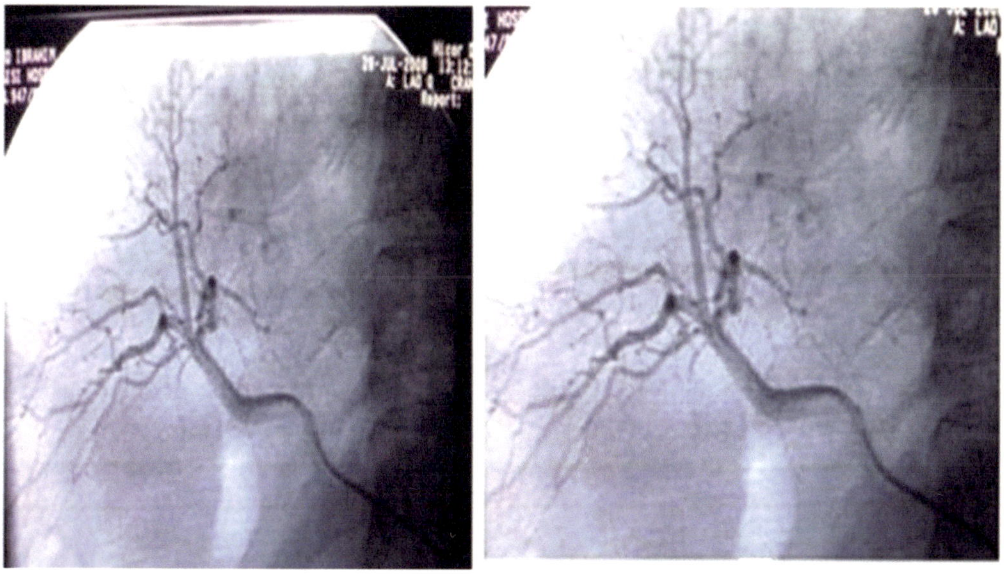

Fig. 3. Comparison of First and second Hepatic arteriogram following administration of Hepatic Progenitor cells to patient with Liver Cirrhosis (Khan et al., 2010)

Several studies have demonstrated the presence of cells of bone marrow origin in the human liver. This was elegantly demonstrated with bone marrow transplantation in female patients who received bone marrow from male donors and were found to be carrying Y-chromosome- and CK8- positive hepatocytes, thus, suggesting that extrahepatic stem cells can engraft in the liver [94]. An improvement of liver function and hepatocyte production was seen in nine patients with liver cirrhosis after the infusion of bone marrow cells through peripheral vein [37]. In another study, 2 patients with alcoholic- induced liver cirrhosis were treated with autologous mobilised HSCs [95]. In a study on 10 patients with chronic end-stage liver disease transplantation of committed progenitor cells and no bone marrow cells via the hepatic artery, showed an improvement in liver function [96]. A significant clinical improvement in the liver function, and Child-Pugh Score was seen in ten patients with advanced liver cirrhosis due to hepatitis B infection following autologous bone marrow infusion [97]. Pai et al., reported a study where autologous expanded mobilised adult bone marrow CD34+ cells were administered via the hepatic artery in 9 patients with alcoholic liver cirrhosis. Significant decrease in serum bilirubin, ALT, and AST levels were observed, whilst the Child-Pugh Scores and radiological ascites improved in 7 and 5 patients, respectively [98].

3.4.4 Tracking and monitoring the fate of transplanted cells

One of the major problems facing the delivery and monitoring of cell transplants is their noninvasive *in vivo* visualization. Tracking of transplanted cells is a very challenging area of cell based therapy. Monitoring transplanted cells becomes imperative in order to understand the route of migration of cells upon injection. It is important that the cells do not enter the pulmonary capillaries as this could lead to complications. In deciding choice of the tracking agent the primary points to be considered are safety of the tracking agent upon administration and the duration that the tracking agent can provide signals to be able to track the cells. The tracking agent must not cause any effects to the viability or multiplication of the transplanted cells. Though multiple studies have used various biomarkers in pre-clinical studies, their feasibility in human clinical studies warrants further validation.

In one of our clinical studies, tracking of the cells labeled with Tc-HM-PAO was done after cells were infused through hepatic artery (Figure 4). Hepatic scintiography showed that transplanted human fetal hepatic progenitor cells homed in the total lobes of liver with a high rate of engraftment, thereby again reiterating the effectiveness of hepatic artery in the cell delivery. No other clinical complications were observed during and after 6 months of follow-up [91].

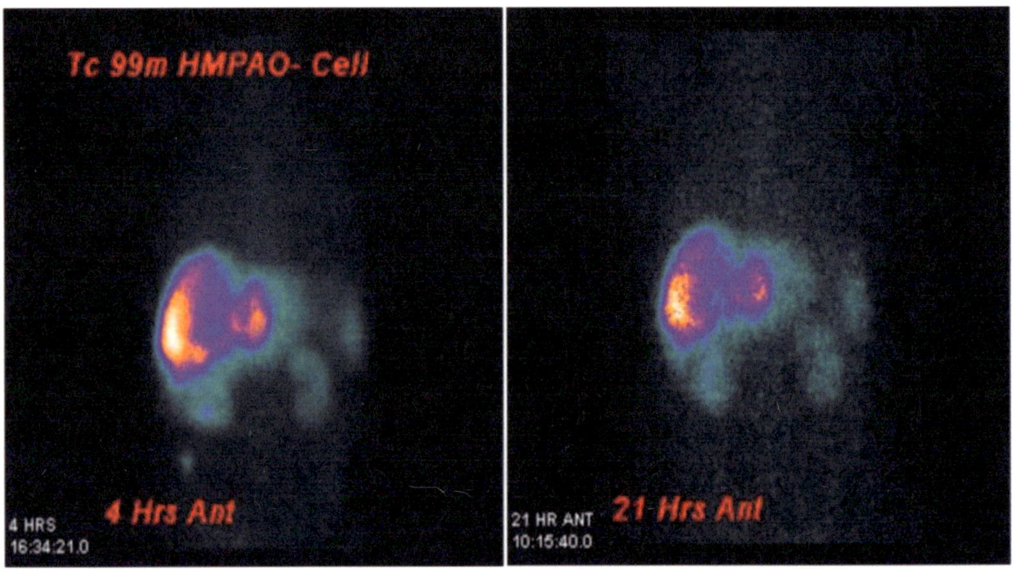

Fig. 4. Tc 99m HMPAO Progenitor cell labeling (Khan et al., 2010)

4. Basic studies in animal models to monitor the progression of the disease and treatment

Over the last decade several advancements have been made in the area of liver disease management. Further research is required to improvise the tissue imaging methods and new technologies are needed to monitor the fate of the cells that are injected. Animal models serve as efficient systems to study these methodologies.

4.1 Non-invasive monitoring of the transplanted cells

Monitoring the fate of transplanted cells over time is important to understand the route of cells migration and also to evaluate the efficacy of the transplanted cells. Several different methods based on magnetic, fluorescence or radio imaging are currently being evaluated for their efficiency in detecting transplanted cells. Though some of the methods are currently being used in clinical settings majority are still in the preclinical stages.

4.1.1 Radionuclide imaging

Radionuclide imaging is a very sensitive method that utilizes radio labeled markers for detection of cells. Single photon emission computed tomography (SPECT) and positron emission tomography (PET), allow the imaging of radio labeled markers and their interaction with biochemical processes in living subjects. SPECT provides the advantages of cell quantification, and lower background signals, but also has disadvantage of a lower spatial resolution compared with MRI and optical imaging [99].

PET is more sensitive than SPECT and permits more accurate quantification of cell numbers. Long term expression of reporter genes such as herpes simplex type 1 thymidine kinase (*HSV1-tk*) for PET imaging is more beneficial than direct and indirect labeling as the cells divide and increase the signal and is also indicative of viability of the cells [100, 101]. The use of reporter gene imaging is currently limited only to animal model studies and needs further validation for its use in clinical settings.

Using radiolabeling the organ bio distribution of human adult hepatocytes and fetal liver cells upon transplantation, was studied in NOD/SCID mice. The cells were labelled with indium-111 (111In)-oxine and technetium-99m (99mTc)-Ultratag or 99mTc-Ceretec and injected via the intrasplenic or intraportal routes. The adult hepatocytes and fetal liver stem/progenitor cells incorporated 111In but not 99mTc labels and the tracking of the cells confirmed that the cells were retained in the liver and spleen without translocating into pulmonary or systemic circulations [93].

4.1.2 Magnetic Resonance Imaging (MRI)

MRI provides a high spatial and temporal resolution in monitoring the distribution, migration, survival and differentiation of the transplanted cells in animals over weeks, *in vivo*. Cells are labeled in advance with contrast agents such as gadolinium-diethylenetriamine penta-acetic acid (Gd-DTPA) for short-term monitoring for upto a week, or superparamagnetic iron oxide (SPIO) for long-term monitoring (reviewed in [102]). Since these contrast agents are exogenous and are often degraded, efforts to develop suitable reporter genes such as the transferrin receptor (TfR) gene that is highly expressed on the target cell membrane have been done for years, however the field is still in its early stage [103, 104]. The major limitation with MRI is the inability to distinguish viable from nonviable cells or proliferating from non-proliferating cell populations. Further it also, cannot distinguish iron-labeled cells from free iron released upon cell death, therefore, iron particle labeling should better be looked as a marker for high-resolution detection of cell location rather than monitoring cell viability in MRI stem cell tracking [105].

4.1.3 Bioluminescence, fluorescence and infra red imaging

Bioluminescence and fluorescence imaging are efficient, noninvasive methods that provide rapid assessments of transgene expression in preclinical models [106]. Bioluminescence imaging is based on transgenic expression of certain bioluminescent genes such as the luciferase gene in transplanted cells and the *in vivo* detection is done by injecting the non-toxic bioluminescent substrate solution and the photons that are generated are captured to provide the images [107-109]. However, bioluminescence is limited by a lower spatial resolution and the inability to produce 3-dimensional and tomographic images [105].

Fluorescent imaging is based on the fluorescence property of a widely used protein such a green fluorescent protein (GFP) that is also introduced into the transplanted cells for stable expression [110]. Due to the low tissue penetrance and non-specific background generated by the autofluorescence of the surrounding tissue GFP reporter cannot be used to reliably track *in vivo* characteristics of transplanted stem cells [109-112].

Imaging in the near-infrared (NIR) wavelength (700–900 nm) spectrum can maximize tissue penetrance in addition to minimizing the autofluorescence from non target tissue. Our lab is currently working on tracking transplanted cells using the fluorescent dye Di-D. EpCAM +ve human fetal liver cells were labeled with Di-D and injected in the liver of nude mice to monitor their survival and engraftment of the liver. Using Di-D the labeled cells were imaged with KODAK Fx PRO animal imaging system at various time points indicating the persistence of fluorescencent labeled EpCAM+ve cells upto 110 days post transplantation (Figure 5). The fluorescence images of animal were overlaid with the X-ray images of the animal. Further studies on the safety and efficacy of the dye needs to be confirmed before clinical trials could be attempted.

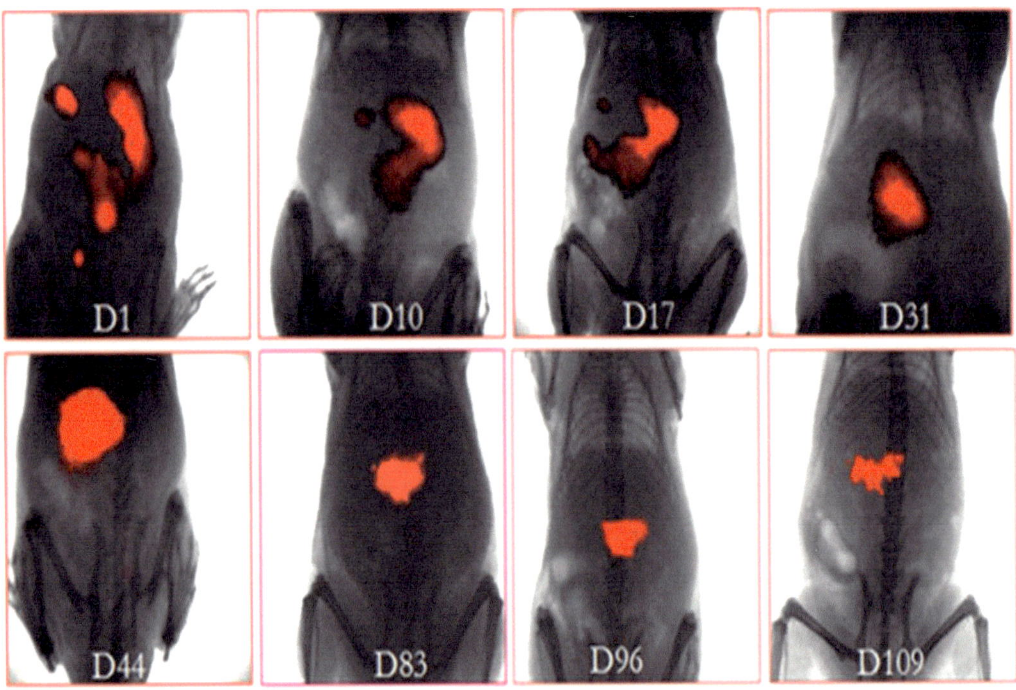

Fig. 5. *In vivo* images of nude mice after intra hepatic transplantation with DiD labeled EpCAM +ve cells at different time points (Unpublished data)

4.2 Need for more non- invasive tissue imaging methods to monitor disease progression

Non-invasive tissue imaging methods to detect fibrosis play an important role in the management of liver disease. Methods to monitor the disease progression are as important as the treatment procedures, as timely detection of fibrosis could help in prevention of the further progression and restore the liver functions with minimal intervention or cell therapy. Imaging methods offer an invaluable diagnostic tool in detection and monitoring the disease progression.

The conventional methods of imaging such as the ultrasonography and computed tomography are routinely used in the evaluation of fibrosis but with limited success in predicting the stage of the disease. The more recent methodologies based on elastography such as the fibroscan (ultrasound elastography) and Magnetic Resonace elastography (based on magnetic resonance) measure the stiffness of the tissue and correlate to the extent of disease progression and are more sensitive to detect fibrosis. However, the sensitivity of these methods are limited to detect fibrosis only in later stage of disease progression and new sensitive methods need to be developed that can predict the early fibrotic changes of the liver. Since fibrosis is characterized by extensive and unorganized accumulation of ECM with the principal component being collagen, our lab has initiated work on this front by targeting collagen to detect fibrosis. We propose to achieve this by using probes that are specific to collagen and combine these to a fluorescent or magnetic reporter molecule that would be detected by imaging methods. The amount of collagen detected would be correlated to the stage of fibrosis. Studies are currently being carried on rat and mouse models and would need extensive confirmation before these can be extended to human studies.

5. Conclusions

Cell based treatment is a promising area which is slowly gaining importance as a therapeutic approach for liver diseases. Preclinical and clinical trials with different types of adult cells have provided evidence for their usefulness in treating cirrhotic liver disease though with certain limitations. Recently, human fetal progenitor cells with their low immunogenicity and a good proliferative capacity are emerging as safe and potential sources for treating liver diseases. However, the efficiency of these cells in treating various types of liver diseases is yet to be established. Long term monitoring of the fate of the transfused cells also remains a challenging area that needs to be addressed before these cells can be used as routine treatment options for liver diseases. Improvised methods of cell isolation and infusion in patients and the development of faster and efficient non-invasive methodologies for detecting fibrosis at early stages would be an important step towards the management of liver diseases.

6. Acknowledgement

The work reported in this review was supported by Grants No. GAP220, from the Department of Science and Technology, New Delhi to GP and GAP338 from Indian Council of Medical Research New Delhi to GP and AAK.

7. References

[1] Mormone E, George J, Nieto, N (2011) Molecular pathogenesis of hepatic fibrosis and current therapeutic approaches. Chem-Biol. Inter. 193: 225–231.

[2] Rojkind M, Martinez-Palomo A (1976) Increase in type I and type III collagens in human alcoholic liver cirrhosis. Proc. Natl. Acad. Sci. USA 73: 539-543.

[3] Seyer JM, Hutcheson ET, Kang AH (1977) Collagen polymorphism in normal and cirrhotic human liver. J. Clin. Invest. 59: 241-248.
[4] Taniyama Y, Morishita R, Nakagam H, Moriguchi A, Sakonjo H, Shokei-Kim MD, Matsumoto K, Nakamura S, Higaki J, Ogihara T (2000) Potential Contribution of a Novel Antifibrotic Factor, Hepatocyte Growth Factor, to Prevention of Myocardial Fibrosis by Angiotensin II Blockade in Cardiomyopathic Hamsters. J. American Heart Association 102: 246-252.
[5] Ueki T, Kaneda Y, Tsutsui H, Nakanishi K, Sawa Y, Morishita R, Matsumoto K, Nakamura T, Takahashi H, Oakamoto E, Fujimoto J (1999) Curative gene therapy of liver cirrhosis by hepatocyte growth factor in rats. Nat Med. 5: 226 -230.
[6] Yaekashiwa M, Nakayama S, Ohnuma K, Sakai T, Abe T, Satoh K, Matsumoto K, Nakamura T, Takahashi T, Nukiwa T (1997) Simultaneous or delayed administration of hepatocyte growth factor equally represses the fibrotic changes in murine lung injury induced by bleomycin: a morphologic study. Am. J. Respir. Crit. Care Med. 156: 1937–1944.
[7] Liu Y (2002) Hepatocyte growth factor and the kidney. Curr. Opin. Nephrol. Hypertens. 11: 23–30.
[8] Matsumoto K, Nakamura T (2001) Hepatocyte growth factor: renotropic role and potential therapeutics for renal diseases. Kidney Int. 59: 2023– 2038.
[9] Janczewska-Kazek E, Marek B, Kajdaniuk D, Ziółkowsk A, Kukla M (2005) Correlation between TGF-beta1, VEGF, HGF, EGF, TGF-alpha and FGF serum levels, necroinflammatory activity and fibrosis in chronic hepatitis C. E&C Hepatology 1: 24-28.
[10] Kallis YN, Robson AJ, Fallowfield JA, Thomas HC, Alison MR, Wright NA, Goldin RD, Iredale JP, Forbes SJ (2011) Remodelling of extracellular matrix is a requirement for the hepatic progenitor cell response. Gut 60: 525–533.
[11] Gehling UM, Willems M, Schlagner K, Benndorf RA, Dandri M, Petersen J, Sterneck M, Pollok JM, Hossfeld DK, Rogiers X (2010) Mobilization of hematopoietic progenitor cells in patients with liver cirrhosis, World J. Gastro. 16: 217–224.
[12] Piscaglia C (2008) Stem cells, a two-edged sword: risks and potentials of regenerative medicine. World J. Gastro. 14: 4273–4279.
[13] Liongue C, Wright C, Russell AP, Ward AC (2009) Granulocyte colony-stimulating factor receptor: stimulating granulopoiesis and much more. Int. J. Biochem. Cell Biol. 41: 2372–2375.
[14] Piscaglia C, Shupe TD, Oh S, Gasbarrini A, Petersen BE (2007) Granulocyte-colony stimulating factor promotes liver repair and induces oval cell migration and proliferation in rats. Gastroenterology. 133: 619–631.
[15] Gaia S, Smedile A, Omed`e P Olivero, A, Sanavio F, Balzola F,Ottobrelli A, Abate ML, Marzano A, Rizzetto M, Tarella C (2006) Feasibility and safety of G-CSF administration to induce bone marrow-derived cells mobilization in patients with end stage liver disease. J. Hep. 45: 13–19.
[16] Khan AA, Parveen N, Mahaboob VS, Rajendraprasad A, Ravindraprakash HR, Venkateswarlu J,. Rao SG, Narusu ML, Khaja MN, Pramila R, Habeeb A, Habibullah CM (2008) Safety and efficacy of autologous bone marrow stem cell transplantation through hepatic artery for the treatment of chronic liver failure: A preliminary study. Transp. Proc. 40: 1140–1144.
[17] Campli CD, Zocco MA, Saulnier N, Grieco A, Rapaccini G, Addolorato G, Rumi C, Santoliquido A, Leone G, Gasbarrini G, Gasbarrini A (2007) Safety and efficacy

profile of G-CSF therapy in patients with acute on chronic liver failure. Dig. Liv. Dis. 39: 1071–1076.

[18]. Lorenzini S, Isidori A, Catani L, Gramenzi A, Talarico S, Bonifazi F, Giudice V, Conte R, Baccarani M, Bernardi M, Forbes SJ, Lemoli RM, Andreone P (2008) Stem cell mobilization and collection in patients with liver cirrhosis. Aliment. Pharmcol. Ther. 27: 932–939.

[19] Spahr L, Lambert JF, Rubbia-Brandt L, Chalandon Y, Frossard JL, Giostra E, Hadengue A. (2008) Granulocyte-colony stimulating factor induces proliferation of hepatic progenitors in alcoholic steatohepatitis: a randomized trial. Hepatology 48: 221–229.

[20] Garg V, Garg H, Khan A, Trehanpati N, Kumar A, Sharma BC, Sakhuja P, Sarin SK (2012) Granulocyte Colony-Stimulating Factor Mobilizes CD34(+) Cells and Improves Survival of Patients With Acute-on-Chronic Liver Failure. Gasteroenterology 142: 505-512

[21] Groth CG, Arborgh B, Bjorken C, Sundberg B, Lundgren G (1977) Correction of hyperbilirubinemia in the glucuronyltransferase deficient rat by intraportal hepatocyte transplantation. Transplant. Proc. 9: 313–316.

[22] Bilir BM, Guinette D, Karrer F, Kumpe DA, Krysl J, Stephens J, McGavran L, Ostrowska A, Durham J (2000) Hepatocyte transplantation in acute liver failure. Liver Transpl. 6: 32–40.

[23] Mito M, Kusano M (1993) Hepatocyte transplantation in man. Cell Transplant 2: 65–74.

[24] Strom SC, Fisher RA, Thompson MT, Sanyal AJ, Cole PE, Ham JM, Posner MP (1997) Hepatocyte transplantation as a bridge to orthotopic liver transplantation in terminal liver failure. Transplantation 63: 559–569.

[25] Kung JW, Forbes SJ (2009) Stem cells and liver repair. Curr. Opin. Biotech. 20: 568–574.

[26] Sancho-Bru P, Najimi M, Caruso M et al Pauwelyn K, Cantz T, Forbes S, Roskams T, Ott M, Gehling U, Sokal E, Verfaillie CM, Muraca M (2009) Stem and progenitor cells for liver repopulation: can we standardise the process from bench to bedside? Gut 58: 594- 603.

[27] Chan C, Berthiaume F, Nath BD, Tilles AW, Toner M, Yarmush ML (2004) Hepatic tissue engineering for adjunct and temporary liver support: critical technologies. Liver Transplant. 10: 1331–1342.

[28] Demetriou AA, Brown Jr RS, Busuttil RW, Fair J, McGuire BM, Rosenthal P, Am Esch JS 2nd, Lerut J, Nyberg SL, Salizzoni M, Fagan EA, de Hemptinne B, Broelsch CE, Muraca M, Salmeron JM, Rabkin JM, Metselaar HJ, Pratt D, De La Mata M, McChesney LP, Everson GT, Lavin PT, Stevens AC, Pitkin Z, Solomon BA (2004) Prospective, randomized, multicenter, controlled trial of a bioartificial liver in treating acute liver failure. Ann. Surg. 239: 660.

[29] Ellis AJ, Hughes RD, Wendon JA, Dunne J, Langley PG, Kelly JH, Gislason GT, Sussman NL, Williams R (1996) Pilot-controlled trial of the extracorporeal liver assist device in acute liver failure. Hepatology 24: 1446–51.

[30] Piscaglia C, Novi M, Campanale M, Gasbarrini A (2008) Stem cell-based therapy in gastroenterology and hepatology. Minimally Invasive Therp. Allied Technol. 17: 100–118.

[31] Duret C, Gerbal-Chaloin S, Ramos J, Fabre JM, Jacquet E, Navarro F, Blanc P, Sa-Cunha A, Maurel P, Daujat-Chavanieu M (2007) Isolation, characterization, and differentiation to hepatocyte-like cells of nonparenchymal epithelial cells from adult human liver. Stem Cells 25: 1779–1790.

[32] Schmelzer E, Wauthier E, Reid LM (2006) The phenotypes of pluripotent human hepatic progenitors. Stem Cells 24: 1852–1858.

[33] Haruna Y, Saito K, Spaulding S, Nalesnik MA, Gerber MA (1996) Identification of bipotential progenitor cells in human liver development. Hepatology 23: 476-481.
[34] Badve S, Logdberg L, Sokhi R, Sigal SH, Botros N, Chae S, Das KM, Gupta S (2000) An antigen reacting with Das-1 monoclonal antibody is ontogenically regulated in diverse organs including liver and indicates sharing of developmental mechanisms among cell lineages. Pathobiology 68: 76-86.
[35] Pittenger MF, Mackay AM, Beck SC, Jaiswal RK, Douglas R, Mosca JD, Moorman MA, Simonetti DW, Craig S, Marshak DR (1999) Multilineage potential of adult human mesenchymal stem cells. Science 284: 143–147.
[36] Bianco P, Riminucci M, Gronthos S, Robey PG (2001) Bone marrow stromal stem cells: nature, biology, and potential applications. Stem Cells 19: 180-192.
[37] Terai S, Ishikawa T, Omori K, Aoyama K, Marumoto Y, Urata Y, Yokoyama Y, Uchida K, Yamasaki T, Fujii Y, Okita K, Sakaida I (2006) Improved liver function in patients with liver cirrhosis after autologous bone marrow cell infusion therapy. Stem Cells 24: 2292–2298.
[38] Ismail, Fouad O, Abdelnasser A, Chowdhury A, Selim A (2010) Stem cell therapy improves the outcome of liver resection in cirrhotics. J. Gastro. Cancer 41: 17–23.
[39] Russo FP, Alison MR, Bigger BW, Amofah E, Florou A, Amin F, Bou-Gharios G, Jeffery R, Iredale JP, Forbes SJ (2006) The bone marrow functionally contributes to liver fibrosis. Gastroenterology 130: 1807–1821.
[40] Thomas JA, Pope C, Wojtacha D, Robson AJ, Gordon-Walker TT, Hartland S, Ramachandran P, Van Deemter M, Hume DA, Iredale JP, Forbes SJ. (2011) Macrophage therapy for murine liver fibrosis recruits host effector cells improving fibrosis, regeneration and function. Hepatology 53: 2003–2015.
[41] Duffield JS, Forbes SJ, Constandinou CM, Clay S, Partolina M, Vuthoori S, Wu S, Lang R, Iredale JP (2005) Selective depletion of macrophages reveals distinct, opposing roles during liver injury and repair. J. Clin. Invest. 115: 56–65.
[42] Wognum AW, Eaves AC, Thomas TE (2003) Identification and isolation of hematopoietic stem cells. Arch. Med. Res. 34: 461-475.
[43] Zuk PA, Zhu M, Mizuno H, Huang J, Futrell JW, Katz AJ, Benhaim P, Lorenz HP, Hedrick MH (2001) Multi lineage cells from human adipose tissue: implications for cell-based therapies. Tissue. Eng. 7: 211–228.
[44] Banas A, Teratani T, Yamamoto Y, Tokuhara M, Takeshita F, Quinn G, Okochi H, Ochiya T (2007) Adipose tissue derived mesenchymal stem cells as a source of human hepatocytes. Hepatology 46: 219–228.
[45] Tong ML, Martina M, Hutmacher DW, Hui JHPO, Eng HL, and Lim B (2007) Identification of common pathways mediating differentiation of bone marrow and adipose tissue-derived human mesenchymal stem cells into three mesenchymal lineages. Stem Cells 25: 750–760.
[46] Kern S, Eichler H, Stoeve J, Kl¨uter H, Bieback K (2006) Comparative analysis of mesenchymal stem cells from bone marrow, umbilical cord blood, or adipose tissue. Stem Cells 24: 1294–1301.
[47] Gluckman E, Rocha V, Boyer-Chammard A, Locatelli F, Arcese W, Pasquini R, Ortega J, Souillet G, Ferreira E, Laporte JP, Fernandez M, Chastang C (1997) Outcome of cord-blood transplantation from related and unrelated donors. N. Engl. J. Med. 337: 373–81.
[48] Grewal SS, Barker JN, Davies SM, Wagner JE (2003) Unrelated donor hematopoietic cell transplantation: marrow or umbilical cord blood. Blood 101: 4233–44.

[49] Bahk JY, Piao Z, Jung JH, Han H (2011) Treatment of the end Stage Liver Cirrhosis by Human Umbilical Cord Blood Stem Cells: Preliminary Results. In: Ali Gholamrezanezhad editor. Stem Cells in Clinic and Research. InTech. pp 469-500.
[50] Bartholomew A, Sturgeon C, Siatskas M, Ferrer K, McIntosh K, Patil S, Hardy W, Devine S, Ucker D, Deans R, Moseley A, Hoffman R. (2002) Mesenchymal stem cells suppress lymphocyte proliferation *in vitro* and prolong skin graft survival *in vivo*. Exp. Hematol. 30: 42-48.
[51] Le Blanc K, Tammik L, Sundberg B, Haynesworth SE, Ringden O (2003) Mesenchymal stem cells inhibit and stimulate mixed lymphocyte cultures and mitogenic responses independently of the major histocompatibility complex. Scand. J. Immunol. 57: 11-20.
[52] Fibbe WE, Noort WA (2003) Mesenchymal stem cells and hematopoietic stem cell transplantation. Ann. N. Y. Acad. Sci. 996: 235-244.
[53] Le Blanc K, Rasmusson I, Sundberg B, Gotherstrom C, Hassan M, Uzunel M, Ringden O (2004) Treatment of severe acute graft-versus-host disease with third party haploidentical mesenchymal stem cells. Lancet 363: 1439-1441.
[54] Maitra B, Szekely E, Gjini K, Laughlin MJ, Dennis J, Haynesworth SE, Koc ON (2004) Human mesenchymal stem cells support unrelated donor hematopoietic stem cells and suppress T-cell activation. Bon. Marr. Transplant. 33: 597-604.
[55] Aggarwal S, Pittenger MF (2005) Human mesenchymal stem cells modulate allogeneic immune cell responses. Blood 105:1815-1822.
[56] Zaret KS, Grompe M (2008) Generation and regeneration of cells of the liver and pancreas. Science 322: 1490–1494.
[57] Cai J, Zhao Y, Liu Y, Ye F, Song Z, Qin H, Meng S, Chen Y, Zhou R, Song X, Guo Y, Ding M, Deng H (2007) Directed differentiation of human embryonic stem cells into functional hepatic cells. Hepatology 45: 1229–39.
[58] Yamanaka S (2009) Elite and stochastic models for induced pluripotent stem cell generation. Nature 460: 49–52.
[59] Sullivan GJ, Hay DC, Park IH, Fletcher J, Hannoun Z, Payne CM, Dalgetty D, Black JR, Ross JA, Samuel K, Wang G, Daley GQ, Lee JH, Church GM, Forbes SJ, Iredale JP, Wilmut I (2010) Generation of functional human hepatic endoderm from human induced pluripotent stem cells. Hepatology 51: 329–335.
[60] Si-Tayeb K, Noto FK, Nagaoka M, Li J, Battle MA, Duris C, North PE, Dalton S, Duncan SA (2010) Highly efficient generation of human hepatocyte-like cells from induced pluripotent stem cells. Hepatology 51: 297– 305.
[61] Jozefczuk J, Prigione A, Chavez L, Adjaye J (2011). Comparative analysis of human embryonic stem cell and induced pluripotent stem cell-derived hepatocyte like cells reveals current drawbacks and possible strategies for improved differentiation. Stem Cells Dev. 20: 1259–1275.
[62] Inamura M, Kawabata K, Takayama K, Tashiro K, Sakurai F, Katayama K, Toyoda M, Akutsu H, Miyagawa Y, Okita H, Kiyokawa N, Umezawa A, Hayakawa T, Furue MK, Mizuguchi H (2011) Efficient generation of hepatoblasts from human ES cells and iPS cells by transient overexpression of homeobox gene HEX. Mol. Ther. 19: 400–407.
[63] SubbaRao M, Khan AA, Parveen N, Habeeb MA, Habibullah CM, Pande G (2008) Characterization of hepatic progenitors from human fetal liver during second trimester. World J. Gastroenterol. 14: 5730-5737.

[64] Nyamath P, Alvi A, Habeeb A, Khosla S, Khan AA, Habibullah CM (2007) Characterization of hepatic progenitors from human fetal liver using CD34 as a hepatic progenitor marker. World J. Gastroenterol. 13: 2319-2323.
[65] Malhi H, Irani AN, Gagandeep S, Gupta S (2002). Isolation of human progenitor liver epithelial cells with extensive replication capacity and differentiation into mature hepatocytes. J. Cell Sci. 115: 2679-2688.
[66] Petersen BE, Goff JP, Greenberger Michalopoulos GK (1998) Hepatic oval cells express the hematopoietic stem cell marker Thy-1 in the rat. Hepatology 27: 433–445.
[67] Fujio K, Evarts RP, Hu Z, Marsden ER, Thorgeirsson SS (1994) Expression of stem cell factor and its receptor, c-kit, during liver regeneration from putative stem cells in adult rat. Lab. Invest. 70: 511–516.
[68] Weiss TS, Lichtenauer M, Kirchner S, Stock P, Aurich H, Christ B, Brockhoff G, Kunz-Schughart LA, Jauch KW, Schlitt HJ, Thasler WE (2008) Hepatic progenitor cells from adult human livers for cell transplantation. Gut. 57:1129-38.
[69] Harrison MF (1953) Composition of the liver cell. Proc R Soc Lond B Biol Sci. 141: 203-16
[70] Anderson NG (1953) The mass isolation of whole cells from rat liver.Science. 117: 627-8.
[71] Castagna M, Chauveau J (1963) Dispersion of hepatic tissue in the state of isolated cells. C R Hebd Seances Acad Sci. 22: 969-977.
[72] Gallai-Hatchard JJ, Gray GM (1971) A method of obtaining a suspension of intact parenchymal cells from adult rat liver. J Cell Sci. 8: 73-86.
[73] Hommes FA, Draisma MI, Molenaar I (1970) Preparation and some properties of isolated rat liver cells. Biochim. Biophys Acta. 222: 361–371.
[74] Gridelli B, Vizzini G, Pietrosi G, Luca A, Spada M, Gruttadauria S, Cintorino D, Amico G, Chinnici C, Miki T, Schmelzer E, Conaldi PG, Triolo F, Gerlach JC (2012) Efficient human fetal liver cell isolation protocol based on vascular perfusion for liver cell-based therapy and case report on cell transplantation. Liver Transpl. 18:226-37.
[75] Yaswen P, Hayner NT, Fausto N (1984) Isolation of oval cells by centrifugal elutriation and compersion with other cell types purified from normal and preneoplastic livers. Cancer Res. 44: 324-331.
[76] Balzar M, Winter MJ, de Boer CJ, Litvinov SV (1999) The biology of the 17-1A antigen (Ep-CAM). J. Mol. Med. 77: 699–712.
[77] Schmelzer E, Wauthier E, Reid LM (2006) The phenotypes of pluripotent human hepatic progenitors. Stem Cells 24:1852-1858.
[78] Goodell MA, Brose K, Paradis G, Conner AS, Mulligan RC (1996) Isolation and functional properties of murine hematopoietic stem cells that are replicating *in vivo*. J Exp Med. 183: 1797-806
[79] Uchida N, Fujisaki T, Eaves AC, Eaves CJ (2001) Transplantable hematopoietic stem cells in human fetal liver have a CD34(+) side population (SP)phenotype. J Clin Invest. 108: 1071-7
[80] Khan AA, Parveen N, Habeeb MA, Habibullah CM (2006) Journey from hepatocyte transplantation to hepatic stem cells: A novel treatment strategy for liver diseases. Indian. J. Med. Res. 123: 601-614.
[81] Liu Y-N, Zhang J, He Q-H, Dai X, and Shen L (2008) Isolation and characterization of epithelial progenitor cells from human fetal liver. Hepatol Res. 38: 103–113.
[82] Habibullah CM, Syed IH, Qamar A, Taher-Uz Z (1994) Human fetal hepatocyte transplantation in patients with fulminant hepatic failure. Transplantation 58: 951-2.

[83] Blei AT (2005) Selection for acute liver failure: have we got it right? Liver Transpl. 11: S30–S34

[84] Khan AA, Habeeb A, Parveen N, Naseem B, Babu RP, Capoor AK, Habibullah CM. (2004) Peritoneal transplantation of human fetal hepatocytes for the treatment of acute fatty liver pregnancy; a case report. Trop. Gastroenterol. 25: 141-3.

[85] Fox IJ, Roy Chowdhary J, Kauffman SS (1998) Treatment of the Crigler-Najjar syndrome type I with hepatocyte transplantation. N Engl. J. Med. 338: 1422-6.

[86] Khan AA, Parveen N, Mahaboob VS, Rajendraprasad A, Ravindraprakash HR, Venkateswarlu J, Rao P, Pande G, Narusu ML, Khaja MN, Pramila R, Habeeb A, Habibullah CM (2008) Treatment of Crigler-Najjar Syndrome type 1 by hepatic progenitor cell transplantation: a simple procedure for management of hyperbilirubinemia. Transplant Proc. 40: 1148-50.

[87] Sokal EM, Smets F, Bourgos A, Van Maldergem L, Buls JP, Reding R (2003) Hepatocyte transplantation in a 4-year girl with peroxisomal biogenesis disease: technique, safety and metabolic follow-up. Transplantation. 76: 735-43.

[88] Sommer BG, Sutherland DE, MatasAJ, SimmonsRL, Najarian, JS (1979) Hepato cellular transplantation for treatment of galactosamine induced acute liver failure in rats. Transplant. Proc. 9: 578-584.

[89] Overturf K, Al-Dhalimy M, Tanguay R, Brantly M, Ou CN (1996) Hepatocytes corrected by gene therapy are selected *in vivo* in a murine model of hereditary tyrosinaemia type I. Nat. Genet. 12: 266–273.

[90] Gunsalus JR, Brady DA, Coulter SM, Gray BM, Edge AS (1997) Reduction of serum cholesterol in Watanabe rabbits by xenogeneic hepatocellular transplantation. Nat. Med. 3: 48–53.

[91] Khan AA, Shaik MV, Parveen N, Rajendraprasad A, Aleem MA, Aejaz Habeeb M, Srinivas G, Avinash Raj T, Tiwari SK, Kumaresan K, Venkateswarlu, Pande G, Habibullah CM (2010) Human Fetal Liver-Derived Stem Cell Transplantation as Supportive Modality in the Management of End-Stage Decompensated Liver Cirrhosis. Cell Transplant. 19: 409–418.

[92] Nagata H, Ito M, Shirota C, Edge A, McCowan TC (2003) Route of hepatocyte delivery affects hepatocyte engraftment in the spleen. Transplantation 76: 732–734.

[93] Cheng K, Benten D, Bhargava K, Inada M, Joseph B, Palestro C, Gupta S (2009) Hepatic Targeting and Biodistribution of Human Fetal Liver Stem/ Progenitor Cells and Adult Hepatocytes in Mice. Hepatology 50: 1194–1203.

[94] Alison MA, Poulsom R, Jeffery R (2000) Hepatocytes from non-hepatic adult stem cells. Nature 406: 257

[95] Yannaki E, Anagnostopoulos A, Kapetanos D, Xagorari A, Iordanidis F, Batsis I, Kaloyannidis P, Athanasiou E, Dourvas G, Kitis G, Fassas A (2006) Lasting amelioration in the clinical course of decompensated alcoholic cirrhosis with boost infusions of mobilized peripheral blood stem cells Exper. Hematol. 34: 1583–1587.

[96] Lyra AC, Soares MB, Silva (2007) Feasiblity and safety of autologous bone marrow mononuclear cell transplantation in patients with advanced chronic liver disease. World J. Gastroenterol. 13: 1067– 1073.

[97] Kim JK, Park YN, Kim JS, Park MS, Paik YH, Seok JY, Chung YE, Kim HO, Kim KS, Ahn SH, Kim do Y, Kim MJ, Lee KS, Chon CY, Kim SJ, Terai S, Sakaida I, Han KH (2010) Autologous bone marrow infusion activates the progenitor cell compartment in patients with advanced liver cirrhosis. Cell Transplant. 19: 1237–1246.

[98] Pai M, Zacharoulis D, Milicevic MN (2008) Autologous infusion of expanded mobilized adult bone marrow-derived CD34+ cells into patients with alcoholic liver cirrhosis. The Am. J. Gastroenterol. 103: 1952– 1958.
[99] Beeres SL, Bengel FM, Bartunek J, Atsma DE, Hill JM, Vanderheyden M, Penicka M, Schalij MJ, Wijns W, Bax JJ (2007) Role of imaging in cardiac stem cell therapy. J. Am. Coll. Cardiol. 49: 1137-48.
[100] Tian M, Perin E, Silva G, et al. (2009) Long-term monitoring of persistence, migration and differentiation of HSV1-tk expressed mesenchymal stem cells in an ischemia-reperfusion porcine model with 18F-FEAU PET/CT and MRI. Presentation at the World Congress of Molecular Imaging, Montreal, Canada.
[101] Arbab AS, Janic B, Haller J, Pawelczyk E, Liu W, Frank JA (2009) In-vivo cellular imaging for translational medical research. Curr. Med. Imaging. Rev.5: 19-38.
[102] Zhao C, Tian M, Zhang H (2010) *In Vivo* Stem Cell Imaging. The Open Nucl. Med. J. 2: 171-177
[103] Kang JH, Chung JK. (2008) Molecular-genetic imaging based on reporter gene expression. J. Nucl. Med. 49: 164S-79S.
[104] Arbab AS, Yocum GT, Wilson LB, Parwana A, Jordan EK, Kalish H, Frank JA (2004) Comparison of transfection agents in forming complexes with ferumoxides, cell labeling efficiency, and cellular viability. Mol. Imaging 3: 24-32.
[105] Li Z, Suzuki Y, Huang M, Cao F, Xie X, Connolly AJ, Yang PC, Wu JC (2008) Comparison of reporter gene and iron particle labeling for tracking fate of human embryonic stem cells and differentiated endothelial cells in living subjects. Stem Cells 26: 864-73.
[106] Iyer M, Sato M, Johnson M, Gambhir SS, Wu L (2005) Applications of molecular imaging in cancer gene therapy. Curr. Gene. Ther. 5: 607-18.
[107] Kim DE, Schellingerhout D, Ishii K, Shah K, Weissleder R (2004) Imaging of stem cell recruitment to ischemic infarcts in a murine model. Stroke 35: 952-7.
[108] Shah K, Hingtgen S, Kasmieh R, Figueiredo JL, Garcia-Garcia E, Martinez-Serrano A, Breakefield X, Weissleder R (2008) Bimodal viral vectors and *in vivo* imaging reveal the fate of human neural stem cells in experimental glioma model. J. Neurosci. 28: 4406-13.
[109] Kang JH, Chung JK (2008) Molecular-genetic imaging based on reporter gene expression. J. Nucl. Med. 49: 164S-79S.
[110] Shah K, Jacobs A, Breakefield XO, Weissleder R (2004) Molecular imaging of gene therapy for cancer. Gene Ther. 11: 1175-87.
[111] van der Bogt KE, Swijnenburg RJ, Cao F, Wu JC (2006) Molecular imaging of human embryonic stem cells: Keeping an eye on differentiation, tumorigenicity and immunogenicity. Cell Cycle 5: 2748-52.
[112] Troy T, Jekic-McMullen D, Sambucetti L, Rice B (2004) Quantitative comparison of the sensitivity of detection of fluorescent and bioluminescent reporters in animal models. Mol. Imaging 3: 9-23.

Liver Regeneration and Bioengineering – The Emergence of Whole Organ Scaffolds

Pedro M. Baptista, Dipen Vyas and Shay Soker
Wake Forest University Health Sciences,
Wake Forest Institute for Regenerative Medicine, Winston-Salem, NC,
USA

1. Introduction

Every year an estimated two million people die of advanced liver disease. The World Health Organization estimates that over six hundred and fifty million people worldwide are affected by some form of liver disease, including thirty million Americans. On a worldwide base, the bleak cenario of one to two million deaths are accounted to liver related diseases annually. From all the countries, China has the world's largest population of Hepatitis B patients (approx. 120 million) with five hundred thousand people dying of the liver disease every year(1, 2). In the US alone, there are around five hundred thousand critical episodes of liver problems every year requiring hospitalization with a huge burden to the patients and an enormous cost to the health care system. In the European Union and United States of America alone, over eighty one thousand and twenty six thousand people died of chronic liver disease in 2006, respectively(1, 3). For these patients, liver transplantation is currently the only therapy proven to extend survival for end-stage liver disease, as it is also the only treatment for severe acute liver failure and the some forms of inborn errors of metabolism. However, the waiting list for liver transplants is extensive and many on the list will not receive an organ due to a dramatic shortage of donors or not being eligible(1).

A good example of this is that in 2007 there were nearly seventeen thousand individuals on the US waiting list for a liver transplant. Only 30% of those in need were transplanted. The average waiting time was more than 400 days. The same year, about one thousand and three hundred people died while waiting for a suitable donor with no available medical option for saving their life. Also, for those patients with fulminant hepatic failure, a severe liver disease with 60-90% mortality, depending on the cause, only 10% received a transplant. Nevertheless, liver transplantation still has a relatively high mortality of 30-40% at 5-8 years with 65% of the deaths occurring in the first 6 months. In addition, patients who have undergone transplantation have to use lifelong immunosuppressive therapy, with sometimes severe side effects(4).

The etiologies of end-stage chronic liver disease that lead to transplantation are numerous and ~80% of people in the liver transplant waiting list have as primary diagnosis a cirrhotic liver. Fortunately, some of the causes of the disease are nowadays preventable. A good example is the successful vaccination programs in many countries in the world against Hepatitis B virus that have considerably reduced the incidence of chronic carriers and viral induced cirrhosis(5). Regrettably, close to 20% of the livers transplanted in the USA and 30% in Europe have a preventable underlying cause, alcoholic liver disease. Also ~45% of deaths due to liver cirrhosis in the USA are related with alcohol abuse(1, 3, 4). Patients with pathologies like hepatic cancer, congenital malformations and metabolic diseases, and acute hepatic necrosis compose the remaining percentage of the list.

2. Transplantation

The success of liver transplantation has resulted in a progressively increasing demand for such treatment. However, as mentioned above, at the same time the availability of donor organs has diminished, resulting in the number of potential recipients for liver transplantation far exceeding organ supply. Given this, several strategies have been explored in the last decade or so with the aim to increase access to liver transplantation. These include obtaining organs from non-heart-beating donors and live donors, and splitting and using livers from expanded donor criteria. Also, the introduction of the Model for End-Stage Liver Disease (MELD) system implemented February 27 of 2002 in the United States helped Organ Procurement Organizations to prioritize patients waiting for a liver transplant. The MELD score is a numerical scale used for adult liver transplant candidates that ranges between 6 (less ill) and 40 (gravely ill). The individual score determines how urgently a patient needs a liver transplant within the next 3 months. The number is calculated using the most recent laboratory tests - table 1(6).

Bilirubin	which measures how effectively the liver excretes bile
INR	formally known as the prothrombin time, measures the liver's ability to make blood clotting factors
Creatinine	which measures kidney function. Impaired kidney function is often associated with severe liver disease

Table 1. Laboratory values used in the MELD score calculation

The MELD score is then distributed in 4 levels according with the severety of the disease. Less than or equal to 10, 11-18, 19-24 and greater than or equal to 25, being the last the level that includes the most severe patients. Nevertheless, MELD score is not the only factor used for organ allocation to a patient.

In general, for organ distribution a donor is matched to a potential recipient on the basis of several factors: ABO blood type, body size, degree of medical urgency and MELD score. Organ Procurement Organizations (e.g. OPTN/UNOS, etc) uses a computerized point system to distribute organs in a fair manner. Recipients are chosen primarily on the basis of medical urgency within each ABO blood group. Waiting time is only a factor when patients have the same MELD score.

Nevertheless, there are four Special Case Exceptions that will be assigned a higher MELD score than that assigned by the patient's laboratory test results:

- Hepatocellular Carcinoma
- Hepatopulmonary Syndrome
- Familial Amyloidosis
- Primary Oxaluria

In addition to the previously mentioned four special case exceptions, a transplant center can apply for a MELD exception for a patient whose medical urgency is not reflected by the MELD score(6).

The implementation of more fair and efficient allocation systems, improvement in the immunosupressive regimens, and the increase of living donation have all helped to increase overall patient survival and graft survival in the past decade in the United States. The number of livers transplanted also increased to a all time high in 2006, with a marked decrease on the waiting time for liver transplantation after MELD implementation, especially for the sickest patients.

An example of the impact of these improvements is the increase of 6% (86% in 2007) and 16% (87% in 2007) of the unadjusted 1-year graft survival for deceased donor and living donor liver recipients between 1998 and 2007, respectively. These accounts also for an improvement of 3% (89% in 2007) and 11% (91% in 2007) of the unadjusted 1-year patient survival for deceased donor and living donor liver recipients for the same period, respectively(7). However, these numbers decrease significantly for the 5-year patient survival. In 2007 it was 74% and 79% for deceased donor and living donor liver recipients, respectively. These numbers decrease even further for the 10-year patient survival, where in 2007 we had 61% and 71% patient survival for deceased donor and living donor liver recipients, respectively. One important note is that patient survival was higher than graft survival ~5%, because of the opportunity for repeat liver transplantation in the event of graft failure(8).

These numbers highlight the need for novel therapies that can increase patient survival, as well as lower costs to the health care systems. Tolerance research and its clinical induction is a good example of this. The identification of molecular signatures in tolerant patients in whom immunosuppression could be stopped, and induction of tolerance, through lymphocyte depletion or T lymphocyte co-stimulation blockade, are the most advanced approaches to reduce complications of immunosuppression(9).

3. Bioartificil liver devices

In the past few decades, due to scarcity of donors, extracorporeal liver support devices have been developed to support the failing liver resulting from different complications. These devices were created initially for the management of patients waiting for a suitable donor for orthotopic liver transplantation. Recent advances in the design of these devices allow now utilizing them in the recovery of the native liver from an acute injury. Thus, these devices can either bridge the patients to liver transplantation or can fully avoid the need for it (10). Liver support devices can be broadly classified into two classes: artificial liver (AL) devices and bioartificial (BAL) devices. The artificial support devices are designed to detoxify the blood or plasma via different methods like hemodialysis, hemofiltration, hemodiafiltration, hemadsorption, plasmapheresis, plasma fractionation and albumin dialysis (10, 11). Bioartificial support devices are targeted towards providing essential metabolic and synthetic functions of liver along with removal of toxins. BAL devices generally utilize primary hepatocytes or hepatoma cell lines incorporated into a bioreactor system to perform the essential liver functions (12). Here, we will discuss the operating principles of several artificial and bioartificial support systems which have been or are currently used in clinical trials.

3.1 Artificial liver devices

Various metabolic functions of the liver are severely impaired during acute or chronic liver failure which leads to accumulation of lethal toxins in the body. ALs were developed as support devices which can efficiently remove these toxins from blood or plasma by using membrane filtration and/or adsorbents. Liver Dialysis device, Molecular Adsorbent Recirculating System (MARS) and Prometheus are the most widely used artificial support systems.

In liver dialysis device the patient's blood is drawn from a central vein and passed through a low-to-medium permeability membrane and at the same time a suspension of powdered activated charcoal and cation exchange resin is pumped through the dialysate side of the dialyser. This result in removal of toxins from the blood as the toxins are adsorbed based on their binding affinities (10, 13). The MARS uses an albumin-impermeable membrane (50 kDa cutoff) which separates the high-flux albumin coated dialyser from albumin filled dialysate. The toxins bound to the albumin are dissolved from the patient's albumin and they pass through the membrane and ultimately bind to the albumin solution in dialysate side. This albumin is then recycled by dialysis or adsorption through charcoal and resin-binding columns (14, 15). Prometheus primarily uses fractional plasma separation and adsorption techniques for removal of toxins. It uses an albumin permeable membrane (250 kDa cutoff) in contrast to MARS. The albumin bound toxins passes through the membrane and passes through special adsorbers which directly remove the toxins from the plasma and delivering the free albumin back to the patient (13, 14).

Liver Dialysis, MARS and Prometheus have been widely used in clinical trials across Europe and Asia, and have showed some benefits to the patients but no major outcome benefits as far as patient survival is concerned.

3.2 Bioartificial liver devices

Although ALs have been shown to provide temporary support to the patients with acute liver failure by detoxifying the blood or plasma, they have major limitations in replacing synthetic and metabolic functions of liver (12). Thus, attempts have been made to develop bioartificial liver (BAL) systems, which can provide both metabolic and synthetic hepatic functions and its detoxification. BALs incorporate primary hepatocytes or hepatoma cell lines as a biological component and hollow fiber or porous matrix membranes on which the functional hepatocytes are coated (12, 16). Many cell types from various sources have been investigated for providing bioartificial liver support. Primary human hepatocytes have been widely studied as an ideal cell source due to their biocompatibility but they are not readily available and their proliferative capacity *in vitro* is limited (13, 16). Animal cell source such as porcine primary hepatocytes are being investigated due to ease of availability and their ability to maintain metabolic functions similar to human hepatocytes. However, concerns regarding immunological reaction to the animal proteins and transmission of disease exist (17). Nonetheless, porcine hepatocytes remain a popular choice as a hepatocyte source in various BAL systems.

A bioreactor is a vital component of BALs and it has a major influence on the efficacy of these systems. In order to be used in BAL systems, the bioreactor should be able to provide a suitable environment for hepatic cells to thrive and remain functional along with an adequate interface between blood and hepatocytes for mass transport (18). The bioreactors should also have a potential for scale up and flexibility as BALs may be required to be customized to the patient's needs. Thus, there is an inevitable need for structural optimization and modifications of the bioreactors even though there have been recent advances in this technology. It should be highlighted that no bioreactor is currently approved for patient use, although some have been used in clinical trials (19). Here is a list of BAL devices currently under clinical investigation:

1. Extracorporeal Liver Assist Device (ELAD)
2. HepatAssist
3. Bioartificial Liver Support System (BLSS)
4. The Academic Medical Center - Bioartificial Liver (AMC-BAL)
5. Modular Extracorporeal Liver Support device (MELS)

So far over 200 patients have been treated with HepatAssist and over 40 patients treated with ELAD, making them the most common BALs used for treatment (15, 20). ELAD is the only BAL system which uses human hepatocyte cell line. The other BAL systems listed above use porcine hepatocytes as a cell source. ELAD uses immortalized C3A cell line derived from human hepatoma cell line HepG2 (21). The cells are located in the extracapillary space of hollow fibre cartridges (200 gram total cells in four cartridges). The membrane is impermeable to immunoglobulins, blood cells and C3A cells. The blood flows through the lumen of cartridges as the ultrafiltrated plasma from the membrane comes in direct contact with hepatocytes (12). HepatAssist incorporates approximately 5-7 billion cryopreserved porcine hepatocytes attached to microcarriers and loaded onto a hollow fibre. The separated plasma passes through a charcoal column and oxygenator prior to entering the hollow fibres in the bioreactor. An upgraded version of HepatAssist contains 15-20 billion porcine hepatocytes. The membrane pores are 0.15μm in size which prevents a physical contact between human cells and porcine hepatocytes (10, 20). Currently, none of the BALs have been approved by FDA for clinical use. All of the above listed BALs are undergoing several clinical trials in USA and Europe.

3.3 Future of liver support devices

Recent developments in artificial and bioartificial devices have shown a promising path towards the management of patients with acute liver failure. However, considerable technical challenges and regulatory issues remain to be tackled in order to efficiently utilize these devices in the clinic. ALs have demonstrated the ease of use and cost effectiveness along with proving to improve biochemical parameters and clinical symptoms by detoxifying the blood/plasma, but it has a major limitation of not replacing critical metabolic and synthetic functions of liver. In order to ameliorate liver injury and subsequently prevent the lethal effects of loss of liver function on other critical organs, liver function needs to be performed by the extracorporeal support devices while the patient awaits the transplantation. BALs have been developed in recent years which utilize hepatocytes in a bioreactor to carry out the metabolic and synthetic functions of liver. For these reasons, BALs hold a promising future as they have shown potential by efficiently treating several patients across different clinical trials. Several challenges exist in BAL technology including the debate on ideal cell source, requirement of large number of cells, maintainance of the functional hepatocytes for longer period of time in a bioreactor, complexity of the design and high cost. These challenges have delayed the entry of BAL systems in the clinic. Nonetheless, plenty of optimized designs of liver support devices are under development and undergoing clinical trials which is a sign of optimism in this area of critical care and management.

4. Cell therapies

Hepatocyte transplantation is certainly in the forefront of new therapeutic strategies. The first successful hepatocyte transplantation into a patient was carried out in June 1992 to a French Canadian woman with familial hypercholesterolaemia. After *ex vivo* transduction with a retrovirus encoding for the human LDL receptor, the patient's hepatocytes were infused through the inferior mesenteric vein into the liver. LDL and HDL levels improved throughout the next 18 months and transgene expression was detected in a liver biopsy(22). Following this first success, other patients followed through. However, not all the patients treated had a clear benefit from the procedure(23). Since then, several other metabolic diseases have been treated with hepatocyte transplantation with different degrees of success(24-28). It has also been used as a support treatment to acute(29-31) and chronic liver diseases(30-33) in bridging severely ill patients to orthotopic liver transplantation (OLT). Low efficacy and lack of long-term therapeutic effect have been common in all these procedures. These failures could be explained by the relatively small number of hepatocytes that engraft in the recipient liver due to quality, quantity and possibly immunosuppresion protocols(34). However, transplantation of a number of hepatocytes corresponding to 1-5% of the total liver mass has been able to show a positive impact in transplanted patients, even if for a limited period of time(34).

Due to the shortage of available human hepatocytes for transplantation, other cell sources have been used. Specifically, bone marrow derived mesenchymal stem cells(35), hematopoietic stem cells(36, 37) and fetal liver progenitor/stem cells(38) have shown to improve, to a certain extent, the condition of cirrhotic patients. The latter cell type holds an enormous potential for cell/regenerative medicine therapies due to their high expansion capabilities and differentiation into hepatocytes and biliary epithelium(39).

Recent data suggests that human embrionic (hES) and induced pluripotent (iPS) stem cells hold great promise to regenerative applications in every medical field. Specifically for the liver, several studies have established the required pathways to differentiate a hES or iPS into a hepatic fate by using defined soluble growth factor signals that mimic embryonic development(40, 41). These cells, once transplanted into rodent livers were able to engraft and express several normal hepatic functions(42). However, more extensive characterization, as well as further safety evaluation, are needed to determine wether these cells will fully function as primary adult hepatocytes.

5. Liver bioengineering

Tissue engineering is one of the most promising fields in regenerative medicine. As described in 1993 by Robert Langer and Joseph Vacanti it is the conjugation of biomaterials (synthetic or naturally derived) with cells, in order to generate tissue constructs that can be implanted into patients to substitute a lost function, maintain or gain new functions(43). The current paradigm is suitable for the engineering of thin constructs like the bladder, skin or blood vessels. Although, in the specific case of the liver, the 3D architecture and dense cellular mass requires novel tissue engineering approaches and the development of vascularized biomaterials, in order to support thick tissue masses and be readily transplantable. Additionally to the vascular support for large tissue masses, hepatocyte function maintenance represents the ultimate aim in any organ engineering or regenerative medicine strategy for liver disease.

Hepatocytes are known to be attachment-dependent cells and lose rather quickly their specific functions without optimal media- and ECM- composition and cell-cell contacts. Also, function and differentiation of liver cells are influenced by the 3D organ architecture(44).

In the last two decades innumerous strategies for the culture of adult hepatocytes in combination with several types of 3D, highly porous polymeric matrices have been attempted(45-49). However, in the absence of vasculature, restriction in cell growth and function is common due to the limitations in nutrient and oxygen diffusion. Some of these problems are being now partially overcome with the development of bioreactors that provide continuous perfusion of culture media and gases allowing a 3D culture configuration and hepatocyte function maintenance(50-52).

The tissue engineering concept has several advantages over the injection of cell suspensions into solid organs. The matrices provide sufficient volume for the transplantation of an adequate cell mass up to whole-organ equivalents[45]. Transplantation efficiency could readily be improved by optimizing the microarchitecture and composition of the matrices as well as by attaching growth factors and extracellular matrix molecules to the polymeric scaffold, helping to recreate the hepatic microenvironment(44). The use of naturally derived matrices has also proved to be very helpful in hepatocyte culture(47). These matrices, besides preserving some of the microarchitecture features of the tissues that they are derived from, also retain bioactive signals (e.g., cell-adhesion peptides and growth factors) required for the retention of tissue-specific gene expression(53, 54). Additionally, cell transplantation into polymeric matrices is, in contrast to cell injection into tissues and organs, a reversible procedure since the cell-matrix-constructs may be removed if necessary.

Finally, heterotopic hepatocyte transplantation in matrices has already been demonstrated in long-term studies(55, 56). Nonetheless, initial engraftment rates are suboptimal. One of the reasons for this is the absolute requirement of the transplanted hepatocytes for hepatotrophic factors that the liver constantly receives through its portal circulation(57). Thus, the development of a tissue engineered liver construct capable of being orthotopically transplanted is essential.

Apart from cellular therapies, other early developments of experimental approaches are not showing results that will indicate clinical translation in the next few years. However, two experimental approaches that show higher level of maturity may have the potential for succesful clinical translation. The first experimental approach is the "cell sheet" technology developed by Okano *et al.* in Japan(58). Its simple configuration and fabrication allows for the stacking of up to four hepatocyte cell sheets that can readily engraft and provide a defined metabolic relief to the recipient(59). This technology has already been applied successfully to one patient with heart failure. Other technology that shows great promise is tissue and organ decellularization. Our lab and others have been able to generate several decellularized scaffolds for tissue engineering applications like tissue engineering of urethra(60), heart valves(61), blood vessel(62). More recentely, Ott *et al.* developed a novel method of perfusion decellularization that is able to generate whole organ scaffolds. The use of this method allowed the decellularization of a whole heart that was later repopulated with neonatal rat cardiomyocytes. This bioengineered heart was able to contract up to 2% of the normal contractile function(63). This approach may have a tremendous potential for the field of organ bioengineering. We have recently used a similar perfusion decellularization technique to liver, pancreas, intestine and kidney generating decellularized organ scaffolds for organ bioengineering(64, 65). These bioscaffolds preserve their tissue microarchitecture and an intact vascular network that can be readily used as a route for recellularization by perfusion of culture medium with different cell populations. In an analogous fashion, Uygun *et al.* decellularized rat livers and repopulated them with rat primary hepatocytes, showing promising hepatic function and the ability of heterotopicaly transplant these bioengineered livers into animals for up to eight hours(66). Baptista *et al.* were able to take this a step further by using human primary liver progenitor/stem and endothelial cells to bioengineer a vascularized liver. These bioengineered livers displayed some of the functions of a native human liver (albumin and urea secretion, drug metabolism enzyme expression, etc), exhibiting also an endothelialized vascular network that prevented platelet adhesion and aggregation, critical for blood vessel patency after transplantation(65). Nonetheless, it is difficult to predict the outcome and the real translational value of this technology in the present days, but the potential is certainly vast. Translation of it into the bioengineering of human size livers might help mitigate the endless hurdle of organ shortage for transplantation.

6. References

[1] CDC. Centers for Disease Control and Prevention Database. In; 2007.

[2] WHO. World Health Organization - Global Burden of Disease: 2004 update (2008). In: WHO publications; 2008.

[3] Eurostat. Eurostat's Harmonised Regional Statistical Database. In; 2007.

[4] OPTN. Transplant Database. In; 2011.

[5] Kao JH, Chen DS. Global control of hepatitis B virus infection. Lancet Infect Dis 2002;2:395-403.

[6] OPTN. MELD score. In; 2011.

[7] Wolfe RA, Merion RM, Roys EC, Port FK. Trends in organ donation and transplantation in the United States, 1998-2007. Am J Transplant 2009;9:869-878.

[8] Thuluvath PJ, Guidinger MK, Fung JJ, Johnson LB, Rayhill SC, Pelletier SJ. Liver transplantation in the United States, 1999-2008. Am J Transplant;10:1003-1019.

[9] Turka LA, Wood K, Bluestone JA. Bringing transplantation tolerance into the clinic: lessons from the ITN and RISET for the Establishment of Tolerance consortia. Curr Opin Organ Transplant;15:441-448.

[10] Carpentier B, Gautier A, Legallais C. Artificial and bioartificial liver devices: present and future. Gut 2009;58:1690-1702.

[11] Phua J, Lee KH. Liver support devices. Curr Opin Crit Care 2008;14:208-215.

[12] Park JK, Lee DH. Bioartificial liver systems: current status and future perspective. J Biosci Bioeng 2005;99:311-319.

[13] Pless G. Artificial and bioartificial liver support. Organogenesis 2007;3:20-24.

[14] Rifai K. Extracorporeal albumin dialysis. Hepatol Res 2008;38:S41-S45.
[15] Brophy CM, Nyberg SL. Extracorporeal treatment of acute liver failure. Hepatol Res 2008;38:S34-S40.
[16] Cao S, Esquivel CO, Keeffe EB. New approaches to supporting the failing liver. Annu Rev Med 1998;49:85-94.
[17] Stange J, Mitzner S. Cell sources for bioartificial liver support. Int J Artif Organs 1996;19:14-17.
[18] Tilles AW, Berthiaume F, Yarmush ML, Tompkins RG, Toner M. Bioengineering of liver assist devices. J Hepatobiliary Pancreat Surg 2002;9:686-696.
[19] Yu CB, Pan XP, Li LJ. Progress in bioreactors of bioartificial livers. Hepatobiliary Pancreat Dis Int 2009;8:134-140.
[20] McKenzie TJ, Lillegard JB, Nyberg SL. Artificial and bioartificial liver support. Semin Liver Dis 2008;28:210-217.
[21] Adham M. Extracorporeal liver support: waiting for the deciding vote. ASAIO J 2003;49:621-632.
[22] Grossman M, Raper SE, Kozarsky K, Stein EA, Engelhardt JF, Muller D, Lupien PJ, et al. Successful ex vivo gene therapy directed to liver in a patient with familial hypercholesterolaemia. Nat Genet 1994;6:335-341.
[23] Grossman M, Rader DJ, Muller DW, Kolansky DM, Kozarsky K, Clark BJ, 3rd, Stein EA, et al. A pilot study of ex vivo gene therapy for homozygous familial hypercholesterolaemia. Nat Med 1995;1:1148-1154.
[24] Fox IJ, Chowdhury JR, Kaufman SS, Goertzen TC, Chowdhury NR, Warkentin PI, Dorko K, et al. Treatment of the Crigler-Najjar syndrome type I with hepatocyte transplantation. N Engl J Med 1998;338:1422-1426.
[25] Horslen SP, McCowan TC, Goertzen TC, Warkentin PI, Cai HB, Strom SC, Fox IJ. Isolated hepatocyte transplantation in an infant with a severe urea cycle disorder. Pediatrics 2003;111:1262-1267.
[26] Ambrosino G, Varotto S, Strom SC, Guariso G, Franchin E, Miotto D, Caenazzo L, et al. Isolated hepatocyte transplantation for Crigler-Najjar syndrome type 1. Cell Transplant 2005;14:151-157.
[27] Muraca M, Gerunda G, Neri D, Vilei MT, Granato A, Feltracco P, Meroni M, et al. Hepatocyte transplantation as a treatment for glycogen storage disease type 1a. Lancet 2002;359:317-318.
[28] Sokal EM, Smets F, Bourgois A, Van Maldergem L, Buts JP, Reding R, Bernard Otte J, et al. Hepatocyte transplantation in a 4-year-old girl with peroxisomal biogenesis disease: technique, safety, and metabolic follow-up. Transplantation 2003;76:735-738.
[29] Strom SC, Fisher RA, Thompson MT, Sanyal AJ, Cole PE, Ham JM, Posner MP. Hepatocyte transplantation as a bridge to orthotopic liver transplantation in terminal liver failure. Transplantation 1997;63:559-569.
[30] Strom SC, Chowdhury JR, Fox IJ. Hepatocyte transplantation for the treatment of human disease. Semin Liver Dis 1999;19:39-48.
[31] Strom SC, Fisher RA, Rubinstein WS, Barranger JA, Towbin RB, Charron M, Mieles L, et al. Transplantation of human hepatocytes. Transplant Proc 1997;29:2103-2106.

[32] Combs C, Brunt EM, Solomon H, Bacon BR, Brantly M, Di Bisceglie AM. Rapid development of hepatic alpha1-antitrypsin globules after liver transplantation for chronic hepatitis C. Gastroenterology 1997;112:1372-1375.
[33] Mito M, Kusano M, Kawaura Y. Hepatocyte transplantation in man. Transplant Proc 1992;24:3052-3053.
[34] Fisher RA, Strom SC. Human hepatocyte transplantation: worldwide results. Transplantation 2006;82:441-449.
[35] Kharaziha P, Hellstrom PM, Noorinayer B, Farzaneh F, Aghajani K, Jafari F, Telkabadi M, et al. Improvement of liver function in liver cirrhosis patients after autologous mesenchymal stem cell injection: a phase I-II clinical trial. Eur J Gastroenterol Hepatol 2009;21:1199-1205.
[36] Salama H, Zekri AR, Zern M, Bahnassy A, Loutfy S, Shalaby S, Vigen C, et al. Autologous hematopoietic stem cell transplantation in 48 patients with end-stage chronic liver diseases. Cell Transplant 2010.
[37] Zacharoulis D, Milicevic MN, Helmy S, Jiao LR, Levicar N, Tait P, Scott M, et al. Autologous infusion of expanded mobilized adult bone marrow-derived CD34+ cells into patients with alcoholic liver cirrhosis. Am J Gastroenterol 2008;103:1952-1958.
[38] Khan AA, Shaik MV, Parveen N, Rajendraprasad A, Aleem MA, Habeeb MA, Srinivas G, et al. Human fetal liver derived stem cell transplantation as supportive modality in the\ management of end stage decompensated liver cirrhosis. Cell Transplantation 2010.
[39] Schmelzer E, Zhang L, Bruce A, Wauthier E, Ludlow J, Yao HL, Moss N, et al. Human hepatic stem cells from fetal and postnatal donors. J Exp Med 2007;204:1973-1987.
[40] Gouon-Evans V, Boussemart L, Gadue P, Nierhoff D, Koehler CI, Kubo A, Shafritz DA, et al. BMP-4 is required for hepatic specification of mouse embryonic stem cell-derived definitive endoderm. Nat Biotechnol 2006;24:1402-1411.
[41] Gadue P, Huber TL, Paddison PJ, Keller GM. Wnt and TGF-beta signaling are required for the induction of an in vitro model of primitive streak formation using embryonic stem cells. Proc Natl Acad Sci U S A 2006;103:16806-16811.
[42] Basma H, Soto-Gutierrez A, Yannam GR, Liu L, Ito R, Yamamoto T, Ellis E, et al. Differentiation and transplantation of human embryonic stem cell-derived hepatocytes. Gastroenterology 2009;136:990-999.
[43] Langer R, Vacanti JP. Tissue engineering. Science 1993;260:920-926.
[44] Mooney D, Hansen L, Vacanti J, Langer R, Farmer S, Ingber D. Switching from differentiation to growth in hepatocytes: control by extracellular matrix. J Cell Physiol 1992;151:497-505.
[45] Fiegel HC, Kaufmann PM, Bruns H, Kluth D, Horch RE, Vacanti JP, Kneser U. Hepatic tissue engineering: from transplantation to customized cell-based liver directed therapies from the laboratory. J Cell Mol Med 2008;12:56-66.
[46] Kim SS, Sundback CA, Kaihara S, Benvenuto MS, Kim BS, Mooney DJ, Vacanti JP. Dynamic seeding and in vitro culture of hepatocytes in a flow perfusion system. Tissue Eng 2000;6:39-44.
[47] Lin P, Chan WC, Badylak SF, Bhatia SN. Assessing porcine liver-derived biomatrix for hepatic tissue engineering. Tissue Eng 2004;10:1046-1053.

[48] Linke K, Schanz J, Hansmann J, Walles T, Brunner H, Mertsching H. Engineered liver-like tissue on a capillarized matrix for applied research. Tissue Eng 2007;13:2699-2707.
[49] Tong JZ, Bernard O, Alvarez F. Long-term culture of rat liver cell spheroids in hormonally defined media. Exp Cell Res 1990;189:87-92.
[50] Gerlach J, Unger J, Hole O, Encke J, Muller C, Neuhaus P. [Bioreactor for long-term maintenance of differentiated hepatic cell functions]. ALTEX 1994;11:207-215.
[51] Torok E, Pollok JM, Ma PX, Kaufmann PM, Dandri M, Petersen J, Burda MR, et al. Optimization of hepatocyte spheroid formation for hepatic tissue engineering on three-dimensional biodegradable polymer within a flow bioreactor prior to implantation. Cells Tissues Organs 2001;169:34-41.
[52] Torok E, Vogel C, Lutgehetmann M, Ma PX, Dandri M, Petersen J, Burda MR, et al. Morphological and functional analysis of rat hepatocyte spheroids generated on poly(L-lactic acid) polymer in a pulsatile flow bioreactor. Tissue Eng 2006;12:1881-1890.
[53] Kim BS, Baez CE, Atala A. Biomaterials for tissue engineering. World J Urol 2000;18:2-9.
[54] Voytik-Harbin SL, Brightman AO, Kraine MR, Waisner B, Badylak SF. Identification of extractable growth factors from small intestinal submucosa. J Cell Biochem 1997;67:478-491.
[55] Kaufmann PM, Kneser U, Fiegel HC, Kluth D, Herbst H, Rogiers X. Long-term hepatocyte transplantation using three-dimensional matrices. Transplant Proc 1999;31:1928-1929.
[56] Johnson LB, Aiken J, Mooney D, Schloo BL, Griffith-Cima L, Langer R, Vacanti JP. The mesentery as a laminated vascular bed for hepatocyte transplantation. Cell Transplant 1994;3:273-281.
[57] Starzl TE, Francavilla A, Halgrimson CG, Francavilla FR, Porter KA, Brown TH, Putnam CW. The origin, hormonal nature, and action of hepatotrophic substances in portal venous blood. Surg Gynecol Obstet 1973;137:179-199.
[58] Yang J, Yamato M, Shimizu T, Sekine H, Ohashi K, Kanzaki M, Ohki T, et al. Reconstruction of functional tissues with cell sheet engineering. Biomaterials 2007;28:5033-5043.
[59] Ohashi K, Yokoyama T, Yamato M, Kuge H, Kanehiro H, Tsutsumi M, Amanuma T, et al. Engineering functional two- and three-dimensional liver systems in vivo using hepatic tissue sheets. Nat Med 2007;13:880-885.
[60] El-Kassaby AW, Retik AB, Yoo JJ, Atala A. Urethral stricture repair with an off-the-shelf collagen matrix. J Urol 2003;169:170-173; discussion 173.
[61] Lee DJ, Steen J, Jordan JE, Kincaid EH, Kon ND, Atala A, Berry J, et al. Endothelialization of heart valve matrix using a computer-assisted pulsatile bioreactor. Tissue Eng Part A 2009;15:807-814.
[62] Amiel GE, Komura M, Shapira O, Yoo JJ, Yazdani S, Berry J, Kaushal S, et al. Engineering of blood vessels from acellular collagen matrices coated with human endothelial cells. Tissue Eng 2006;12:2355-2365.
[63] Ott HC, Matthiesen TS, Goh SK, Black LD, Kren SM, Netoff TI, Taylor DA. Perfusion-decellularized matrix: using nature's platform to engineer a bioartificial heart. Nat Med 2008;14:213-221.

[64] Baptista PM, Orlando G, Mirmalek-Sani SH, Siddiqui M, Atala A, Soker S. Whole organ decellularization - a tool for bioscaffold fabrication and organ bioengineering. Conf Proc IEEE Eng Med Biol Soc 2009;2009:6526-6529.
[65] Baptista PM, Siddiqui MM, Lozier G, Rodriguez SR, Atala A, Soker S. The use of whole organ decellularization for the generation of a vascularized liver organoid. Hepatology 2011;53:604-617.
[66] Uygun BE, Soto-Gutierrez A, Yagi H, Izamis ML, Guzzardi MA, Shulman C, Milwid J, et al. Organ reengineering through development of a transplantable recellularized liver graft using decellularized liver matrix. Nat Med 2010.